Lectures on Elementary Particles and Quantum Field Theory
Volume 1

Edited by
Stanley Deser, Marc Grisaru, and Hugh Pendleton

Lectures by Stephen L. Adler, Stanley Mandelstam,
Steven Weinberg, and Wolfhart Zimmermann

Lectures on Elementary Particles and Quantum Field Theory

1970 Brandeis University Summer Institute in Theoretical Physics, Volume 1

THE M.I.T. PRESS
Cambridge, Massachusetts, and London, England

ISBN 0 262 04031 X (hardcover)
ISBN 0 262 54013 4 (paperback)

Library of Congress catalog card number: 70-138840

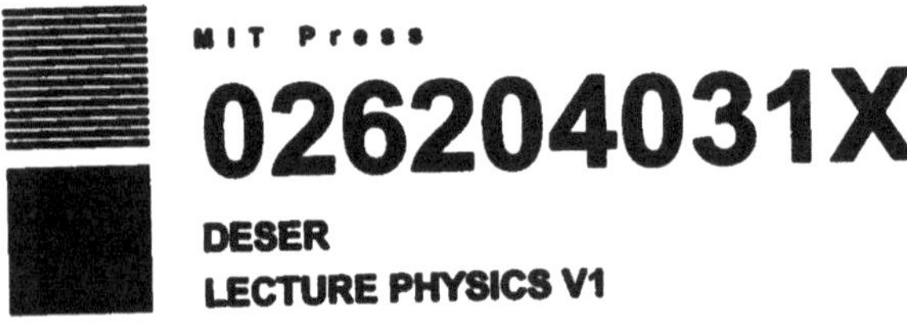

CONTENTS

FOREWORD

The notes contained in this volume, the first of two, are based on the lecture series delivered by the authors at the 1970 Brandeis Summer Institute in Theoretical Physics. In order to permit rapid publication and availability to the physics community, they are printed directly from typewritten copy. We are grateful to the lecturers for their assistance in the preparation of the typescripts, and to Betty Griffin, Sylvia Pendleton, and Mary Sider for their typing. We thank Geraldine Prentice, secretary of the Brandeis Summer Institute, for her continuing helpfulness.

The 1970 Brandeis Summer Institute in Theoretical Physics was made possible by the generous support of the National Science Foundation and of the North Atlantic Treaty Organization.

The second volume of notes contains lectures by Rudolf Haag on observables and fields, by Maurice Jacob on duality, by Henry Primakoff on weak interactions, by Michael Reed on non-Fock representations, and by Bruno Zumino on effective Lagrangians and broken symmetries.

Stanley Deser
Marc Grisaru
Hugh Pendleton
—Editors

Perturbation Theory Anomalies

Stephen L. Adler
Institute for Advanced Study
Princeton, New Jersey

CONTENTS

1. INTRODUCTION AND REVIEW OF PERTURBATION THEORY

The purpose of these lectures is to study some recent developments in renormalized perturbation theory. The subject of renormalized perturbation theory is, of course, old and well developed; for large classes of Lagrangian field theory models, renormalization procedures and proofs of renormalizability to all orders of perturbation theory have been given.[1] The new aspects which we will discuss here have been brought into focus by recent work on current algebras, which makes heavy use of Ward identities and of Bjorken limits involving currents, both of which are statements about time-ordered products of the type which commonly occur in Lagrangian field theories. Since some of these statements are obtained by rather naive formal manipulations of highly divergent quantities, it is natural to ask whether the manipulations are indeed correct. There is at present no general way to answer this question, even in specific Lagrangian field theory models, since no general methods for calculation in field theory exist. However, it is possible, and is quite illuminating, to examine the question within the framework of renormalized perturbation theory, where definite methods of calculation exist and concrete answers can be obtained. This track has been pursued by a number of authors during the past several years, and will be described below in detail. The results show that, in perturbation theory, there are many cases in which the usual naive manipulations break down, leading to modifications, or anomalies, in Ward identities and Bjorken limits. These anomalies and their properties are an interesting mathematical physics question in their own right, as well as having important implications for certain current algebra calculations.

1.1 Review of Quantum Electrodynamics and Renormalization Theory

We will begin our study of perturbation theory anomalies by reviewing the usual renormalization theory, in the familiar case of quantum electrodynamics (QED). We will find that QED exhibits many of the anomalies which will interest us, and the generalizations to other cases of physical interest, such as the quark model with massive vector gluon and the σ-model, involve questions of detail rather than of general principle. The Lagrangian for QED is[2]

$$\mathcal{L}(x) = \bar{\psi}(x)(i\gamma\cdot\Box - m_0)\psi(x) - \tfrac{1}{4}F_{\mu\nu}(x)\,F^{\mu\nu}(x) \qquad (1)$$
$$e_0\bar{\psi}(x)\,\gamma_\mu\,\psi(x)\,A^\mu(x)$$

with $\psi(x)$ the electron field, $A^\mu(x)$ the photon field, $F_{\mu\nu}(x) = \partial A_\mu(x)/\partial x^\nu - \partial A_\nu(x)/\partial x^\mu$ the electromagnetic field strength tensor, $\gamma\cdot\Box = \gamma^\mu\partial/\partial x^\mu$ and with $-e_0$ and m_0, respectively, the electron bare charge and bare mass. From the Lagrangian, we find the equations of motion for the fields,

$$(i\gamma\cdot\Box - m_0)\,\psi(x) = e_0\gamma_\mu A^\mu(x)\,\psi(x), \qquad (2)$$
$$\partial F^{\mu\nu}(x)/\partial x^\nu = e_0\, j^\mu(x),$$
$$j^\mu(x) = \bar{\psi}(x)\,\gamma^\mu\psi(x).$$

Using the equation of motion for $\psi(x)$ to calculate $\partial j^\mu(x)/\partial x^\mu$ and assuming that no subtlety arises when we apply the chain rule of differentiation to the product of two field operators at the same space time point, we find

$$\partial j^\mu(x)/\partial x^\mu = \partial\bar{\psi}(x)/\partial x^\mu\,\gamma^\mu\,\psi(x) + \bar{\psi}(x)\gamma^\mu\,\partial\psi(x)/\partial x^\mu \qquad (3)$$
$$= \bar{\psi}(x)[\,im_0 + ie_0\gamma_\mu A^\mu(x)\,]\,\psi(x) + \bar{\psi}(x)[\,-im_0 - ie_0\gamma_\mu A^\mu(x)\,]\psi(x)$$
$$= 0,$$

which is the usual equation of current conservation. Finally, the canonical anticommutation relations of the spinor fields are

$$\{\psi_\alpha(\underset{\sim}{x}, t),\ \psi^\dagger_\beta(\underset{\sim}{y}, t)\} = \delta_{\alpha\beta}\,\delta^3(\underset{\sim}{x} - \underset{\sim}{y})\,, \qquad (4)$$

with $\alpha, \beta = 1, \ldots, 4$ the labels of the spinor components, and, if we use the Feynman gauge, the canonical commu-

tation relations of the photon fields are

$$[A_\mu(\underset{\sim}{x}, t), \partial A_\nu(\underset{\sim}{y}, t)/\partial t] = -ig_{\mu\nu}\delta^3(\underset{\sim}{x}-\underset{\sim}{y}). \qquad (5)$$

Eqs. (1) - (5) are the basic equationsof QED.

The usual method for dealing with these equations is, of course, to work in momentum space and to expand in a perturbation series in powers of e_0. This leads to the familiar Feynman rules, which we summarize as follows:

(i) For each internal electron line with momentum p we include a factor $i(\not{p}-m_0+i\varepsilon)^{-1}$ and for each vertex a factor $-ie_0\gamma_\mu$. For each internal photon line of momentum q we include a factor $-ig_{\mu\nu}(q^2+i\varepsilon)^{-1}$.

(ii) There is a factor $\int d^4\ell/(2\pi)^4$ for each internal integration over loop variable ℓ and a factor -1 for each fermion loop.

(iii) For each external photon line there is a factor $\varepsilon_\mu\sqrt{Z_3}$, where ε_μ is the photon polarization four-vector and Z_3 is the photon wave-function renormalization. For each external electron line entering (leaving) the graph there is a factor $\sqrt{Z_2}\, u(p,s)\,[\sqrt{Z_2}\,\bar{u}(p,s)]$, and similar factors for external positron lines, with Z_2 the electron wave-function renormalization. Disconnected bubbles and self-energy insertions on external electron and photon lines are excluded.

Using these rules, we can construct the invariant Feynman amplitude $\mathcal{M}$ for any process, to any order of perturbation theory. As is well known, divergences are encountered which must be removed by a renormalization procedure, which we will now briefly sketch. Let us define the electron propagator $S'_F(p)$, the photon propagator $D'_F(q)^{\mu\nu}$ and the vertex part $\Gamma_\mu(p,p')$ by

$$iS'_F(p) = \int d^4x\, e^{ip\cdot x} < 0|\, T(\psi(x)\bar{\psi}(0))|\, 0>, \tag{6a}$$

$$iD'_F(q)^{\mu\nu} = \int d^4x\, e^{iq\cdot x} <0|\, T(A^\mu(x)A^\nu(0))|\, 0>, \tag{6b}$$

$$S'_F(p)\,\Gamma_\mu(p,p')\,S'_F(p') \tag{6c}$$
$$= -\int d^4x\, d^4y\, e^{ip\cdot x} e^{-ip'\cdot y} <0|\, T(\psi(x) j_\mu(0)\bar{\psi}(y))|\, 0>.$$

They have the following diagrammatic representations:

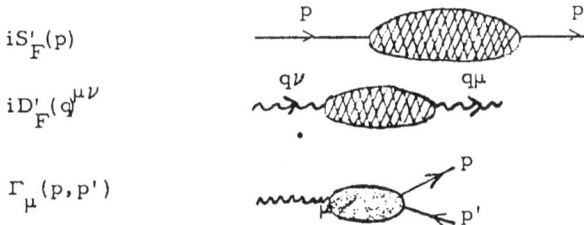

$iS'_F(p)$ p p

$iD'_F(q)^{\mu\nu}$ $q\nu$ $q\mu$

$\Gamma_\mu(p,p')$ μ p p'

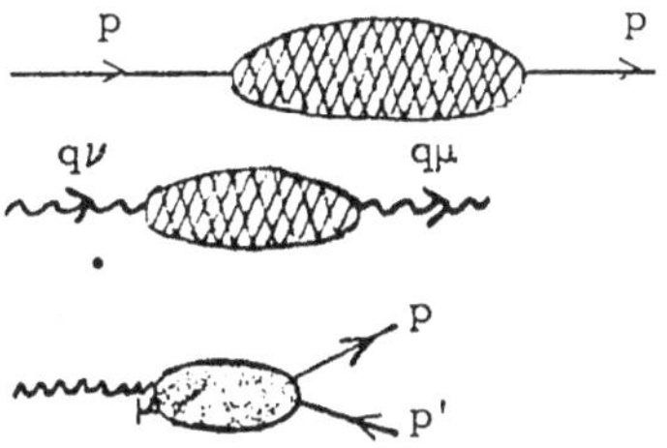

The vertex part is proper; that is, it cannot be divided into two disjoint graphs by cutting a single line. The electron and photon propagators can be expressed in terms of the <u>proper</u> electron and photon self-energy parts, defined by

$-i\,\Sigma(p) =$ (the blob includes proper diagrams only with the two electron ends removed)

$$ie_0^2\, \Pi(q)^{\mu\nu} = \text{[diagram]}$$ (the blob includes proper diagrams only with the two photon ends removed)

For the electron propagator, we find

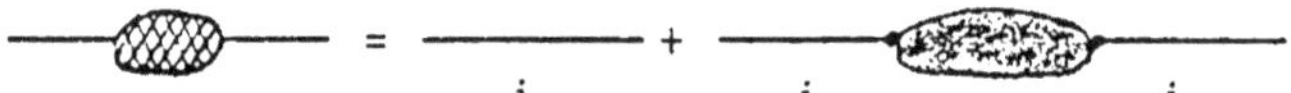

$$iS_F'(p) = \frac{i}{\not p - m_0} + \frac{i}{\not p - m_0}[-i\Sigma(p)]\frac{i}{\not p - m_0}$$

$$+ \frac{i}{\not p - m_0}[-i\Sigma(p)]\frac{i}{\not p - m_0}[-i\Sigma(p)]\frac{i}{\not p - m_0} + \ldots$$

which sums to

$$S_F'(p) = \frac{1}{\not p - m_0 - \Sigma(p)} \tag{7}$$

Similarly, writing

$$D_F'(q)^{\mu\nu} = -g^{\mu\nu}\, D_F'(q) + \text{longitudinal terms} \tag{8a}$$

and using the fact that current conservation requires $\Pi(q)^{\mu\nu}$ to have the form

$$\Pi(q)^{\mu\nu} = (-q^2 g^{\mu\nu} + q^\mu q^\nu)\, \Pi(q^2) \tag{8b}$$

we find that a similar summation gives us

$$D_F'(q) = \frac{1}{q^2\, [1 + e_0^2\, \Pi(q^2)]} \,. \tag{9}$$

The reason for defining the propagators and vertex part is that <u>all divergences reside in these quantities</u>. To see this, we note that the superficial degree of divergence of a graph is given by

$$D = 4k - 2b - f, \tag{10a}$$

b = number of internal photon lines,

f = number of internal electron lines,

k = number of internal-momentum integrations.

Letting

n = number of vertices

B = number of external boson lines, (10b)

F = number of external fermion lines,

and using the topological relations

$$F + 2f = 2n, \tag{10c}$$

$$B + 2b = n,$$

we can rewrite D in terms of the numbers of external lines above,

$$D = 4 - \frac{3}{2}F - B. \tag{11}$$

The condition for a graph to converge is that $D < 0$ for the graph itself and for all subgraphs contained inside the graph. From Eq. (11), we learn that the potentially dangerous types of graphs are (a) the electron proper self energy part $\Sigma(p)$ ($D=1$), (b) the proper vertex part $\Gamma_\mu(p, p')$ ($D=0$), (c) the photon proper self-energy part $\Pi(q^2)$ ($D=2$), (d) the proper vertex of three photons

and the proper photon-photon scattering amplitude

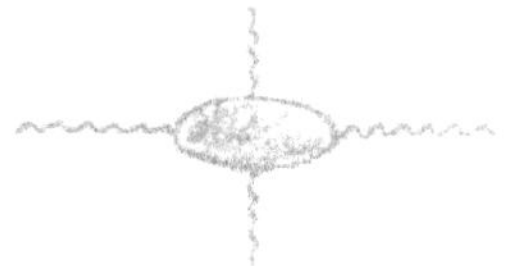

The three photon vertex vanishes because the photon is odd under charge conjugation. (Furry's theorem.) Although D=2 for the photon proper-self-energy part, two powers of momentum are used up to form the $q^2 g_{\mu\nu} - q_\mu q_\nu$ term in Eq. (8b), and thus $\Pi(q^2)$ has only an effective divergence $D_{eff} = 0$. Similarly, in the case of photon-photon scattering, four powers of internal momenta are used up in forming electromagnetic field strengths $F_{\mu\nu}(q) \sim (q_\mu \epsilon_\nu - q_\nu \epsilon_\mu)$ for each of the four external photons. Hence this graph has $D_{eff} = -4$ and is highly convergent.

Having shown that divergences are always associated with self-energy and vertex parts, we can now state a procedure for studying, and then removing, the divergences of an arbitrary graph. Given an arbitrary graph, let us define the skeleton as the new graph obtained by contracting all self-energy and vertex parts into points. [In doing this, we write $m_0 = m - \delta m$, with m the physical mass, and treat δm as part of the electron-self-energy part. Hence the electron propagators appearing in the skeleton are all of the form $(\not{p} - m)^{-1}$.] Clearly, the skeleton graph is always finite.

If we can devise a procedure for making the self-energy and vertex parts finite, we can then obtain a finite value for our original graph by making appropriate insertions of the finite self-energy and vertex parts on the skeleton. For skeletons with three legs or more, the insertions can always be made in a non-overlapping way, and the recipe is simple to implement. For skeletons with two legs (electron and photon self-energy parts themselves) there are situations which necessarily involve overlapping vertex insertions, such as

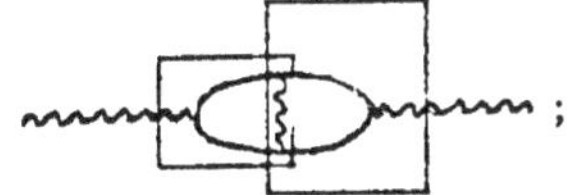

this is what makes proofs of renormalizability so difficult.

The recipe for making the self-energy and vertex parts finite is the following. First, we choose $\delta m = m - m_0$ such that

$$\delta m - \Sigma(p) = 0 \quad \text{at} \quad \not{p} = m, \tag{12}$$

guaranteeing that S'_F has a pole when $\not{p}$ is equal to the physical electron mass m,

$$S'_F \rightarrow \frac{Z_2}{\not{p} - m} \quad \text{for} \quad \not{p} \rightarrow m. \tag{13}$$

Z_2 is the electron wave-function renormalization constant. Similarly, we define a photon wave function renormalization constant by examining the behavior of the photon propagator near mass shell,

$$D'_F(q)^{\mu\nu} \to \frac{-Z_3\, g^{\mu\nu}}{q^2} + \text{longitudinal terms, for } q^2 \to 0, \quad (14)$$

and a vertex renormalization constant by examining the zero momentum transfer limit of the vertex with the electron lines on mass shell,

$$\Gamma_\mu(p,p)\big|_{\not p=m} = Z_1^{-1}\,\gamma_\mu . \quad (15)$$

The renormalization recipe consists of <u>rescaling</u> the self-energy and vertex functions and the bare charge so as to define renormalized functions $\tilde{S}'_F(p)$, $\tilde{D}'_F(q)^{\mu\nu}$ and $\tilde{\Gamma}_\mu(p,p')$ and a renormalized (physical) charge e,

$$\begin{aligned} S'_F(p) &= Z_2\, \tilde{S}'_F(p) \\ D'_F(q)^{\mu\nu} &= Z_3\, \tilde{D}'_F(q)^{\mu\nu}, \\ \Gamma_\mu(p,p') &= Z_1^{-1}\, \tilde{\Gamma}_\mu(p,p') \end{aligned} \quad (16)$$

$$e_0 = \frac{Z_1\, e}{Z_2\sqrt{Z_3}} .$$

The remarkable fact is that Eqs. (16), when used to express the tilde functions in terms of the physical charge and mass e and m, lead to finite values of the tilde functions for all values of the four-momenta p and p'. That is, all of the infinities can be removed into the renormalization <u>constants</u> Z_1, Z_2 and Z_3. It further turns out that, as a consequence of current conservation, one has

$$Z_1 = Z_2 \ . \tag{17}$$

When Eqs. (16) are substituted into a skeleton graph with external legs removed [item (iii) of our Feynman rules specifies that self-energy insertions in external lines are to be omitted], they give the graph obtained from the skeleton by making renormalized self-energy and vertex insertions, times a product of renormalization factors, which is clearly

$$\left(\frac{Z_1}{Z_2\sqrt{Z_3}}\right)^n Z_1^{-n} Z_2^{f} Z_3^{b} = Z_2^{f-n} Z_3^{b-\frac{1}{2}n} . \tag{18}$$

Upon using the topological relations of Eq. (10c), Eq. (18) reduces to

$$Z_2^{-\frac{1}{2}F} Z_3^{-\frac{1}{2}B} , \tag{19}$$

which is exactly canceled by the product of external line factors $\sqrt{Z_2}$ and $\sqrt{Z_3}$ specified in item (iii) of our Feynman rules. Thus, our Feynman rules, with the rescalings of Eq. (16), always lead to a finite renormalized matrix element $\mathcal{M}$.

The proof that the rescalings do really make the self-energy and vertex parts finite is based on mathematical induction: one assumes that the procedure makes all graphs of order $n-2$ in e finite, and then demonstrates the convergence of the rescaled graphs of order n. We will briefly sketch the proof in the cases of the vertex and electron-self-energy parts, since the arguments involved are simple and involve concepts which will be useful to us later on in our discussion of anomalies. The proof in the case of the photon self-energy part is made complicated by the overlapping divergence problem, and will be omitted.[3] In order to prove renormalizability of the vertex part, we first formulate an integral equation which it satisfies. To do this, let us define an electron-positron scattering kernel $K(p', p, q)_{\alpha\beta, \gamma\delta}$, which represents all diagrams of the form

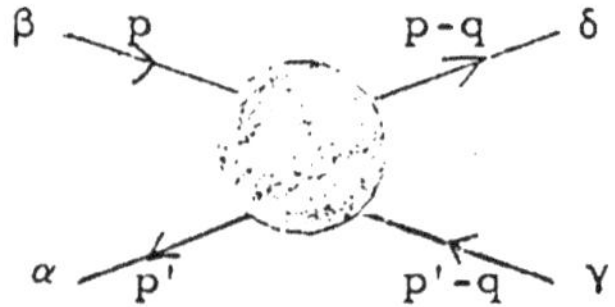

with external leg propagators removed and with disconnected diagrams and diagrams of the following two classes omitted,

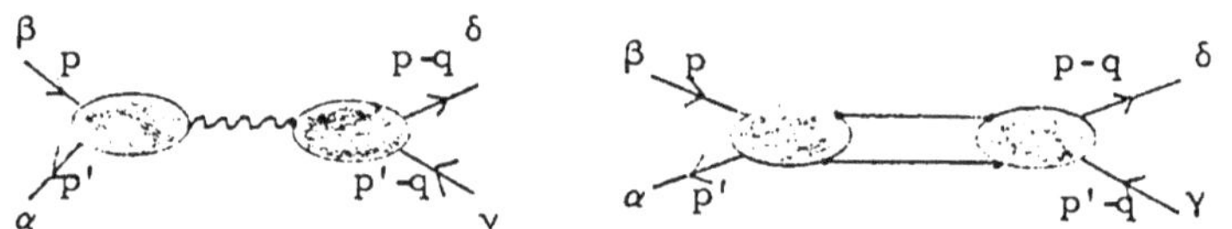

The lowest order contribution to K is

$$K^{(0)}(p', p, q)_{\alpha\beta, \gamma\delta} = \frac{ie_0^2}{q^2} (\gamma_\mu)_{\alpha\gamma} (\gamma^\mu)_{\delta\beta}, \tag{20}$$

coming from the diagram

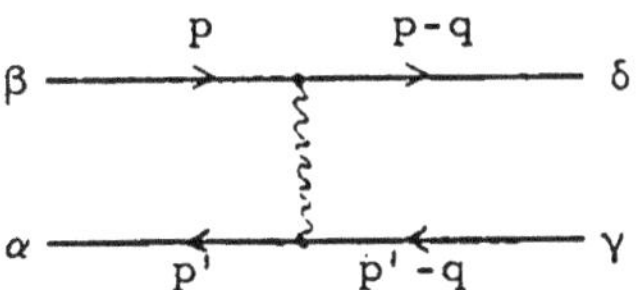

In terms of this kernel, we can write an integral equation for the vertex part, according to the following diagrammatic representation:

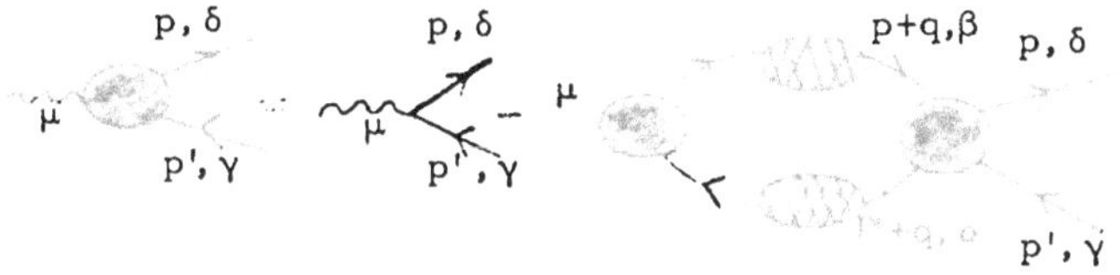

We get

$$\Gamma_\mu(p,p')_{\delta\gamma} = (\gamma_\mu)_{\delta\gamma} + \int \frac{d^4q}{(2\pi)^4} [\, i S'_F(p+q)\, \Gamma_\mu(p+q, p'+q) \times i S'_F(p'+q)\,]_{\beta\alpha} K(p'+q, p+q, q)_{\alpha\beta,\gamma\delta}, \tag{21}$$

or, in a condensed notation,

$$\Gamma = \gamma - \int \Gamma\, S'_F\, S'_F\, K. \tag{22}$$

If we define a rescaled kernel $\tilde{K}$ by

$$\tilde{K} \equiv Z_2^2 K, \tag{23}$$

then Eq. (22) can be rewritten in terms of rescaled quantities as

$$\tilde{\Gamma}_\mu = Z_1 \gamma_\mu - \int \tilde{\Gamma}_\mu \tilde{S}'_F \tilde{S}'_F \tilde{K}. \tag{24}$$

We wish to show that when the theory has been made finite to order n-2, the renormalization constant defined by Eq. (15) makes $\tilde{\Gamma}$ finite in order n. Since $\tilde{K}$ begins in order e^2 [see Eq. (20)], we can get the order n contribution to $\tilde{\Gamma}$ by substituting the order n-2 contributions to $\tilde{\Gamma}$, $\tilde{S}'_F$ and $\tilde{K}$ into the right hand side of Eq. (24):

$$\tilde{\Gamma}_\mu^{(n)} = Z_1^{(n)} \gamma_\mu - \int \tilde{\Gamma}_\mu^{(n-2)} \tilde{S}_F'^{(n-2)} \tilde{S}_F'^{(n-2)} \tilde{K}^{(n-2)} \qquad (25)$$

The definitions of the rescaled vertex, Eq. (15) and (16), tell us that

$$\tilde{\Gamma}_\mu^{(n)}(p,p)\big|_{\not p = m} = 0 \qquad (26)$$

which implies that

$$Z_1^{(n)} \gamma_\mu = \int \tilde{\Gamma}_\mu^{(n-2)} \tilde{S}_F'^{(n-2)} \tilde{S}_F'^{(n-2)} \tilde{K}^{(n-2)} \big|_{\not p = \not p' = m}, \qquad (27)$$

determining $Z_1^{(n)}$ in terms of known quantities.

Note that the statement that the right hand side of Eq. (26) is proportional to a (divergent) constant times γ_μ involves no assumptions: the right hand side is a dimensionless Lorentz vector function of m, and therefore must be proportional to γ_μ, the only Lorentz vector in the problem when $\not p$ and $\not p'$ have both been replaced by m. Substituting Eq. (27) into Eq. (25), we get

$$\tilde{\Gamma}_\mu^{(n)} = \int \tilde{\Gamma}_\mu^{(n-2)} \tilde{S}_F'^{(n-2)} \tilde{S}_F'^{(n-2)} \tilde{K}^{(n-2)} \big|_{\not p = \not p' = m} \qquad (28)$$

$$- \int \tilde{\Gamma}_\mu^{(n-2)} \tilde{S}_F'^{(n-2)} \tilde{S}_F'^{(n-2)} \tilde{K}^{(n-2)}.$$

Now, using the induction hypothesis, it is an easy matter to show that all subintegrations in Eq. (24) involving some, but not all of the internal lines, are finite. Thus the only divergence is a logarithmic one connected with the overall subintegration in which the four-momenta passing through all internal lines become large simultaneously. This overall divergence is, however made finite by the single differencing of Eq. (28), and hence $\tilde{\Gamma}_{\mu}^{(n)}$ is seen to be convergent. We have therefore established that the vertex part Γ_{μ} is multiplicatively renormalizable. The important thing to note about our argument is that it has proceeded entirely from the integral equation of Eq. (22), but in no way depended on specific properties of the vector-vertex γ_{μ}. Since the pseudoscalar and axial-vector vertices in quantum electrodynamics satisfy similar integral equations,

$$\begin{aligned} \Gamma_{\mu}^{5} &= \gamma_{\mu}\gamma_{5} - \int \Gamma_{\mu}^{5} S_{F}' S_{F}' K, \\ \Gamma^{5} &= \gamma_{5} - \int \Gamma^{5} S_{F}' S_{F}' K, \end{aligned} \tag{29}$$

it follows by an inductive argument identical to the one given above that these vertices are also multiplicatively renormalizable,

$$\begin{aligned} \Gamma_{\mu}^{5}(p,p') &= Z_{A}^{-1} \tilde{\Gamma}_{\mu}^{5}(p,p'), \\ \Gamma^{5}(p,p') &= Z_{D}^{-1} \tilde{\Gamma}^{5}(p,p'), \end{aligned} \tag{30}$$

with Z_A and Z_D divergent constants.

Next, let us turn our attention to the electron propagator $S_F'(p)$. Although overlapping divergences are present here, the overlapping divergence problem can be circumvented by using an important connection between the propagator and the vertex part known as the Ward identity. To obtain the Ward identity, we multiply Eq. (6c) by $(p-p')^\mu$, giving

$$
\begin{aligned}
&(p-p')^\mu S_F'(p)\,\Gamma_\mu(p,p')S_F'(p') \qquad (31)\\
&= -(p-p')^\mu \int d^4x d^4y\, e^{ip\cdot x} e^{-ip'\cdot y} <0|T(\psi(x) j_\mu(0)\overline{\psi}(y))|0>\\
&= -(p-p')^\mu \int d^4x d^4y\, e^{i(p'-p)\cdot x} e^{-ip'\cdot y} <0|T(\psi(0) j_\mu(x)\overline{\psi}(y))|0>\\
&= \int d^4x\, d^4y\, e^{i(p'-p)\cdot x} e^{-ip'\cdot y}\, i\frac{\partial}{\partial x_\mu} <0|T(\psi(0) j_\mu(x)\overline{\psi}(y))|0>\\
&= i\int d^4x d^4y\, e^{i(p'-p)\cdot x} e^{-ip'\cdot y} <0|T(\psi(0)\frac{\partial}{\partial x_\mu} j_\mu(x)\,\overline{\psi}(y))\\
&+ \delta(x_0)(T[j_0(x),\psi(0)]\overline{\psi}(y)) + \delta(x_0-y_0)T(\psi(0)[j_0(x),\overline{\psi}(y)])|0>.
\end{aligned}
$$

[In going from the first line to the second line on the right-hand side of Eq. (31), we have set $x \to -x$, $y \to y-x$, and used translation invariance of the T-product. In going to the next line we have integrated by parts, and in the final line we have used the standard formula for the time derivative of a T-product.] Using Eq. (3) to evaluate the first term in the final line, using the canonical commutation relations to evaluate the two remaining terms, and comparing with Eq. (6a), we get

$$(p-p')^\mu S_F'(p)\,\Gamma_\mu(p,p')S_F'(p') = S_F'(p')-S_F'(p), \qquad (32)$$

or multiplying by $S_F'(p)^{-1}\ldots S_F'(p')^{-1}$,

$$(p-p')^\mu \Gamma_\mu(p,p') = S_F'(p)^{-1}-S_F'(p')^{-1} \qquad (33)$$

which is the Ward identity. Now, when $\not p' = m$, Eqs. (7) and (12) tell us that $S_F'(p')^{-1} = 0$. Hence

$$S_F'(p)^{-1} = (p-p')^\mu \Gamma_\mu(p,p')\big|_{\not p' = m}, \tag{34}$$

which immediately implies that $Z_1 S_F'(p)^{-1}$ is finite to order n, since $Z_1 \Gamma_\mu$ is. Furthermore, examining Eq. (34) in the neighborhood of $\not p = m$, we have

$$\begin{aligned} Z_2^{-1}(\not p - m) \approx S_F'(p)^{-1}\big|_{\not p \approx m} &= (p-p')^\mu \Gamma_\mu(p,p')\big|_{\substack{\not p' = m \\ \not p \approx m}} \\ &\approx Z_1^{-1}(\not p - m), \end{aligned} \tag{35}$$

which tells us that $Z_1 = Z_2$, as claimed. We see, then, that the Ward identity plays a very useful role in discussing the renormalization of QED. In deriving the current conservation condition in Eq. (3) and the Ward identity in Eq. (16), we have made with impunity the type of dangerous manipulations referred to in the opening paragraphs. It turns out, however, that these manipulations are justified, and the Ward identity and other consequences of vector current conservation are valid in all orders of perturbation theory. In other words, in QED with only the vector current $j_\mu(x)$ considered, there are no Ward identity anomalies.

2. THE VVA TRIANGLE ANOMALY

To get our first example of a Ward identity anomaly, we will consider the axial-vector current $j_\mu^5(x)$ in quantum electrodynamics,

$$j_\mu^5(x) = \bar{\psi}(x)\,\gamma_\mu\gamma_5\psi(x). \tag{36}$$

This current plays no role in pure quantum electrodynamics, but appears when the weak interaction between electrons and neutrinos is taken into account within the framework of the local current-current theory. {To see this, we note that in the current-current theory without intermediate boson, the leptonic weak interactions are described by the effective Lagrangian

$$\mathcal{L}_{\rm eff} = (G/\sqrt{2})\, j_\lambda^\dagger\, j^\lambda, \tag{37}$$

where $G \approx 10^{-5}/M_{\rm proton}^{\;2}$ is the Fermi constant and where

$$j^\lambda = \bar{\nu}_\mu\gamma^\lambda(1-\gamma_5)\mu + \bar{\nu}_e\gamma^\lambda(1-\gamma_5)e \tag{38}$$

is the leptonic current. In addition to the usual terms describing muon decay, Eq. (37) contains the terms

$$(G/\sqrt{2})[\bar{\mu}\gamma_\lambda(1-\gamma_5)\nu_\mu\bar{\nu}_\mu\gamma^\lambda(1-\gamma_5)\mu + \bar{e}\gamma_\lambda(1-\gamma_5)\,\nu_e\bar{\nu}_e\gamma^\lambda(1-\gamma_5)e], \tag{39}$$

which describe elastic neutrino-lepton scattering. It is frequently convenient to rewrite Eq. (39), by means of a Fierz transformation,[4] in the form (the so-called charge retention ordering)

$$(G/\sqrt{2})[\bar{\mu}\gamma_\lambda(1-\gamma_5)\mu\bar{\nu}_\mu\gamma^\lambda(1-\gamma_5)\nu_\mu + \bar{e}\gamma_\lambda(1-\gamma_5)e\bar{\nu}_e\gamma^\lambda(1-\gamma_5)\nu_e], \tag{40}$$

which clearly involves the muon and electron axial-vector currents as well as the corresponding vector currents.} Proceeding in analogy with our treatment of the vector current in the last section, we use the equations of motion to calculate the divergence of the axial-vector current,

$$\begin{aligned}\partial j_\mu^5(x)/\partial x_\mu &= \bar{\psi}(x)[im_0+ie_0\gamma_\mu A^\mu(x)]\gamma_5\psi(x) \\ &+\bar{\psi}(x)\gamma_5[im_0+ie_0\gamma_\mu A^\mu(x)]\psi(x) \\ &= 2im_0 j^5(x),\end{aligned} \tag{41}$$

with

$$j^5(x) = \bar{\psi}(x)\gamma_5\psi(x) \tag{42}$$

the pseudoscalar current. Defining axial-vector and pseudoscalar vertex parts by analogy with Eq. (6c);

$$\begin{aligned}&S_F'(p)\,\Gamma_\mu^5(p,p')\,S_F'(p') \\ &\quad = -\int d^4x\,d^4y\,e^{ip\cdot x}e^{-ip'\cdot y}\langle 0|\,T(\psi(x)j_\mu^5(0)\bar{\psi}(y))\,|0\rangle, \\ &S_F'(p)\,\Gamma^5(p,p')\,S_F'(p') \\ &\quad = -\int d^4x\,d^4y\,e^{ip\cdot x}e^{-ip'\cdot y}\langle 0|\,T(\psi(x)j^5(0)\bar{\psi}(y))\,|0\rangle,\end{aligned} \tag{43}$$

a derivation precisely analogous to that of Eq. (31) gives the naive axial-vector Ward identity

$$(p-p')^\mu\,\Gamma_\mu^5(p,p') = 2m_0\,\Gamma^5(p,p') + S_F'(p)^{-1}\gamma_5 + \gamma_5 S_F'(p')^{-1}. \tag{44}$$

2.1 Ward Identity in Perturbation Theory

We wish to examine whether the naive formal manipulations which led to Eq. (44) are actually valid in perturbation theory. Defining vertex corrections Λ_μ^5 and Λ^5 by

$$\begin{aligned}\Gamma_\mu^5 &= \gamma_\mu \gamma_5 + \Lambda_\mu^5, \\ \Gamma^5 &= \gamma_5 + \Lambda^5\end{aligned} \tag{45}$$

and using Eq. (7), we may rewrite Eq. (44) as

$$(p-p')^\mu \Lambda_\mu^5(p,p') = 2m_0 \Lambda^5(p,p') - \Sigma(p)\gamma_5 - \gamma_5 \Sigma(p'). \tag{46}$$

In order to derive Eq. (46), let us divide the diagrams contributing to $\Lambda_\mu^5(p,p')$ into two types: (a) diagrams in which the axial-vector vertex $\gamma_\mu \gamma_5$ is attached to the fermion line beginning with external four-momentum p' and ending with external four-momentum p; (b) diagrams in which the axial-vector vertex $\gamma_\mu \gamma_5$ is attached to an internal closed loop. Because the axial-vector current is charge conjugation <u>even</u>,[5] Furry's theorem tells us that the number of photon lines emerging from the closed loop must be even

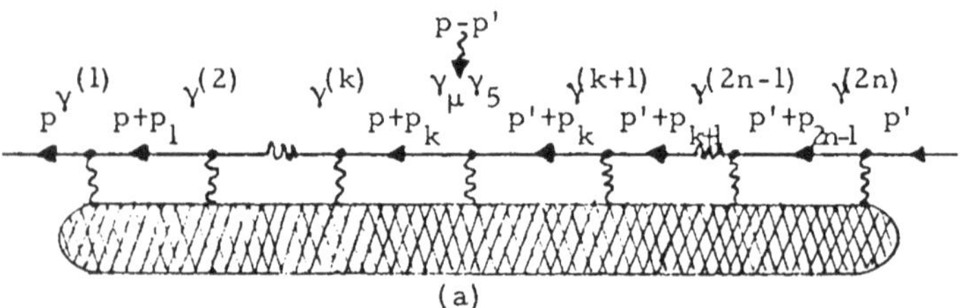

(a)

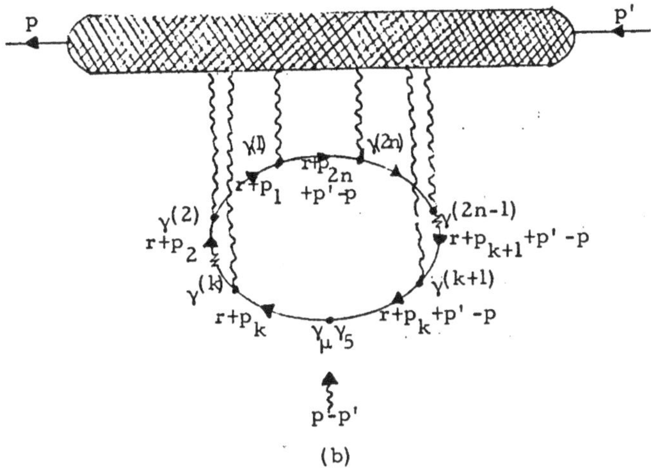

(b)

A typical contribution of type (a) has the form

$$\sum_{k=1}^{2n-1}\prod_{j=1}^{k-1}\left[\gamma^{(j)}\frac{1}{\not p+\not p_j-m_0}\right]\gamma^{(k)}\frac{1}{\not p+\not p_k-m_0}\gamma_\mu\gamma_5\frac{1}{\not p'+\not p_k-m_0}$$

$$\times\prod_{j=k+1}^{2n-1}\left[\gamma^{(j)}\frac{1}{\not p'+\not p_j-m_0}\right]\gamma^{(2n)}(\dots), \tag{47}$$

where we have focused our attention on the line to which the $\gamma_\mu\gamma_5$ vertex is attached and have denoted the remainder of the diagram by $(\dots)$. Multiplying Eq. (47) by $(p-p')^\mu$ and making use of the identity

$$\frac{1}{\not p+\not p_k-m_0}(\not p-\not p')\gamma_5\frac{1}{\not p'+\not p_k-m_0}=\frac{1}{\not p+\not p_k-m_0}(2m_0\gamma_5)$$

$$\times\frac{1}{\not p'+\not p_k-m_0}+\frac{1}{\not p+\not p_k-m_0}\gamma_5+\gamma_5\frac{1}{\not p'+\not p_k-m_0} \tag{48}$$

gives, after a little algebraic rearrangement,

$$\sum_{k=1}^{2n-1}\prod_{j=1}^{k-1}\left[\gamma^{(j)}\frac{1}{\not p+\not p_j-m_0}\right]\gamma^{(k)}\frac{1}{\not p+\not p_k-m_0}2m_0\gamma_5$$

$$\times \frac{1}{\not p' + \not p_k - m_0} \prod_{j=k+1}^{2n-1} \left[\gamma^{(j)} \frac{1}{\not p' + \not p_j - m_0} \right] \gamma^{(2n)} (\ldots)$$

$$-(\ldots) \prod_{j=1}^{2n-1} \left[\gamma^{(j)} \frac{1}{\not p + \not p_j - m_0} \right] \gamma^{(2n)} \gamma_5$$

$$-\gamma_5 \prod_{j=1}^{2n-1} \left[\gamma^{(j)} \frac{1}{\not p' + \not p_j - m_0} \right] \gamma^{(2n)} (\ldots). \tag{49}$$

The first, second, and third terms in Eq. (49) are, respectively, the type-(a) piece of Λ^5 and the pieces of $-\Sigma(p)\gamma_5$ and $-\gamma_5\Sigma(p')$ corresponding to the type-(a) piece of Λ_μ^5 in Eq. (47). Summing over all type-(a) contributions to Λ_μ^5, we get

$$(p-p')^\mu \Lambda_\mu^{5(a)}(p,p')$$
$$= 2m_0 \Lambda^{5(a)}(p,p') - \Sigma(p)\gamma_5 - \gamma_5 \Sigma(p'). \tag{50}$$

We turn next to contributions to Λ_μ^5 of type (b). A typical term is

$$\int d^4r \,\mathrm{Tr}\{ \sum_{k=1}^{2n} \prod_{j=1}^{k-1} \left[\gamma^{(j)} \frac{1}{r + \not p_j - m_0} \right] \gamma^{(k)} \frac{1}{r + \not p_k - m_0} \gamma_\mu \gamma_5$$
$$\times \frac{1}{r + \not p_k + \not p' - \not p - m_0} \prod_{j=k+1}^{2n} \left[\gamma^{(j)} \frac{1}{r + \not p_j + \not p' - \not p - m_0} \right] \} (\ldots). \tag{51}$$

Multiplying by $(p-p')^\mu$ and using Eq. (48) gives

$$\int d^4r \,\mathrm{Tr}\{ \sum_{k=1}^{2n} \prod_{j=1}^{k-1} \left[\gamma^{(j)} \frac{1}{r + \not p_j - m_0} \right] \gamma^{(k)} \frac{1}{r + \not p_k - m_0} 2m_0 \gamma_5$$
$$\times \frac{1}{r + \not p_k + \not p' - p - m_0} \prod_{j=k+1}^{2n} \left[\gamma^{(j)} \frac{1}{r + \not p_j + \not p' - \not p - m_0} \right] \}$$
$$\times (\ldots) + \int d^4r \,\mathrm{Tr}\{ \gamma^5 \prod_{j=1}^{2n} \left[\gamma^{(j)} \frac{1}{r + \not p_j - m_0} \right]$$
$$-\gamma_5 \prod_{j=1}^{2n} \left[\gamma^{(j)} \frac{1}{r + \not p_j + \not p' - \not p - m_0} \right] \} (\ldots). \tag{52}$$

The first term in Eq. (52) is the type-(b) contribution to Λ^5 corresponding to Eq. (51), while making the change of variable $r \to r + p' - p$ in the integration in the second term causes the second and third terms to cancel. This gives, when we sum over all type-(b) contributions,

$$(p-p')^{\mu} \Lambda_{\mu}^{5(b)}(p,p') = 2m_0 \Lambda^{5(b)}(p,p'). \qquad (53)$$

The Ward identity of Eq. (44) is finally obtained by adding Eqs. (50) and (53).

Clearly, the only step in the above derivation which is not simply an algebraic rearrangement is the change of integration variable in the second term of Eq. (52). This will be a valid operation provided that the integral is at worst superficially logarithmically divergent, a condition that is satisfied by loops with four or more photons, that is, loops with $n \geq 2$. However, when the loop is a triangle graph with only two photons emerging,

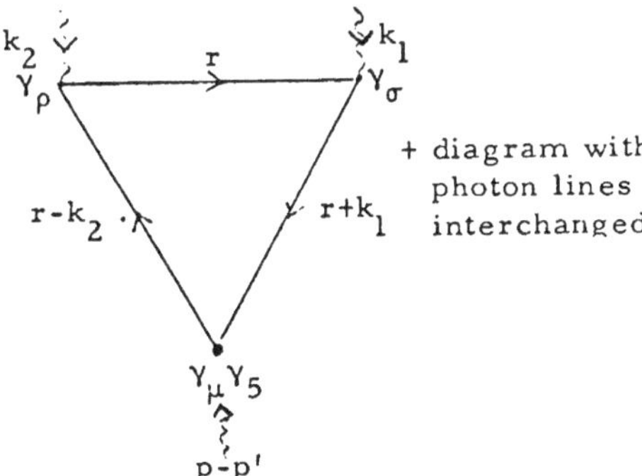

we have $n = 1$, and the integral in Eq. (52) appears to be quadratically divergent. Actually, since $\mathrm{tr}\{\gamma_5 \gamma^{(1)} \not{r}\, \gamma^{(2)} \not{r}\} = 0$, the integral in the $n = 1$ case is superficially linearly divergent. Since it is well known that translation of a linearly divergent Feynman integral is not necessarily a valid operation,[6] we suspect that Eq. (53) breaks down for the triangle graph.

To see that this really does happen, we make use of an explicit expression for the triangle graph calculated by Rosenberg.[7] The sum of the triangle illustrated above and the corresponding graph with the two photons interchanged is

$$\frac{-ie_0^2}{(2\pi)^4} R_{\sigma\rho\mu} \equiv 2\int \frac{d^4 r}{(2\pi)^4}(-1)\,\mathrm{Tr}\{\frac{i}{\not{r}+\not{k}_1-m_0}(-ie_0\gamma_\sigma) \times \frac{i}{\not{r}-m_0}(-ie_0\gamma_\rho)\frac{i}{\not{r}-\not{k}_2-m_0}\gamma_\mu\gamma_5\} . \tag{54}$$

This expression is linearly divergent, but as in the case of the photon self-energy part, current conservation requires that the photons couple through their field-strength tensors $k_2^\xi \varepsilon_2^\rho - k_2^\rho \varepsilon_2^\xi$, $k_1^\eta \varepsilon_1^\sigma - k_1^\sigma \varepsilon_1^\eta$, using up two powers of momentum and leaving a convergent integral with $D_{eff} = -1$. The simplest way to make use of current conservation in practice is to first write down the most general form for the axial-tensor $R_{\sigma\rho\mu}$, consistent with the requirements of parity and Lorentz invariance. A little thought shows that this is

$$\begin{aligned} R_{\sigma\rho\mu}(k_1, k_2) = & A_1 k_1^\tau \epsilon_{\tau\sigma\rho\mu} + A_2 k_2^\tau \epsilon_{\tau\sigma\rho\mu} \\ & + A_3 k_{1\rho} k_1^\xi k_2^\tau \epsilon_{\xi\tau\sigma\mu} + A_4 k_{2\rho} k_1^\xi k_2^\tau \epsilon_{\xi\tau\sigma\mu} \\ & + A_5 k_{1\sigma} k_1^\xi k_2^\tau \epsilon_{\xi\tau\rho\mu} + A_6 k_{2\sigma} k_1^\xi k_2^\tau \epsilon_{\xi\tau\rho\mu}, \end{aligned} \tag{55}$$

with the $A_j (j=1, \ldots, 6)$ Lorentz scalar functions of k_1 and k_2. The requirement of Bose symmetry, $R_{\sigma\rho\mu}(k_1, k_2) = R_{\rho\sigma\mu}(k_2, k_1)$ implies that

$$\begin{aligned} A_1(k_1, k_2) &= -A_2(k_2, k_1) \\ A_3(k_1, k_2) &= -A_6(k_2, k_1) \\ A_4(k_1, k_2) &= -A_5(k_2, k_1). \end{aligned} \tag{56}$$

Imposing the condition of current conservation, $k_1^\sigma R_{\sigma\rho\mu} = k_2^\rho R_{\sigma\rho\mu} = 0$ gives us relations between A_1, A_2 and the remaining A' s,

$$\begin{aligned} A_1 &= k_1 \cdot k_2 A_3 + k_2^2 A_4, \\ A_2 &= k_1^2 A_5 + k_1 \cdot k_2 A_6. \end{aligned} \tag{57}$$

Now $A_{3,4,5,6}$ each appear in Eq. (55) multiplied by three powers of external photon four-momentum, and therefore will each involve a highly convergent Feynman integral with $D_{\text{eff}} = 1-3 = -2$. On the other hand, the scalars A_1 and A_2 each multiply just one power of momentum, and therefore are represented by formally logarithmically divergent Feynman integrals with $D_{\text{eff}} = 1-1=0$. But current conservation saves the day, since it allows us to calculate A_1 and A_2 directly from the convergent quantities $A_{3,4,5,6}$. Introducing Feynman parameters and doing the r integration in the standard manner, we find

$$\begin{aligned} A_3(k_1,k_2) &= -16\pi^2 I_{11}(k_1,k_2), \\ A_4(k_1,k_2) &= 16\pi^2 [I_{20}(k_1,k_2)-I_{10}(k_1,k_2)], \end{aligned} \tag{58}$$

with

$$\begin{aligned} I_{st}(k_1,k_2) = \int_0^1 dx \int_0^{1-x} dy\, x^s y^t [\, y(1-y)k_1^2 \\ +x(1-x)k_2^2+2xyk_1\cdot k_2-m_0^2]^{-1}. \end{aligned} \tag{59}$$

In order to check the Ward identity, we will also need an expression for the triangle graph with $\gamma_\mu\gamma_5$ replaced by $2m_0\gamma_5$. Defining

$$\begin{aligned} \frac{-ie_0^2}{(2\pi)^4} 2m_0 R_{\sigma\rho} \equiv 2\int \frac{d^4r}{(2\pi)^4}(-1)\mathrm{Tr}\{\frac{i}{\not r+\not k_1-m_0}(-ie_0\gamma_\sigma) \\ \times \frac{i}{\not r-m_0}(-ie_0\gamma_\rho)\frac{i}{\not r-\not k_2-m_0} 2m_0\gamma_5\}, \end{aligned} \tag{60}$$

we find by straightforward calculation that

$$\begin{aligned} R_{\sigma\rho} &= k_1^\xi k_2^\tau \epsilon_{\xi\tau\sigma\rho} B_1 \\ B_1 &= 8\pi^2 m_0 I_{00}(k_1,k_2). \end{aligned} \tag{61}$$

We are now ready to calculate the divergence of the axial-vector triangle diagram. If the Ward identity holds, we should find

$$-(k_1+k_2)^\mu R_{\sigma\rho\mu} = 2m_0 R_{\sigma\rho}, \tag{62}$$

but from Eqs. (55) - (61) we find, instead,

$$-(k_1+k_2)^{\mu} R_{\sigma\rho\mu} = 2m_0 R_{\sigma\rho} + 8\pi^2 k_1^{\xi} k_2^{\tau} \epsilon_{\xi\tau\sigma\rho}. \tag{63}$$

We see that the <u>axial-vector Ward identity fails in the case of the triangle graph.</u> The failure is a result of the fact that the integration variable in a linearly divergent Feynman integral cannot be freely translated.

2.2 Impossibility of Eliminating the Anomaly by a Subtraction

The question now immediately arises, whether it is possible to redefine $R_{\sigma\rho\mu}$ by a subtraction or in some other manner, so as to eliminate the Ward identity anomaly of Eq. (63), but without introducing any new types of anomalous behavior? A subtraction term, in order to preserve the expected behavior of $R_{\sigma\rho\mu}$, must have the following properties: (i) It must be a three-index Lorentz axial-tensor. (ii) It must be symmetric under interchange of the photon variables (k_1, σ) and (k_2, ρ). (iii) It must be a polynomial in the momentum variables k_1 and k_2. This requirement follows from generalized unitarity, which says that discontinuities of $R_{\sigma\rho\mu}$ with respect to external variables are related to Feynman amplitudes for intermediate state processes by the Cutkosky rules. Since taking a discontinuity renders the Feynman integrals in Eq. (54) convergent, the discontinuities of $R_{\sigma\rho\mu}$ have no anomalies, and $R_{\sigma\rho\mu}$ satisfies generalized unitarity by itself. Thus the subtraction term must have vanishing discontinuities, i.e., it must be a polynomial. (iv) If we set $k_1 = \xi Q$, $k_2 = -\xi Q + \xi R + S$, with Q, R, S arbitrary, and let $\xi \to \infty$, the subtraction must diverge at worst as ξ times a power of $\ln \xi$. This requirement follows from Weinberg's theorem,[8] which states that when the external momenta of a Feynman graph approach infinity as above, the largest power appearing is $\xi^{D_{MAX}}$, where D_{MAX} is the maximum of the superficial divergences of the graph and of its subgraphs. For the triangle, $D_{MAX} = 1$. Since $R_{\sigma\rho\mu}$ already has asymptotic behavior consistent with Weinberg's theorem, so must the subtraction. We note for future reference that when R = 0,

$$R_{\sigma\rho\mu}(k_1 = \xi Q,\ k_2 = -\xi Q + p' - p) \to -8\pi^2 \xi Q^\tau \epsilon_{\tau\sigma\rho\mu} + O(\ln \xi). \quad (64)$$

(v) The subtraction must have the dimensionality of a mass. (vi) The subtraction must satisfy the requirements of vector current conservation.

It is easy to see that it is in fact impossible to find a subtraction satisfying these six conditions. The first five conditions are only satisfied by a term of the form

$\epsilon_{\tau\sigma\rho\mu}(k_1-k_2)^\tau$, but this term does not satisfy condition (vi)! Thus, while it is possible to define a subtracted triangle diagram $R'_{\sigma\rho\mu}$ which has a normal axial-vector Ward identity,

$$R'_{\sigma\rho\mu} = R_{\sigma\rho\mu} + 4\pi^2 \epsilon_{\tau\sigma\rho\mu}(k_1-k_2)^\tau ,$$
$$-(k_1+k_2)^\mu R'_{\sigma\rho\mu} = 2m_0 R_{\sigma\rho}, \tag{65}$$

it violates vector current conservation,

$$k_1^\sigma R'_{\sigma\rho\mu} = -4\pi^2 k_1^\sigma k_2^\tau \epsilon_{\tau\sigma\rho\mu},$$
$$k_2^\rho R'_{\sigma\rho\mu} = 4\pi^2 k_2^\rho k_1^\tau \epsilon_{\tau\sigma\rho\mu} . \tag{66}$$

All we have succeeded in doing is substituting one divergence anomaly for another. Similarly, if we introduce projection operators in order to force all divergences to have the correct values, as in

$$R''_{\sigma\rho\mu} = (g_{\mu\nu} - \frac{q_\mu q_\nu}{q^2})R_{\sigma\rho}{}^{\nu} + \frac{q_\mu}{q^2} 2m_0 R_{\sigma\rho} , q=-(k_1+k_2),$$
$$R'''_{\sigma\rho\mu} = (g_{\sigma\xi} - \frac{k_{1\sigma}k_{1\xi}}{k_1^2})(g_{\rho\eta} - \frac{k_{2\rho}k_{2\eta}}{k_2^2})R^{\xi\eta}{}_{\mu}{}', \tag{67}$$

we introduce spurious kinematic singularities, in violation of condition (iii). We see that the situation is like the proverbial square peg being inserted in a round hole--we can fix the triangle diagram in one respect only by making it anomalous in some other respect. The anomaly really consists, not of Eq. (63) by itself, but <u>rather of the impossibility of finding a redefined triangle diagram which simul-</u>

taneously satisfies all of the six requirements above. Failure to appreciate this chameleon-like quality of the anomaly has resulted in erroneous claims in the literature that the anomaly can be eliminated.

In the remainder of these lectures, we will always use the expression $R_{\sigma\rho\mu}$ for the triangle, rather than the subtracted expression $R'_{\sigma\rho\mu}$ which has normal axial-vector but anomalous vector Ward identities. Since, *a priori*, it would appear that the vector Ward identity is no more sacred than the axial-vector Ward identity, the choice requires some words of justification. We note, first of all, that enforcing vector current conservation is essential if we want the triangle to describe the physical coupling between a neutrino-antineutrino pair and two photons. The reason is that since two photons can never be in a state with $J = 1$, the coupling of the $J=1$ state of the $\nu\bar{\nu}$ pair to two photons must vanish. Expressed in terms of $R_{\sigma\rho\mu}$, this requirement states that if ℓ_μ is an arbitrary spin-one polarization vector satisfying $\ell\cdot(k_1+k_2)=0$, and if (ϵ_1, k_1) and (ϵ_2, k_2) are photon variables satisfying $\epsilon_1\cdot k_1 = \epsilon_2\cdot k_2 = k_1^2 = k_2^2 = 0$, we must have $\ell^\mu \epsilon_1^\sigma \epsilon_2^\rho R_{\sigma\rho\mu} \equiv 0$. It has been shown by Rosenberg[7] that Eqs. (55) - (58) do satisfy this condition. On the other hand, the subtraction term in

Eq. (65) does not satisfy this condition, and hence $R'_{\sigma\rho\mu}$ does not, and so cannot describe a physical $\nu\nu$-pair--two photon coupling. Secondly, we will see below that the relations between the Ward identity anomaly and commutator anomalies take a particularly simple form when vector current conservation (and hence gauge invariance) are maintained. Finally, the most interesting application of the triangle anomaly, the derivation of a low energy theorem for π^0 decay, is independent of which definition of the triangle is used. So again, it proves convenient (although not essential in this case) to maintain gauge invariance.

2.3 Anomaly for General Axial-Vector Current Matrix Element

Let us now return to the diagrammatic analysis which we left off at Eq. (63). Clearly, the breakdown of the Ward identity for the basic triangle will also cause failure of the Ward identity for any graph of the type illustrated below, in which the two photon lines coming out of the triangle graph join onto a "blob" from which 2F fermion and B boson lines emerge.

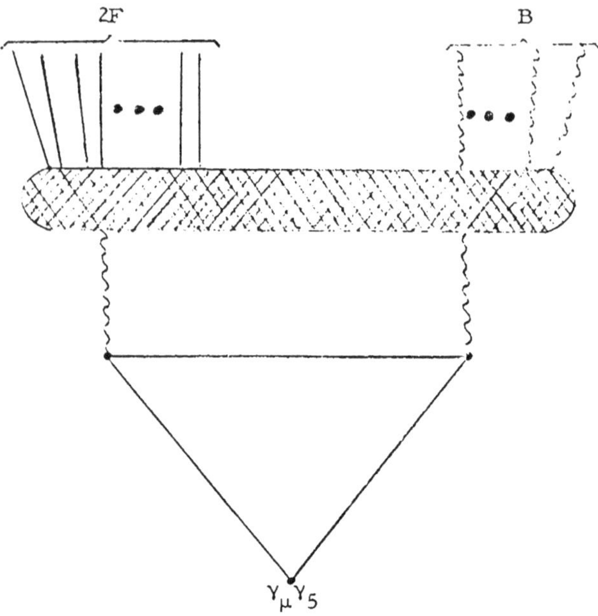

From Eq. (63) for the divergence of the basic triangle graph, it is possible to show that the breakdown of the axial-vector Ward identity in the general case is simply described by replacing Eq. (41) for the axial-vector-current divergence (which we have shown to be incorrect) by[9]

$$\frac{\partial}{\partial x_\mu} j_\mu^5(x) = 2im_0 j^5(x) + \frac{\alpha_0}{4\pi} F^{\xi\sigma}(x)\, F^{\tau\rho}(x)\epsilon_{\xi\sigma\tau\rho}. \qquad (68)$$

Eq. (68) is easily verified by using the following Feynman rules for the vertices of j_μ^5, j^5 and $(\alpha_0/4\pi)\, F^{\xi\sigma}\, F^{\tau\rho}\epsilon_{\xi\sigma\tau\rho}$,

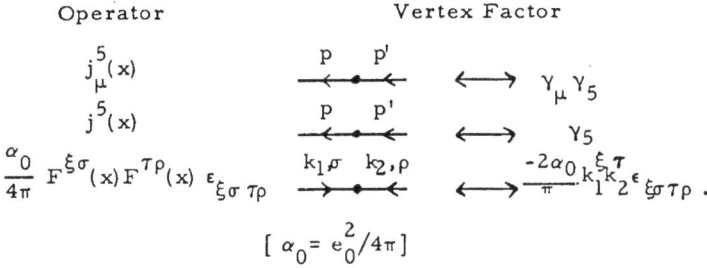

Operator	Vertex Factor		
$j^5_\mu(x)$	$p \quad p'$	$\longleftrightarrow$	$\gamma_\mu\gamma_5$
$j^5(x)$	$p \quad p'$	$\longleftrightarrow$	γ_5
$\frac{\alpha_0}{4\pi} F^{\xi\sigma}(x) F^{\tau\rho}(x)\, \epsilon_{\xi\sigma\tau\rho}$	$k_1\sigma \quad k_2,\rho$	$\longleftrightarrow$	$\frac{-2\alpha_0}{\pi} k_1^\xi k_2^\tau \epsilon_{\xi\sigma\tau\rho}$.

$$[\alpha_0 = e_0^2/4\pi]$$

Using Eq. (68) we can easily see how the Ward identity for the axial-vector vertex is modified. Defining $\overline{F}(p,p')$ by

$$S_F'(p)\,\overline{F}(p,p')S_F'(p') = -\int d^4x\, d^4y\, e^{ip\cdot x} e^{-ip'\cdot y} \times \langle 0|\, T(\psi(x)\, F^{\xi\sigma}(0)\, F^{\tau\rho}(0)\, \epsilon_{\xi\sigma\tau\rho}\overline{\psi}(y))\,|0\rangle, \tag{69}$$

then we find

$$(p-p')^\mu\, \Gamma^5_\mu(p,p') = 2m_0\Gamma^5(p,p') + S_F'(p)^{-1}\gamma_5 + \gamma_5 S_F'(p')^{-1} - i(\alpha_0/4\pi)\,\overline{F}(p,p'), \tag{70}$$

which replaces Eq. (44).

2.4 Coordinate Space Calculation

So far we have worked exclusively in momentum space. However, the fact that Eq. (68) shows the anomaly to have a simple form in coordinate space suggests that a coordinate space derivation should be possible. To proceed in coordinate space, let us confine ourselves to the case of a c-number electromagnetic field and let us regard the axial-vector current j^5_μ as the limit of a nonlocal current in which the fields $\overline{\psi}$ and ψ are evaluated at separated space-time points,

$$j^5_\mu(x) = \lim_{\varepsilon\to 0} j^5_\mu(x,\varepsilon), \qquad (71a)$$

$$j^5_\mu(x,\varepsilon) = \overline{\psi}(x+\tfrac{\varepsilon}{2})\gamma_\mu\gamma_5\psi(x-\tfrac{\varepsilon}{2})\exp[-ie_0\int_{x-\frac{\varepsilon}{2}}^{x+\frac{\varepsilon}{2}} d\ell\cdot A(\ell)]. \qquad (71b)$$

The line integral in Eq. (71b) is necessary in order to insure invariance of $j^5_\mu(x,\varepsilon)$ under the gauge transformation

$$\psi(x) \to e^{-ie_0 v(x)}\psi(x),$$

$$A^\mu(x) \to A^\mu(x) + \frac{\partial v(x)}{\partial x_\mu}. \qquad (72)$$

Expanding the exponential out to first order in ε and then using the equations of motion to calculate the divergence gives

$$j^5_\mu(x,\varepsilon) = \overline{\psi}(x+\tfrac{\varepsilon}{2})\gamma_\mu\gamma_5\psi(x-\tfrac{\varepsilon}{2})[1-ie_0\varepsilon_\lambda A^\lambda(x)] + \text{higher order},$$

$$\frac{\partial}{\partial x_\mu} j^5_\mu(x,\varepsilon) = \overline{\psi}(x+\tfrac{\varepsilon}{2})\gamma_\mu\gamma_5\psi(x-\tfrac{\varepsilon}{2})\{-ie_0\varepsilon_\lambda\frac{\partial}{\partial x_\mu}A^\lambda(x)$$

$$\underbrace{-ie_0[A^\mu(x-\tfrac{\varepsilon}{2})-A^\mu(x+\tfrac{\varepsilon}{2})]}_{ie_0\varepsilon_\lambda\frac{\partial}{\partial x_\lambda}A^\mu(x)}\} + 2im_0 j^5(x,\varepsilon) + \text{higher order}$$

$$= j^5_\mu(x,\varepsilon)\, ie_0\varepsilon_\lambda F^{\mu\lambda} + 2im_0 j^5(x,\varepsilon) + \text{higher order}. \qquad (73)$$

Taking the vacuum expectation of Eq. (73), we get the divergence equation for the generating functional describing the coupling, through a single closed loop, of the axial-vector current to an arbitrary number of external c-number photons,

$$\frac{\partial}{\partial x_\mu} \langle 0 \,|\, j^5_\mu(x,\varepsilon)\,|\, 0\rangle \qquad (74)$$

$$= ie_0\langle 0|j^5_\mu(x,\varepsilon)\,\varepsilon_\lambda|0\rangle F^{\mu\lambda} + 2im_0\langle 0|j^5(x,\varepsilon)|0\rangle + \text{higher order}.$$

The first term on the right hand side of Eq. (74) is formally of order ε, and is neglected in the naive derivation of Eq. (41). A careful calculation shows, however, that $<0|j^5_\mu(x,\varepsilon)|0>$ is of order ε^{-1} as $\varepsilon \to 0$, so that in fact the first term makes a finite contribution as $\varepsilon \to 0$. Carrying out the details[10] gives

$$\begin{aligned} & ie_0 <0|j^5_\mu(x,\varepsilon)\,\varepsilon_\lambda|0> F^{\mu\lambda}(x) \\ & = \frac{\alpha_0}{4\pi}\varepsilon_{\mu\lambda\xi\eta} F^{\mu\lambda}(x)\, F^{\xi\eta}(x) \;+\; O(\varepsilon), \end{aligned} \tag{75}$$

in agreement with the vacuum expectation of Eq. (68). Thus, the anomaly can be obtained by the equation of motion approach, provided that one is careful in handling the singular operator products which appear in the axial-vector current.

3. CONSEQUENCES OF THE TRIANGLE ANOMALY

Let us now examine some of the consequences of the anomalous axial-vector divergence which we found in the previous section. We will see that the anomaly produces changes in certain standard results having to do with renormalization of the axial-vector vertex and with γ_5-symmetry. We will also find that Eq. (68) leads to a low energy theorem for the vacuum to two photon matrix element of the naive divergence $2im_0 j^5$, the generalizations of which have interesting physical implications in π^0 decay.

3.1 Renormalization of the Axial-Vector Vertex

Let us begin with an analysis of the behavior of the axial-vector vertex under renormalization. As we recall, according to Eq. (30) both the axial-vector vertex and the pseudoscalar vertex are multiplicatively renormalizable, with respective renormalization constants Z_A and Z_D. Let us now ask whether these renormalization constants are the same as, or are simply related to, the electron-wave function renormalization Z_3. We will first find the answer which follows from the naive axial-vector Ward identity of Eq. (44), and then see how it is changed when the anomaly is taken into account, as in Eq.(70).

In order to talk in a precise way about the infinite renormalization constants m_0, Z_2, Z_A and Z_D, we will follow the standard procedure of introducing a cutoff Λ into our Feynman rules, so that the renormalization constants become finite functions of Λ which diverge as $\Lambda \to \infty$. There are many different ways of introducing a cutoff which accomplish this. [One particular way is specified in detail in the next section.] As long as we deal only with low-energy theorem type questions, in which all external momenta remain small compared with the cutoff, the precise details of how the cutoff is introduced are irrelevant. In particular, no ambiguities in order of limit are involved in a calculation in which external momenta are allowed to approach zero while the cutoff approaches infinity. On the other hand, we will see that in Bjorken limit calculations, in which external momenta approach infinity, the question of whether the external momenta remain much smaller than the cutoff, or become much larger than the cutoff, as both approach infinity, becomes of crucial importance.

To proceed, we start from the naive axial-vector Ward identity of Eq. (44) and set $p = p'$, so that the axial-vector vertex term on the left-hand side vanishes, and then multiply through by the electron wave function renormalization Z_2. This gives

$$2m_0 Z_2 \Gamma^5(p,p) = -[Z_2 S'_F(p)^{-1}\gamma_5 + \gamma_5 Z_2 S_F(p)^{-1}] \quad (76)$$
$$= -[\tilde{S}'_F(p)^{-1}\gamma_5 + \gamma_5 \tilde{S}'_F(p)^{-1}],$$

and since the right-hand side of Eq. (76) is finite (i.e. Λ-independent in the limit of large Λ) , we see that the left hand side, $2m_0 Z_2 \Gamma^5(p,p)$, is also finite. In Section 1 we saw that, for general p and p', $\Gamma^5(p,p')$ is always made finite by multiplication by a renormalization constant Z_D. Hence we conclude that Z_D and $2m_0 Z_2$ are the same, up to a finite factor,

$$2m_0 Z_2 / Z_D = \text{finite}. \quad (77)$$

Next, we substitute into the naive axial-vector Ward identity the expressions of Eqs. (16) and (30) for the renormalized electron propagator, axial-vector vertex and pseudoscalar vertex, and multiply through by Z_A, giving

$$(p-p')^\mu \tilde{\Gamma}^5_\mu(p,p') = \frac{Z_A}{Z_2}\left[\left(\frac{2m_0 Z_2}{Z_D}\right)\tilde{\Gamma}^5(p,p') + \tilde{S}'_F(p)^{-1}\gamma_5 + \gamma_5 \tilde{S}'_F(p')^{-1}\right] . \quad (78)$$

Let us now differentiate with respect to the cutoff Λ. The tilde quantities, by construction, are Λ-independent in the limit of large Λ, as is the ratio of renormalization constants in Eq. (77) and hence the entire square bracket in Eq. (78) . So we get simply

$$0=\frac{\partial}{\partial\Lambda}\left(\frac{Z_A}{Z_2}\right)\left[\left(\frac{2m_0Z_2}{Z_D}\right)\tilde{\Gamma}^5(p,p')+\tilde{S}_F'(p)^{-1}\gamma_5+\gamma_5\tilde{S}_F'(p')^{-1}\right], \quad (79a)$$

which implies that

$$0=\frac{\partial}{\partial\Lambda}\left(\frac{Z_A}{Z_2}\right),$$
$$Z_A/Z_2=\text{finite}. \quad (80)$$

Eqs. (77) and (80) tell us that, up to arbitrary finite factors, the axial-vector and pseudoscalar vertex renormalizations are just Z_2 and $2m_0Z_2$, respectively.

When we replace the incorrect, naive Ward identity of Eq. (44) by the corrected Ward identity of Eq. (70), part of this conclusion must be modified. Referring back to the Feynman rules for the vertex of $F^{\xi\sigma}F^{\tau\rho}\epsilon_{\xi\sigma\tau\rho}$, we see that when there is no net momentum transfer into the vertex, so that $k_1=-k_2$, the antisymmetric tensor factor $k_1^\xi k_2^\tau \epsilon_{\xi\sigma\tau\rho}$ vanishes. Consequently, when $p = p'$, the additional term $\bar{F}(p,p')$ in Eq. (70) vanishes, and so Eqs. (76) and (77) are still valid. That is, even in the presence of the triangle anomaly, $2m_0Z_2\Gamma^5(p,p')$ is still finite.

On the other hand, the presence of the term $\bar{F}$ in Eq. (70) changes Eq. (79a) to read

$$0=\frac{\partial}{\partial\Lambda}\left(\frac{Z_A}{Z_2}\right)\left[\left(\frac{2m_0Z_2}{Z_D}\right)\tilde{\Gamma}^5(p,p')+\tilde{S}_F'(p)^{-1}\gamma_5+\gamma_5\tilde{S}_F'(p')^{-1}\right]$$
$$+\frac{\partial}{\partial\Lambda}\left[\left(\frac{-i\alpha_0}{4\pi}\right)Z_A\bar{F}(p,p')\right]=0\ . \quad (79b)$$

The presence of the extra term proportional to $\overline{F}$ in Eq. (79b) prevents us from drawing our previous conclusion of Eq. (80), that Z_A/Z_2 or $Z_2\Gamma_\mu^5(p,p')$ are finite. We expect that even after multiplication by Z_2, there will still be divergent terms in the axial-vector vertex. Such terms first appear in order α_0^2 of perturbation theory, as a result of the diagram

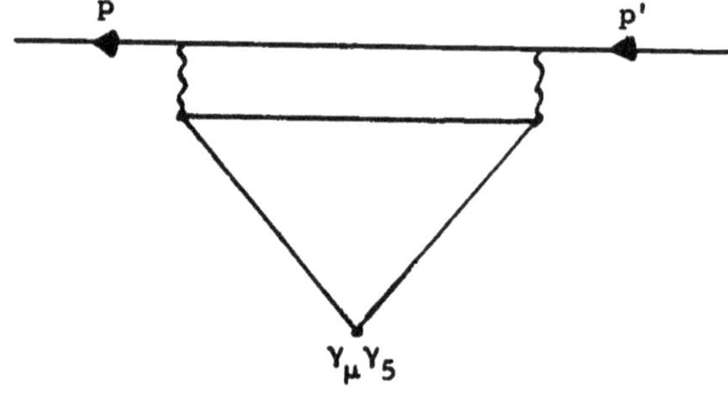

which, by use of Eq. (64), is easily seen to be logarithmically divergent. In heuristic terms, this divergence is not removed by multiplication by Z_2 because Z_2 is obtained from the theory with only vector currents present, and does not "know" about the existence of the axial-vector triangle anomaly. Introducing a cutoff by replacing the photon propagator $-ig_{\mu\nu}(q^2+i\epsilon)^{-1}$ by $-ig_{\mu\nu}[(q^2+i\epsilon)^{-1}-(q^2-\Lambda^2+i\epsilon)^{-1}]$, we find that

$$Z_2\,\Gamma^5_\mu(p,p') = \gamma_\mu\gamma_5[1-\tfrac{3}{4}(\alpha_0/\pi)^2\,\ell n(\Lambda^2/m^2)] + \alpha_0\times\text{finite} + \alpha_0^2\times\text{finite} + O(\alpha_0^3), \tag{81}$$

or equivalently (up to an unspecified finite factor)

$$Z_A = Z_2[1+\tfrac{3}{4}(\alpha_0/\pi)^2\ell n(\Lambda^2/(m^2)+O(\alpha_0^3)]. \tag{82}$$

3.2 Radiative Corrections to $\nu_\ell\,\ell$ Scattering

As an application of Eq. (81), let us consider the radiative corrections to $\nu_\ell\,\ell$ scattering, where ℓ is a μ or an e. As we saw in Eq. (40), after Fierz transformation the terms in the local current-current Lagrangian which describe $\nu_\ell\ell$ scattering become

$$(G/\sqrt{2})[\bar{\mu}\gamma_\lambda(1-\gamma_5)\mu\,\bar{\nu}_\mu\gamma^\lambda(1-\gamma_5)\nu_\mu + \bar{e}\gamma_\lambda(1-\gamma_5)\,e\,\bar{\nu}_e\gamma^\lambda(1-\gamma_5)\nu_e] \tag{83}$$

The radiative corrections to Eq. (83) are simply obtained by calculating the radiative corrections to the charged-lepton currents $\bar{\mu}\gamma_\lambda(1-\gamma_5)\mu$ and $\bar{e}\gamma_\lambda(1-\gamma_5)e$, without any reference to the neutrino currents. Application of our Feynman rules shows that the effect of radiative corrections is to replace the matrix elements $\bar{u}_{(\mu)}\gamma_\lambda(1-\gamma_5)u_{(\mu)}$, $\bar{u}_{(e)}\gamma_\lambda(1-\gamma_5)u_{(e)}$ by

$$\begin{aligned}&\bar{u}_{(\mu)}Z_2^{(\mu)}[\Gamma_\lambda^{(\mu)}-\Gamma_\lambda^{5(\mu)}]u_{(\mu)},\\&\bar{u}_{(e)}Z_2^{(e)}[\Gamma_\lambda^{(e)}-\Gamma_\lambda^{5(e)}]u_{(e)},\end{aligned} \tag{84}$$

with $\Gamma_\lambda^{(\mu,e)}$ and $\Gamma_\lambda^{5(\mu,e)}$ the proper vector and axial-vector vertices and with $Z_2^{(\mu,e)}$ the wave-function renormalization factors coming from item (iii) of the Feynman rules. From the usual vector-current Ward identity, we know that $Z_2^{(\mu)}\Gamma_\lambda^{(\mu)}$ and $Z_2^{(e)}\Gamma_\lambda^{(e)}$ are finite. On the other hand, Eq. (81) tells us that

$$\begin{aligned}Z_2^{(\mu,e)}\Gamma_\lambda^{5(\mu,e)}=&\gamma_\lambda\gamma_5[1-\tfrac{3}{4}(\alpha_0/\pi)^2\ln(\Lambda^2/m^2)]\\&+\alpha_0\times\text{finite}+\alpha_0^2\times\text{finite}+O(\alpha_0^3),\end{aligned} \tag{85}$$

which means that, on account of the presence of axial-vector triangle diagrams, <u>the radiative corrections to $\nu_e e$ and $\nu_\mu\mu$ scattering diverge in the fourth order of perturbation theory</u>. This result contrasts sharply with the fact that the radiative corrections to muon decay or to the scattering reaction $\nu_\mu + e \to \nu_e + \mu$ are finite to all orders in perturbation theory.[11] The crucial difference between the two cases, of course, is that because of separate muon and electron-number conservation, the current $\bar{\mu}\gamma_\lambda(1-\gamma_5)e$ cannot couple into closed electron or muon loops, and thus the troublesome triangle diagram is not present.

Two points of view can be taken towards the divergent radiative corrections in $\nu_\ell \ell$ scattering. One viewpoint is that we know, in any case, that the local current-current theory of leptonic weak interactions cannot be correct, since this theory leads at high energies to non-unitary matrix elements, and since it gives divergent results for higher-order weak-interaction effects.[12] Thus, it is entirely possible that the modifications in Eq. (83) necessary to give a satisfactory weak-interaction theory will also cure the disease of infinite radiative corrections in $\nu_\ell \ell$ scattering. The other viewpoint is that we should try to make the radiative corrections to $\nu_\ell \ell$ scattering finite, within the framework of a local weak-interaction theory. It turns out that this is possible, if we introduce $\nu_e \mu$ and $\nu_\mu e$ scattering terms into the effective Lagrangian so that Eq. (83) is replaced by

$$(G/\sqrt{2})[\bar{\mu}\gamma_\lambda(1-\gamma_5)\mu - \bar{e}\gamma_\lambda(1-\gamma_5)e] \times [\bar{\nu}_\mu\gamma^\lambda(1-\gamma_5)\nu_\mu - \bar{\nu}_e\gamma^\lambda(1-\gamma_5)\nu_e] . \tag{86}$$

This works because the troublesome extra term in Eq. (68) is independent of the bare mass m_0, so that it cancels between the muon and electron terms in Eq. (86), giving

$$\frac{\partial}{\partial x_\lambda}[\bar{\mu}\gamma_\lambda\gamma_5\mu - \bar{e}\gamma_\lambda\gamma_5 e] = 2im_0^{(\mu)}\bar{\mu}\gamma_5\mu - 2im_0^{(e)}\bar{e}\gamma_5 e \ . \qquad (87)$$

Application of the argument of Eqs. (76)-(79) then shows that the radiative corrections to Eq. (87) are finite. What has happened is that the e-triangle and μ-triangle contributions to the total $\nu_e e$ scattering amplitude contribute with opposite sign and regulate each other,

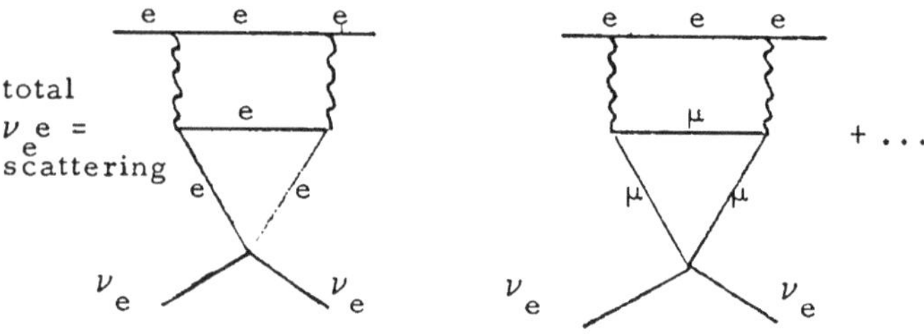

Experimentally, it will be possible to distinguish Eq. (86) from Eq. (83) by looking for elastic scattering of muon neutrinos from electrons; the present upper bound is still consistent with Eq. (86), but is getting very close.[13]

3.3 Connection Between γ_5 Invariance and a Conserved Axial-Vector Current in Massless Electrodynamics

Next, let us discuss the effects of the axial-vector triangle diagram in the case of massless spinor electrodynamics [Eq. (1) with $m_0=0$]. We will find that the triangle diagram leads to a breakdown of the usual connection between symmetries of the Lagrangian and conserved currents. As in our previous discussions, we begin by describing the standard theory, which holds in the absence of singular phenomena. Let $\{\Phi(x)\} = \{\Phi_1(x), \Phi_2(x), \ldots\}$ and $\{\partial_\lambda \Phi\}$ be a set of canonical fields and their space-time derivatives, and let us consider the field theory described by the Lagrangian density

$$\mathcal{L}(x) \equiv \mathcal{L}[\{\Phi\},\{\partial_\lambda \Phi\}]. \tag{88}$$

To establish the connection between invariance properties of $\mathcal{L}$ and conserved currents, we make the infinitesimal, local gauge transformation on the fields,

$$\Phi_j(x) \rightarrow \Phi_j(x) + v(x) G_j[\{\Phi(x)\}], \tag{89}$$

and define the associated current J^α by

$$J^\alpha = -\delta \mathcal{L} / \delta(\partial_\alpha v). \tag{90}$$

Then, by using the Euler-Lagrange equations of motion of the fields, we easily find[14] that the divergence of the current is given by

$$\partial_\alpha J^\alpha = -\delta \mathcal{L} / \delta v . \tag{91}$$

In particular, if the gauge transformation of Eq. (89) with <u>constant gauge</u> function v, leaves the Lagrangian invariant, then $\delta \mathcal{L}/\delta v = 0$ and the current J^α is conserved. Thus, to any continuous invariance transformation of the

Lagrangian there is associated a conserved current. It is also easily verified that the charge $Q(t) = \int d^3x\, J^0(x,t)$ associated with the current J^α has the properties

$$dQ(t)/dt=0, \tag{92a}$$

$$[\,Q, \Phi_j(x)\,] = iG_j(x). \tag{92b}$$

Equation (92b) states that Q is the generator of the gauge transformation in Eq. (89), for constant v.

Let us now specialize to the case of massless electrodynamics, with Eq. (89) the gauge transformation

$$\psi(x) \rightarrow [1 + i\gamma_5 v(x)]\psi(x). \tag{93}$$

When v is a constant and $m_0=0$, this transformation leaves the Lagrangian of Eq. (1) invariant, so that according to Eq. (91), the associated current J^α should be conserved. But calculating J^α, we find

$$J^\alpha = -\delta\mathcal{L}/\delta(\partial_\alpha v) = \bar{\psi}\gamma^\alpha\gamma_5\psi\,, \tag{94}$$

which according to Eq. (68) has the divergence

$$\partial_\alpha J^\alpha = (\alpha_0/4\pi) F^{\xi\sigma}(x) F^{\tau\rho}(x) \epsilon_{\xi\sigma\tau\rho}. \tag{95}$$

Thus, Eq. (91), which was obtained by formal calculation using the equations of motion, breaks down in this case. We see that because of the presence of the axial-vector triangle diagram, even though the Lagrangian (and all orders of perturbation theory) of massless electrodynamics are γ_5

invariant, the axial-vector current associated with the γ_5 transformation is not conserved.

However, it is amusing that even though there is no conserved current connected with the γ_5 transformation, there is still a generator $\bar{Q}^5$ with the properties of Eq. (92). To see this, let us consider the quantity $\bar{j}^5$ defined by

$$\bar{j}^5_\mu(x) = j^5_\mu(x) - \frac{\alpha_0}{\pi} A^\xi(x) \frac{A^\tau(x)}{\partial x_\rho} \epsilon_{\xi\mu\tau\rho}; \tag{96}$$

referring to Eq. (68), we see that

$$\frac{\partial}{\partial x_\mu} \bar{j}^5_\mu(x) = 0. \tag{97}$$

Although $\bar{j}^5_\mu$ is conserved, it is explicitly gauge-dependent and therefore is not an observable current operator. But the associated charge

$$\bar{Q}^5 = \int d^3x\, \bar{j}^5_0(x) = \int d^3x [\psi^\dagger(x)\gamma_5\psi(x) + \frac{\alpha_0}{\pi} \underset{\sim}{A} \cdot \underset{\sim}{\nabla} \times \underset{\sim}{A}] \tag{98}$$

is gauge invariant and therefore observable. According to Eq. (97), $\bar{Q}^5$ is time-independent, and its commutator with $\psi(x)$ (calculated formally by use of the canonical commutation relations) is

$$[\bar{Q}^5, \psi(x)] = -\gamma_5\psi(x) = i[i\gamma_5\psi(x)]. \tag{99}$$

Also, as we will see below, because of an implicit photon field dependence of j^5_0 implied by Eq. (68), $\bar{Q}^5$ does commute with all the photon field variables. Thus, $\bar{Q}^5$ is the conserved generator of the γ_5 transformations.

3.4 Low Energy Theorem for $2im_0 j^5(x)$

Finally, we will show that the anomalous axial-vector divergence equation, Eq. (68), leads to an interesting low energy theorem for the vacuum to two photon matrix element of the naive divergence $2im_0 j^5$. First let us note that we have derived Eq. (68) by considering the triangle without radiative corrections, but have omitted the contributions of diagrams such as the ones shown:

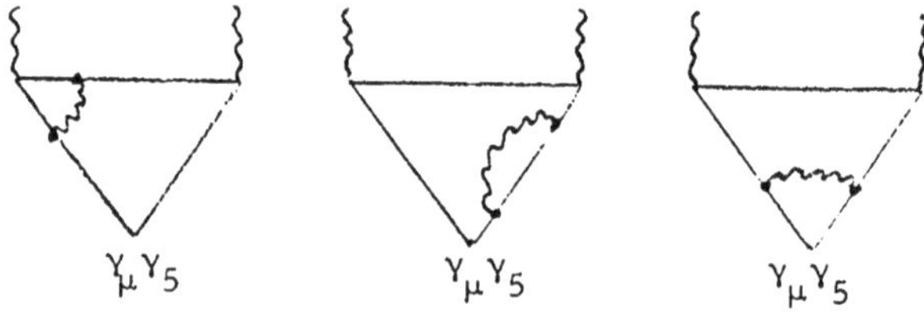

These diagrams are also linearly divergent and hence may also have divergence anomalies of their own. Since the anomalous terms must be Lorentz pseudoscalars satisfying conditions analagous to the six conditions on possible subtraction terms listed in Subsection 2.2, one easily sees that they must have the same form as the lowest order triangle anomaly in Eq. (68). We take into account the possibility of divergence anomalies coming from radiative corrections to the triangle diagram by replacing Eq. (68) by

$$\frac{\partial}{\partial x_\mu} j_\mu^5(x) = 2im_0 j^5(x) + \frac{\alpha_0}{4\pi}(1+C)F^{\xi\sigma}F^{\tau\rho}\epsilon_{\xi\sigma\tau\rho}. \quad (100)$$

We will use Eq. (100) as the basis of our derivation of the low-energy theorem.

To proceed, we take the matrix element of Eq. (68) between the vacuum $|0\rangle$ and the two photon state $\langle\gamma(k_1,\varepsilon_1)\ \gamma(k_2,\varepsilon_2)|$. Since the only pseudoscalar which can be formed from the four-momenta k_1, k_2 and the polarizations ε_1^*, ε_2^* of the two photons is $k_1^{\xi}\ k_2^{\tau}\ \varepsilon_1^{*\sigma}\varepsilon_2^{*\rho}\varepsilon_{\xi\tau\sigma\rho}$, the matrix element of each term in Eq. (100) contains this expression as a factor,

$$\langle\gamma(k_1,\varepsilon_1)\gamma(k_2,\varepsilon_2)|\begin{Bmatrix} \partial\mu j_\mu^5 \\ 2im_0 j^5 \\ \frac{\alpha_0}{4\pi} F^{\xi\sigma}\ F^{\tau\rho}\ \varepsilon_{\xi\sigma\tau\rho} \end{Bmatrix}|0\rangle \qquad (101)$$

$$= (4k_{10}k_{20})^{-\frac{1}{2}} k_1^{\xi} k_2^{\tau}\ \varepsilon_1^{*\sigma}\ \varepsilon_2^{*\rho}\ \varepsilon_{\xi\tau\sigma\rho}\begin{Bmatrix} F(k_1\cdot k_2) \\ G(k_1\cdot k_2) \\ H(k_1\cdot k_2) \end{Bmatrix}$$

The matrix element of Eq. (100) can be rewritten in terms of the amplitudes F, G, H as

$$F(k_1\cdot k_2) = G(k_1\cdot k_2) + (1+C)\ H(k_1\cdot k_2). \qquad (102)$$

To derive the low energy theorem from Eq. (102), we make use of a remarkable kinematic property of the matrix element $\mathcal{M}_\mu = (4k_{10}k_{20})^{\frac{1}{2}}\langle\gamma(k_1,\varepsilon_1)\gamma(k_2,\varepsilon_2)|j_\mu^5|0\rangle$. As we have noted in the previous section, the requirements of Lorentz invariance, gauge invariance and Bose statistics require this matrix element to have the general form

$$\mathcal{M}_\mu = \varepsilon_1^{*\rho}\,\varepsilon_2^{*\sigma}[\,C_1 k_1^\tau \varepsilon_{\tau\sigma\rho\mu} + C_2 k_2^\tau \varepsilon_{\tau\sigma\rho\mu}$$
$$+C_3 k_{1\rho} k_1^\xi k_2^\tau \varepsilon_{\xi\tau\sigma\mu} + C_4 k_{2\rho} k_1^\xi k_2^\tau \varepsilon_{\xi\tau\sigma\mu} \qquad (103)$$
$$+C_5 k_{1\sigma} k_1^\xi k_2^\tau \varepsilon_{\xi\tau\rho\mu} + C_6 k_{2\sigma} k_1^\xi k_2^\tau \varepsilon_{\xi\tau\rho\mu}]\,,$$

with

$$C_1 = k_1\cdot k_2 C_3 + k_2^2 C_4,$$
$$C_2 = k_1^2 C_5 + k_1\cdot k_2 C_6, \qquad (104)$$
$$C_3 \quad (k_1, k_2) = -C_6(k_2, k_1),$$
$$C_4 \quad (k_1, k_2) = -C_5(k_2, k_1).$$

The matrix element of the divergence of the axial-vector current is proportional to $(k_1+k_2)^\mu \mathcal{M}_\mu$. Using the algebraic identity satisfied by the six four-vectors $a, \ldots, f$,

$$(af)\,|bcde| + (bf)\,|cdea| + (cf)\,|deab|$$
$$+ (df)\;|eabc| + (ef)\,|abcd| = 0, \qquad (105)$$
$$(af) \equiv a\cdot f, \quad |abcd| \equiv a^\xi b^\tau c^\sigma d^\eta\; \varepsilon_{\xi\tau\sigma\eta},$$

with $a=f=k_1$, $b=k_2$, $c=k_1+k_2$, $d=e_1^*$ and $e=e_2^*$, we find that $(k_1+k_2)^\mu \mathcal{M}_\mu$ can be rearranged into the form

$$(k_1+k_2)^\mu \mathcal{M}_\mu = [\,C_3 - C_6\,]\, k_1\cdot k_2\; k_1^\xi\, k_2^\tau\, \varepsilon_1^{*\sigma}\; \varepsilon_2^{*\rho} \varepsilon_{\xi\tau\sigma\rho}. \qquad (106)$$

[In obtaining Eq. (106) we have used the fact that for on-shell photons $k_1^2 = k_2^2 = 0$; in the off-shell case the additional terms

$$(k_2^2 C_4 - k_1^2\, C_5)\; k_1^\xi\, k_2^\tau \varepsilon_1^{*\sigma}\;\; \varepsilon_2^{*\rho}\; \varepsilon_{\xi\tau\sigma\rho} \qquad (107)$$

are present.] Comparing Eq. (106) with Eq. (101), we see that

$$F(k_1 \cdot k_2) \propto k_1 \cdot k_2, \qquad (108)$$

$$F(0) = 0$$

which gives us a low energy theorem relating the vacuum to two photon matrix element G of the naive divergence to the corresponding matrix element H of the operator $(\alpha_0/4\pi)\, F^{\xi\sigma} F^{\tau\rho}\, \varepsilon_{\xi\sigma\tau\rho}$,

$$G(0) = -(1+C)\, H(0). \qquad (109)$$

To lowest non-vanishing order in perturbation theory, C can be neglected (it represented possible radiative corrections to the triangle) and H(0) can be evaluated from the Feynman rules preceding Eq. (69), giving

$$H(0) = 2\alpha/\pi, \qquad (110)$$

$$G(0) = -2\alpha/\pi.$$

This result for G(0) could, of course, have been derived without all the fuss directly from the lowest order expression for G given in Eq. (61),

$$G(0) = \left[i\varepsilon_1^{*\sigma} \varepsilon_2^{*\rho} \left(\frac{-ie_0^2}{(2\pi)^4}\right) 2m_0 R_{\sigma\rho} / k_1^{\xi} k_2^{\tau} \varepsilon_1^{*\sigma} \varepsilon_2^{*\rho} \varepsilon_{\xi\tau\sigma\rho} \right]_{k_1=k_2=0}$$

$$\mp \frac{e_0^2}{(2\pi)^4} 2m_0 B_1 \Big|_{k_1=k_2=0} = \frac{-e_0^2}{2\pi^2} \underset{\text{(to lowest order)}}{=} -\frac{2\alpha}{\pi}. \qquad (111)$$

However, as we shall see in detail in the next section, the

real significance of the low energy theorem is that <u>Eq. (110) for</u> G(0) <u>is exact, even when all radiative corrections are carefully taken into account.</u>

4. ABSENCE OF RADIATIVE CORRECTIONS

We must now deal with the question, raised in the last section, of whether radiative corrections to the triangle modify the anomalous axial-vector divergence equation. That is, what is the value of the constant C in Eq. (100), and how is the low energy theorem of Eq. (110) modified by radiative corrections? We will find the remarkable result that $C = 0$ and that Eq. (110) is exact to all orders of perturbation theory. This conclusion can be understood heuristically by noting that radiative corrections to the basic triangle involve axial-vector loops with at least five vertices, which, unlike the lowest order axial-vector triangle, do satisfy the usual axial-vector Ward identities. Thus, when virtual photon momenta are held fixed, the complicated radiative correction diagrams have no divergence anomalies. Since the virtual photon four-momenta appear essentially as parameters on both sides of these Ward identities, one expects that as long as the virtual photon integrations are not too badly divergent, the Ward identities will continue to hold even after the integrations have been performed. The purpose of the present section is to support this heuristic argument with more detailed calculations, and, in particular, to show that no problems are caused by the usual renormalizable infinities in the radiative corrections to the triangle.[15]

4.1 General Argument

We begin by developing a general argument, valid to any order of perturbation theory, which shows that Eqs. (68) and (110) are exact. The basic idea is this: As we have seen in the preceding section, the multiplicative factor Z_2, which makes matrix elements of the naive divergence term in Eq. (68) finite, does not remove the divergences from matrix elements of the axial-vector current term on the left-hand side of Eq. (68). Thus, there is no simple rescaling which simultaneously makes all terms in Eq. (68) finite, and so it is simplest to deal with Eq. (68) directly, even though it involves unrenormalized (and hence divergent) fields, masses and coupling constants. In order to make our manipulations of these divergent quantities well defined, we construct a cutoff version of quantum electrodynamics by introducing a photon regulator field of mass Λ. The cutoff prescription allows the usual renormalization program to be carried out, so that the electron bare mass m_0 and wave function renormalization Z_2, and the axial-vector vertex renormalization Z_A, become specified functions of the renormalized charge and mass, and of the cutoff Λ. In the cutoff field theory it is straightforward to prove the validity of Eq. (68) for the unrenormalized quantities. We then derive the low energy theorem for the matrix element $\langle 2\gamma | 2im_0 j^5 | 0 \rangle$, with the cutoff still present, and finally let the cutoff approach infinity to get a low energy theorem for the renormalized matrix element of the naive divergence.

We introduce the cutoff by modifying the usual Feynman rules for quantum electrodynamics which were stated in Section 1. Our new rules read as follows:

(i) For each internal electron line with momentum p we include a factor $i(\not{p}-m_0+i\varepsilon)^{-1}$ and for each vertex a factor $-i\hat{e}_0\gamma_\mu$. For each internal photon line of momentum q, we replace the usual propagator $-ig_{\mu\nu}(q^2+i\varepsilon)^{-1}$ by the regulated propagator

$$-ig_{\mu\nu}\left(\frac{1}{q^2+i\varepsilon}-\frac{1}{q^2-\Lambda^2+i\varepsilon}\right)=\frac{-ig_{\mu\nu}}{q^2+i\varepsilon}\,\frac{-\Lambda^2}{q^2-\Lambda^2+i\varepsilon}\,. \qquad (112)$$

(ii) Let $\Pi^{(2)}(q)_{\mu\nu}$ denote the two-vertex vacuum polarization loop illustrated below,

given by

$$\Pi^{(2)}(q)_{\mu\nu}=i\int\frac{d^4k}{(2\pi)^4}\mathrm{Tr}\left[\gamma_\mu\frac{1}{\not{k}-m_0+i\varepsilon}\gamma_\nu\frac{1}{\not{k}+\not{q}-m_0+i\varepsilon}\right]. \qquad (113)$$

Wherever $\Pi^{(2)}(q)_{\mu\nu}$ appears, we use its gauge-invariant, subtracted evaluation

$$\Pi^{(2)}(q)_{\mu\nu}=(-q^2 g_{\mu\nu}+q_\mu q_\nu)\,[\Pi^{(2)}(q^2)-\Pi^{(2)}(0)]\,. \tag{114}$$

All vacuum polarization loops with four or more vertices, such as

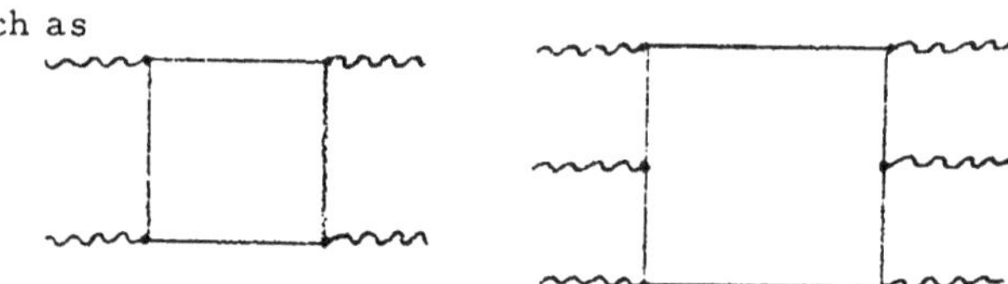

are calculated by imposing the current conservation condition; as we have seen, this suffices to make them finite without need for further subtractions.

(iii) As usual, there is a factor $\int d^4\ell/(2\pi)^4$ for each internal integration over loop variable ℓ and a factor -1 for each fermion loop.

(iv) We use the standard, iterative renormalization procedure outlined in Section 1 to fix the coupling $\hat{e}_0$ and the electron bare mass and wave function renormalization m_0 and Z_2, as functions of the renormalized charge and mass e and m and the cutoff Λ. For finite Λ, the quantities $\hat{e}_0$, m_0 and Z_2 will all be <u>finite</u>. The reason is that regulating the photon propagator (plus gauge invariance for loops) renders finite all vertex and electron self energy parts and all photon self-energy parts other than $\Pi^{(2)}_{\mu\nu}$, such as

The self-energy part $\Pi^{(2)}_{\mu\nu}$ has already been made finite by explicit subtraction. Note that the coupling $\hat{e}_0$ is not the same as the "bare charge" e_0 in Eq. (1), but rather is related to it by

$$\hat{e}_0^2 = \frac{e_0^2}{1 + e_0^2 \Pi^{(2)}(0)} \tag{115}$$

That is, $\hat{e}_0$ is a so-called "intermediate renormalized" charge, obtained from the bare charge by removing <u>only</u> those divergences associated with the lowest order vacuum polarization loop and its iterations.

(v) We include wave-function renormalization factors $Z_2^{\frac{1}{2}}$ and $Z_3^{\frac{1}{2}} = e/\hat{e}_0$ for each external electron and photon line.

This simple set of rules makes all ordinary electrodynamics matrix elements finite. We may summarize the rules compactly by noting that they are the Feynman rules for the following <u>regulated</u> Lagrangian density:

$$\begin{aligned}
\mathcal{L}^R(x) &= \mathcal{L}_0^R(x) + \mathcal{L}_I^R(x), \\
\mathcal{L}_0^R(x) &= \overline{\psi}(x)(i\gamma\cdot\Box - m_0)\psi(x) - \tfrac{1}{4}F_{\mu\nu}(x)F^{\mu\nu}(x) \\
&\quad + \tfrac{1}{4}F^R_{\mu\nu}(x)\, F^{R\mu\nu}(x) - \tfrac{1}{2}\Lambda^2 A^R_\mu(x)\, A^{R\mu}(x), \\
\mathcal{L}_I^R(x) &= -\hat{e}_0\overline{\psi}(x)\gamma_\mu\psi(x)[A^\mu(x) + A^{R\mu}(x)]
\end{aligned} \tag{116}$$

$$+C^{(2)}[F_{\mu\nu}(x)+F^R_{\mu\nu}(x)][F^{\mu\nu}(x)+F^{R\mu\nu}(x)],$$

where A^R_μ is the field of the regulator vector meson of mass Λ, and $F^R_{\mu\nu}(x) = \partial_\nu A^R_\mu(x) - \partial_\mu A^R_\nu(x)$ is the regulator field-strength tensor. The term containing $C^{(2)} \propto \Pi^{(2)}(0)$ is a logarithmically infinite counter term which performs the explicit subtraction of the two vertex vacuum polarization loop in Eq. (114). The regulator free-field Lagrangian density is included in Eq. (116) with the opposite sign from normal; hence, according to the canonical formalism, the regulator field is quantized with the opposite sign from normal. That is, we have

$$[A^R_\mu(\underset{\sim}{x},t), \partial A^R_\nu(\underset{\sim}{y},t)/\partial t] = i\, g_{\mu\nu}\delta^3(\underset{\sim}{x}-\underset{\sim}{y}), \qquad (117)$$

in contrast to Eq. (5). Since the sign of the bare propagator follows directly from the sign of the commutator in Eq. (117), the regulator bare propagator is opposite in sign from the photon bare propagator, as required by Eq. (112).

Having specified our cutoff procedure, we are now ready to introduce the axial-vector and pseudoscalar currents $j^5_\mu(x)$ and $j^5(x)$, and to study their properties. First we must check whether all matrix elements of these currents are finite when calculated in our cutoff theory. The answer is yes, that they are finite, and follows immediately from the fact that all of the basic fermion loops involving one axial-vector or one pseudoscalar vertex,

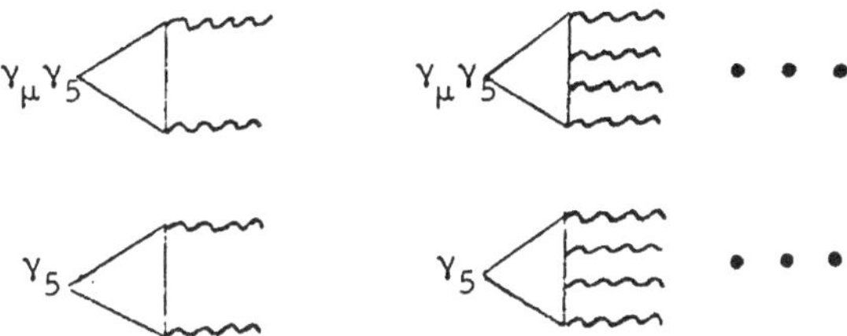

are made finite by the imposition of gauge invariance on the photon vertices, without the need for explicit subtractions. Thus, we can turn immediately to the problem of showing that Eq. (68) is exactly satisfied in our cutoff theory.

To do this, we consider an arbitrary Feynman amplitude involving j_μ^5, with 2F external fermion and B external boson lines

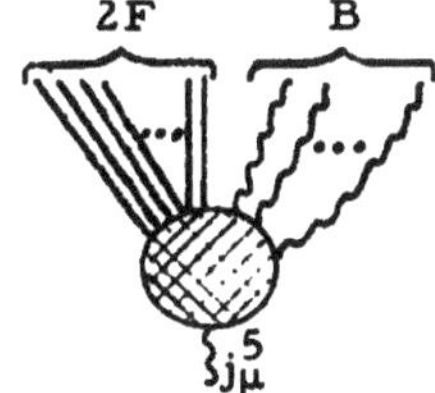

Proceeding as in Section 2, **we divide the diagrams into** two categories, type (a) and type (b), according to whether the axial-vector vertex $\gamma_\mu\gamma_5$ is attached to one of the F fermion lines running through the diagram, or is attached to an internal closed loop; respectively. Typical type-(a) and type-(b) diagrams are drawn below:

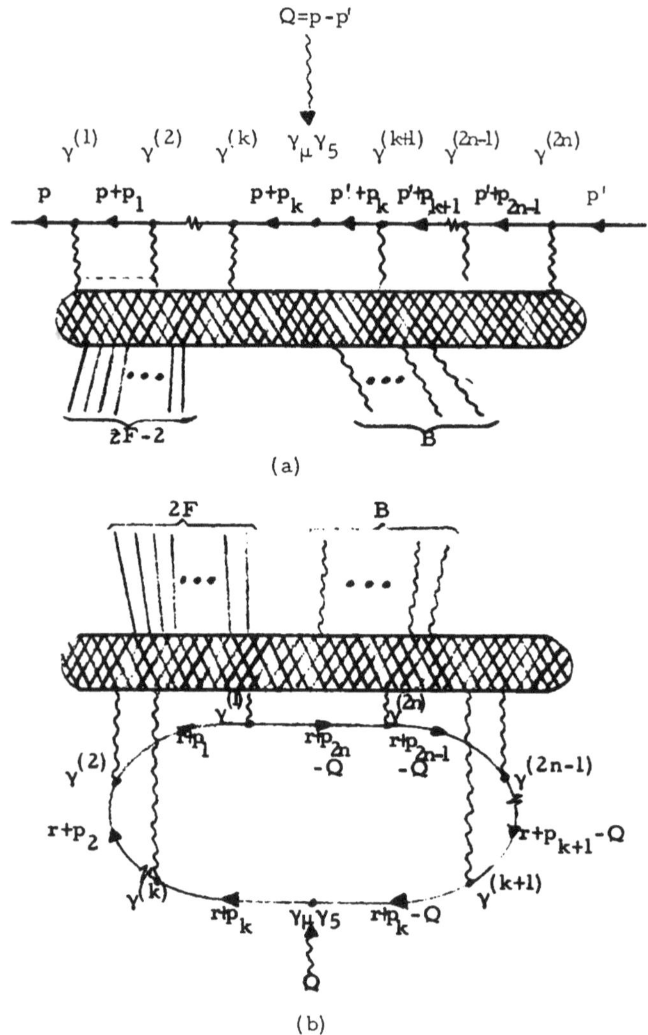

(a)

(b)

For the type-(a) diagrams we find that, just as in Eqs. (47) -(49) of Section 2, the derivation of the Ward identity involves purely <u>algebraic</u> manipulations of the string of fermion propagators on the fermion line containing $\gamma_\mu \gamma_5$. Since the integrals over the four-momenta of the photon propagators joining the fermion propagator string to the shaded "blob" (i.e. the integrals over $p_1, \ldots, p_{2n-1}$) are all <u>convergent</u> in our regulated field theory, it is safe to do these algebraic manipulations <u>inside</u> the integrals. The first term on the right-hand side of Eq. (49) gives the type-(a) contribution to the Feynman amplitude for $2im_0 j^5$ corresponding to the type-(a) diagram for j^5_μ which we started with. The two remaining terms in Eq. (49) give the usual "surface terms" which arise in Ward identities from the equal-time commutator of j^5_0 with the fields of the external fermions of momenta p and p'. Thus, as far as the type-(a) contributions to the Feynman amplitude are concerned, the divergence of j^5_μ is simply $2im_0 j^5$, with no extra terms present. We next turn to a typical type-(b) contribution, which we may write as

$$L(Q;\gamma_\mu\gamma_5;p_1,\ldots,p_{2n-1})(\ldots),$$

$$\begin{aligned} L(Q;\Gamma;p_1,\ldots,p_{2n-1}) = \int d^4r\, \mathrm{Tr}\Big\{ \sum_{k=1}^{2n} \prod_{j=1}^{k-1} \left[\gamma^{(j)} \frac{1}{\not r + \not p_j - m_0} \right] \\ \times \gamma^{(k)} \frac{1}{\not r + \not p_k - m_0} \Gamma \frac{1}{\not r + \not p_k - \not Q - m_0} \prod_{j=k+1}^{2n} \left[\gamma^{(j)} \frac{1}{\not r + \not p_j - \not Q - m_0} \right] \Big\}, \end{aligned} \tag{118}$$

where we have focused our attention on the closed loop and have again denoted the remainder of the diagram by $(\ldots)$.

The same straightforward algebra as before shows that the divergence of Eq. (118) can be rewritten as

$$L(Q;iQ^{\mu}\gamma_{\mu}\gamma_5;p_1,\ldots,p_{2n-1})$$
$$= L(Q;2im_0\gamma_5;p_1,\ldots,p_{2n-1}) \qquad (119)$$
$$+ i\int d^4r\, Tr\{\gamma_5 \prod_{j=1}^{2n}\left[\gamma^{(j)}\frac{1}{\not r+\not p_j-m_0}\right]-\gamma_5\prod_{j=1}^{2n}\left[\gamma^{(j)}\frac{1}{\not r+\not p_j-\not Q-m_0}\right]\}.$$

As we have seen, for loops with $n\geq 2$, the residual integrals in Eq. (119) cancel and we get the Ward identity

$$L(Q;iQ^{\mu}\gamma_{\mu}\gamma_5;p_1,\ldots,p_{2n-1})=L(Q;2im_0\gamma_5;p_1,\ldots,p_{2n-1}). \quad (120)$$

Again, since the integrals over $p_1,\ldots p_{2n-1}$ are all <u>convergent</u> in the regulated field theory, the manipulations leading to Eq. (120) can all be performed <u>inside</u> these integrals. This means that the type-(b) pieces containing loops with $n \geq 2$ all agree with the usual divergence equation $\partial^{\mu} j_{\mu}^5(x) = 2im_0 j^5(x)$. Finally, we must consider the case of the axial-vector triangle, with $n = 1$. As is now familiar, this diagram has an anomalous Ward identity, which in our regulated electrodynamics adds to the normal axial-vector divergence equation the term

$$(\hat{\alpha}_0/4\pi)[F^{\xi\sigma}(x)+F^{R\xi\sigma}(x)][F^{\tau\rho}(x)+F^{R\tau\rho}(x)]\epsilon_{\xi\sigma\tau\rho}, \qquad (121)$$
$$\hat{\alpha}_0 = \hat{e}_0^2/4\pi.$$

To summarize, our diagrammatic analysis shows that the axial-vector divergence equation in the regulated field theory is

$$\partial^{\mu} j_{\mu}^{5}(x)=2im_0 j^5(x)+(\hat{\alpha}_0/4\pi)F^{\xi\sigma}(x)F^{\tau\rho}(x)\epsilon_{\xi\sigma\tau\rho}$$
$$+(\hat{\alpha}_0/4\pi)[F^{\xi\sigma}(x)F^{R\tau\rho}(x)+F^{R\xi\sigma}(x)F^{\tau\rho}(x)+F^{R\xi\sigma}(x)F^{R\tau\rho}(x)]$$
$$\times \epsilon_{\xi\sigma\tau\rho}. \tag{122}$$

Equation (122) is identical with Eq. (68), apart from the terms involving F^R which arise from our explicit inclusion of a regulator field and apart from the fact that Eq. (122) is written in terms of the intermediate renormalized charge and field strength. To see the full equivalence with Eq. (68), we note that the intermediate renormalized quantities used in this section are related to the unrenormalized ones used in Section 2 by

$$\hat{e}_0(F^{\xi\sigma})_{\substack{\text{intermediate}\\ \text{renormalized}}} = e_0(F^{\xi\sigma})_{\text{unrenormalized}}. \tag{123}$$

The crucial point is that the coefficient of the anomalous term is exactly $\hat{\alpha}_0/4\pi$ and <u>does not involve an unknown power series in the coupling constant coming from higher orders in perturbation theory</u>.

The diagrammatic analysis which we have just given may be rephrased succinctly as follows: If we use the regulated Lagrangian density in Eq. (116) to calculate equations of motion, and then use the equations of motion to naively calculate the axial divergence, we find

$$\partial^{\mu} j_{\mu}^{5}(x) = 2im_0 j^5(x). \tag{124}$$

Extra terms on the right-hand side of Eq. (124) can only come from singular diagrams where the naive derivation breaks down. In the regulated field theory, all virtual photon integrations <u>converge</u> and therefore cannot lead to singularities giving additional terms in Eq. (124). Hence breakdown in Eq. (124), if it occurs at all, must be associated with the basic axial-vector loops, without internal radiative corrections. But, as we have seen, the axial-vector loops with four or more photons satisfy Eq. (124), so the basic triangle diagram is the <u>only</u> possible source of an anomaly.

Having derived our basic result, we turn next to the low-energy theorem for $2im_0 j^5(x)$ which is implied by Eq. (122). Taking the matrix element of Eq. (122) between a state containing two physical photons and the vacuum state, using the definition

$$\langle \gamma(k_1,\epsilon_1)\, \gamma(k_2,\epsilon_2) | \left\{ \begin{array}{l} \partial_\mu j_\mu^5 \\ 2im_0 j^5 \\ \frac{\alpha_0}{4\pi}(F^{\xi\sigma}+F^{R\xi\sigma})(F^{\tau\rho}+F^{R\tau\rho})\,\epsilon_{\xi\sigma\tau\rho} \end{array} \right\} |0\rangle_{\Lambda \text{ finite}} \tag{125}$$

$$= (4k_{10}k_{20})^{-\frac{1}{2}} k_1^{\xi} k_2^{\tau} \epsilon_1^{*\sigma} \epsilon_2^{*\rho} \epsilon_{\xi\tau\sigma\rho} \left\{ \begin{array}{l} F_\Lambda(k_1\cdot k_2) \\ G_\Lambda(k_1\cdot k_2) \\ H_\Lambda(k_1\cdot k_2) \end{array} \right\},$$

and proceeding exactly as in Eqs. (101)-(109), we find the low energy theorem

$$G_{\Lambda}(0) = -H_{\Lambda}(0). \tag{126}$$

We wish this time to calculate $H_{\Lambda}(0)$ to all orders in perturbation theory. There are two types of diagrams which contribute to $H_{\Lambda}(k_1 \cdot k_2)$, as illustrated below, where we have used the symbol ⊗ to denote the action of the operator $(\alpha_0/4\pi)(F^{\xi\sigma} + F^{R\xi\sigma})(F^{\tau\rho} + F^{R\tau\rho})\,\epsilon_{\xi\sigma\tau\rho}$:

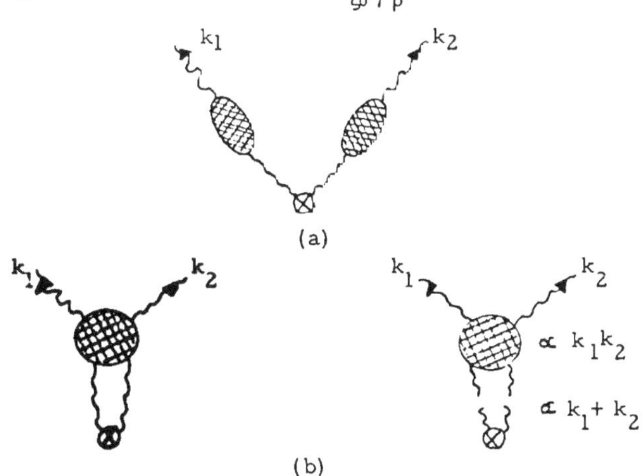

In the diagrams (a), the field strength operators attach directly onto the external photon lines, without photon-photon scattering. The effect of the vacuum polarization parts and the external-line wave-function renormalizations is to change $\hat{\alpha}_0$ to α, giving

$$H_{\Lambda}(0)^{(a)} = 2\alpha/\pi\ . \tag{127}$$

In diagrams (b), there is a photon-photon scattering between ⊗ and the free photons. As a result of the antisymmetric tensor structure of the anomalous divergence term, the vertex ⊗ is proportional to k_1+k_2. Also, the diagram for the scattering of light by light is itself proportional to k_1k_2, since photon gauge invariance implies that the external photons couple through their field strength tensors. Thus, the diagrams (b) are proportional to $k_1k_2(k_1+k_2)$ and are of higher order than the terms which contribute to the low-energy theorem, giving us

$$H_\Lambda(0)^{(b)} = 0. \tag{128}$$

Combining Eqs. (126)-(128), we get an <u>exact low-energy theorem</u> for the operator $2im_0j^5$

$$G_\Lambda(0) = -2\alpha/\pi\ . \tag{129}$$

So far in our discussion we have kept the cutoff Λ finite, so that $G_\Lambda(0)$ is a matrix element calculated with our modified Feynman rules. However, we have seen that all matrix elements of $2im_0j^5$ become cutoff-independent in the limit $\Lambda \to \infty$. Defining a renormalized vacuum to two photon matrix element of the naive axial vector divergence, $\tilde{G}(k_1 \cdot k_2)$, by taking the limit

$$\tilde{G}(k_1 \cdot k_2) = \lim_{\Lambda \to \infty} G_\Lambda(k_1 \cdot k_2), \tag{130}$$

we get from Eq. (129) the low energy theorem

$$\widetilde{G}(0) = -2\alpha/\pi. \tag{131}$$

The observant reader will notice that our definition of the renormalized matrix element in Eq. (130) appears to differ from the skeleton diagram construction described in Section 1. According to the skeleton expansion, an arbitrary renormalized matrix element of the naive divergence (and the vacuum to two photon matrix element in particular) can be constructed by writing down the appropriate skeleton graphs and inserting the renormalized propagators and vertex functions $\widetilde{S}'_F$, $\widetilde{D}'_F$, $\widetilde{\Gamma}_\mu$ and $\widetilde{\Gamma}^5$. These quantities are defined as the Λ-independent limits

$$\begin{aligned} \widetilde{S}'_F &= \lim_{\Lambda\to\infty} Z_2^{-1} S'_F, \\ \widetilde{D}'_F &= \lim_{\Lambda\to\infty} Z_3^{-1} D'_F, \\ \widetilde{\Gamma}_\mu &= \lim_{\Lambda\to\infty} Z_2 \Gamma_\mu, \\ m\widetilde{\Gamma}^5 &= \lim_{\Lambda\to\infty} m_0 Z_2 \Gamma^5; \end{aligned} \tag{132}$$

that is, the skeleton expansion construction consists of taking the $\Lambda\to\infty$ limit <u>first</u> in vertex parts and propagators, and then substituting onto the skeleton. In Eq. (130), however, these operations are performed in the reverse order; the cutoff dependent vertex parts and propagators are substituted onto the skeleton, all integrations are carried out, and then finally the $\Lambda\to\infty$ limit is taken. Can this interchange of order make any difference in the final value of the renormalized matrix element which is obtained? A simple inductive argument shows that the answer to this question is in the negative. Let us suppose that the two procedures give the same answer for all matrix elements of j^5 of order n-2 in perturbation theory. For all matrix elements of order n which have convergent skeletons, the two procedures must obviously agree. According to Eq. (11), the only cases which have potentially divergent skeletons are the pseudoscalar vertex part itself, and the vacuum to two photon matrix element. For the pseudoscalar vertex part, the two constructions agree, by definition. For the vacuum to two photon matrix element, a possible difference $\Delta\tilde{G}$ between the two constructions must have the following properties: (i) $\Delta\tilde{G}$ must be a polynomial in the photon momentum variables k_1 and k_2. This restriction follows from generalized unitarity, which relates discontinuities in the nth order diagram to lower order matrix elements, for which the two constructions agree, by hypothesis. (ii) $\Delta\tilde{G}$ must satisfy the requirements of Weinberg's theorem, since both constructions do. This again means that if we set $k_1 = \xi Q$, $k_2 = -\xi Q + \xi R + S$ and let $\xi \to \infty$, $\Delta\tilde{G}$ must diverge at most as ξ times a power of $\ln \xi$. Together with gauge invariance, the property (i) implies that $\Delta\tilde{G}$ has the form

$$\Delta\tilde{G} = k_1^{\xi} k_2^{\tau} \qquad \epsilon_2^{*\rho} \epsilon_{\xi\tau\sigma\rho} \times \text{ polynomial in } k_1, k_2, \tag{133}$$

but this diverges at least as ξ^2 in the Weinberg limit and violates property (ii). Thus we must have $\Delta\tilde{G} = 0$. We conclude that the two constructions agree in nth order, and by induction, in all orders. Consequently, the low energy theorem of Eq. (131) applies to the renormalized matrix element $\tilde{G}$ obtained from the usual skeleton expansion and makes the remarkable statement that all order α^2, α^3, ... contributions to $\tilde{G}(k_1 \cdot k_2)$ vanish at $k_1 \cdot k_2 = 0$.

4.2 Explicit Second Order Calculation

Let us now briefly outline a calculation which explicitly checks Eq. (131) to second order in perturbation theory. We wish to calculate the sum of the six radiative correction diagrams to the γ_5-γ_σ- γ_ρ triangle,

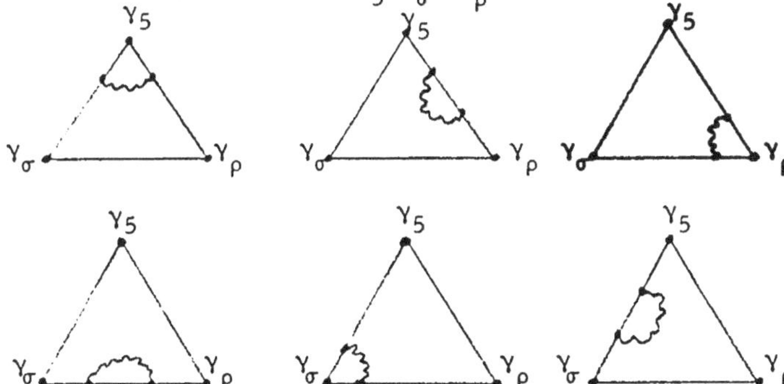

and to verify that they cancel to zero. The first step is to calculate the renormalized quantities $\tilde{\Gamma}_\mu(p,p')$, $\tilde{\Gamma}^5(p,p')$ and $\tilde{S}'_F(p)$. The most straightforward way of doing this is (a) to calculate the unrenormalized quantities Γ_μ, Γ^5 and S'_F using the cutoff Feynman rules, (b) to use Eqs. (12) and (13) to compute the renormalization constants m_0 and Z_2, and (c) to use the recipe of Eq. (132) to find the tilde quantities by taking the limit $\Lambda \to \infty$. This procedure gives, to second order,

$$\tilde{\Gamma}^{(2)}_\lambda(p,p') = \gamma_\lambda + \frac{e^2}{16\pi^2}\int_0^1 z\,dz\int_0^1 dy\left\{2\gamma_\lambda \ln\left[\frac{z^2m^2+(1-z)\mu^2}{D}\right] - \frac{N_\lambda}{D} - \frac{2m^2\gamma_\lambda P_1}{z^2m^2+(1-z)\mu^2}\right\},$$

$$\tilde{\Gamma}^{5(2)}(p,p') = \gamma_5 + \frac{e^2}{16\pi^2}\int_0^1 z\,dz\int_0^1 dy\left\{8\gamma_5 \ln\left[\frac{z^2m^2+(1-z)\mu^2}{D}\right] + \frac{\gamma_5 N}{D} + \frac{4m^2\gamma_5 P_2}{z^2m^2+(1-z)\mu^2}\right\},$$

$$\tilde{S}^{(2)}_F(p) = [\,\not{p} - m - \tilde{\Sigma}^{(2)}(\not{p})\,]^{-1}, \tag{134}$$

$$\tilde{\Sigma}^{(2)}(\not{p}) = \frac{e^2}{16\pi^2}\int_0^1 z\,dz\left\{2g_1 \ln\left[\frac{z^2m^2+(1-z)\mu^2}{-p^2z(1-z)+zm^2+(1-z)\mu^2}\right]\right.$$
$$\left. + g_2\,\frac{m^2-p^2(1-z)^2}{-p^2z(1-z)+zm^2+(1-z)\mu^2} + \frac{2m^2\not{p}\,P_1+4m^3P_2}{z^2m^2+(1-z)\mu^2}\right\},$$

$$D=(y^2z^2-yz)p^2+[(1-y)^2z^2-(1-y)z]\,p'^2+2y(1-y)z^2p\cdot p'+zm^2+(1-z)\mu^2,$$

with

$$N_\lambda = -2m^2\gamma_\lambda - 2[\,(1-z+yz)\not{p}' - yz\not{p}\,] \tag{135}$$
$$\times\gamma_\lambda[\,(1-yz)\not{p}-(1-y)z\not{p}'\,] + 4m[\,(1-2yz)p_\lambda + (1-2z+2yz)p'_\lambda\,],$$
$$N = 4m^2 - 4[\,(1-yz)p-(1-y)zp'\,]\cdot[\,(1-z+yz)p'-yzp\,] + 2m(\not{p}-\not{p}'),$$
$$P_1 = z^2+2z-2,\quad P_2 = 1-2z,$$

$$g_1 = 4m - \not{p}, \quad g_2 = 4m - 2\not{p}.$$

The quantity μ^2 is a fictitious virtual photon mass supplied to avoid logarithmic infrared singularities in the individual radiative correction diagrams. (The sum of the six radiative correction diagrams, however, has no infrared divergences because of cancellations of the troublesome terms.) As a check on our arithmetic, we note that Eqs. (134) and (135) satisfy the vector Ward identity

$$(p-p')^{\lambda}\ \tilde{\Gamma}_{\lambda}(p,p') = \tilde{S}'_F(p)^{-1} - \tilde{S}'_F(p')^{-1} \tag{136}$$

as well as the additional relation

$$0 = 2m\ \tilde{\Gamma}^5(p,p) + \tilde{S}'_F(p)^{-1}\gamma_5 + \gamma_5\ \tilde{S}'_F(p)^{-1}, \tag{137}$$

which follows from the $p=p'$ case of Eq. (70).

The next step is to substitute Eqs. (134) and (135) into the skeleton diagram

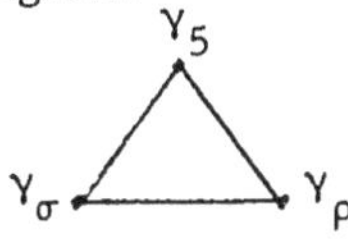

giving the lowest order triangle plus the six radiative correction diagrams illustrated above. The final step is to Taylor expand around the neighborhood $k_1 = k_2 = 0$, since $\tilde{G}(0)$ is the coefficient of the first nonvanishing term in this expansion. Although the integrals for general k_1 and k_2 are very formidible, the leading Taylor coefficient is not very complicated. Some straightforward algebra and integrations then show that the contributions of the radiative corrections to $\widetilde{G}(0)$ do indeed cancel, as required by Eq. (131).[15]

5. GENERALIZATIONS OF OUR RESULTS: π^0 DECAY; OTHER WARD IDENTITY ANOMALIES

Our discussion so far has dealt exclusively with the VVA triangle anomaly in QED. Let us now generalize our results in two directions. First, we will study the consequences of the VVA anomaly in other field theory models, especially in the so-called σ-models, which satisfy the partially-conserved axial-vector current (PCAC) condition as an exact operator identity. We will find that extension to this class of models of the low-energy theorem derived above leads to a prediction of the $\pi^0 \rightarrow 2\gamma$ decay rate. Comparison with experiment provides evidence _against_ the quark model with fractional quark charges. Second, we will briefly examine other triangle, square and pentagon diagrams to see which have anomalous, and which have normal, Ward identities.

5.1 The σ-Models

As we have just noted, the σ-models are a special class of field theory models in which PCAC holds as an operator relation.[16] Since we are interested primarily in the neutral axial-vector current (which can couple to two photons through the triangle diagram), we consider a truncated version of the σ-model in which the charged axial-vector currents do not appear. This simplified model contains only a proton (ψ), a neutral pion (π) and a scalar meson (σ), with Lagrangian density

$$\begin{aligned}\mathcal{L} = \bar{\psi}[i\gamma\cdot\Box - G_0(g_0^{-1} + \sigma + i\pi\gamma_5)]\psi \\ +\lambda_0[4\sigma^2 + 4g_0\sigma(\sigma^2+\pi^2) + g_0^2(\sigma^2+\pi^2)^2] \\ +\tfrac{1}{2}\mu_0^2[2g_0^{-1}\sigma + \sigma^2 + \pi^2] \\ +\tfrac{1}{2}[(\partial\pi)^2 + (\partial\sigma)^2] - \tfrac{1}{2}\mu_1^2(\pi^2+\sigma^2).\end{aligned} \tag{138}$$

In writing Eq. (138) we have chosen the fully translated form of the σ-model, with

$$\langle 0 \mid \sigma \mid 0 \rangle = 0 \tag{139}$$

to all orders of perturbation theory. The neutral axial-vector current is generated by making a chiral gauge transformation on the fields with position dependent gauge parameter $v(x)$,

$$\begin{aligned}\psi &\to (1 + \tfrac{1}{2} i\gamma_5 v)\psi , \\ \pi &\to \pi - v(g_0^{-1} + \sigma) , \\ g_0^{-1} + \sigma &\to g_0^{-1} + \sigma + v\pi .\end{aligned} \tag{140}$$

Using the recipe of Eq. (90), we find

$$j_\mu^5 = -\delta\mathcal{L}/\delta(\partial^\mu v) = \bar{\psi}\tfrac{1}{2}\gamma_\mu\gamma_5\psi + \sigma\partial_\mu\pi - \pi\partial_\mu\sigma + g_0^{-1}\partial_\mu\pi, \tag{141a}$$

$$\partial^\mu j_\mu^5 = -\delta\mathcal{L}/\delta v = -(\mu_1^2/g_0)\pi . \tag{141b}$$

Thus, as claimed, the divergence of the axial-vector current is proportional to the canonical pion field. The various parameters appearing in Eq. (138) have the following significance:

(i) G_0 is the unrenormalized meson-nucleon coupling constant;

(ii) g_0 is related to the bare nucleon mass m_0 by $G_0/g_0 = m_0$;

(iii) μ_1^2 is the bare meson mass which appears in the bare σ and π propagators $(q^2-\mu_1^2+i\varepsilon)^{-1}$;

(iv) the term $\lambda_0[4\sigma^2+4g_0\sigma(\sigma^2+\pi^2)+g_0^2(\sigma^2+\pi^2)^2]$ is a chiral-invariant meson-meson scattering interaction;

(v) the term $\frac{1}{2}\mu_0^2(2g_0^{-1}\sigma+\sigma^2+\pi^2)$ is a chiral-invariant <u>counter term</u> which is necessary to guarantee that

$$<0|\delta\mathcal{L}/\delta\sigma|0> = \partial_\lambda <0|\delta\mathcal{L}/\delta(\partial_\lambda\sigma)|0> = 0\,, \qquad (142)$$

as is required by the Euler-Lagrange equations of motion and translation invariance. Eqs. (139) and (142) fix μ_0^2 to have the value

$$\mu_0^2 = <0|G_0g_0\bar{\psi}\psi - \lambda_0[4g_0^2(3\sigma^2+\pi^2)+4g_0^3\sigma(\sigma^2+\pi^2)]|0>. \qquad (143)$$

The effect of μ_0^2, which is formally quadratically divergent, is to remove the "tadpole" diagrams of the type

so that the condition $<0|\sigma|0> = 0$ is maintained in each order of perturbation theory. It is easily seen that the μ_0^2 counter term simultaneously removes the quadratically divergent parts of the π- and σ-meson self-energies. Consequently, the remaining bare quantities appearing in the Lagrangian (G_0, g_0, μ_1), as well as the wave-function renormalizations, are at most logarithmically divergent, and the theory is no more singular than is QED.[17]

For our future work, it will prove convenient to rewrite the PCAC equation, Eq. (141b), as follows. First, we introduce the pion wave function renormalization constant Z_3^π, which enables us to express Eq. (141b) in terms of the renormalized pion field π^r,

$$\partial^\mu j_\mu^5 = -(\mu_1^2/g_0)(Z_3^\pi)^{\frac{1}{2}} \pi^r. \tag{144}$$

Next, we define the pion weak decay amplitude f_π by writing

$$<\pi(q)|j_\lambda^5|0> = (2q_0)^{-\frac{1}{2}}(-iq_\lambda/\mu^2)\, f_\pi/\sqrt{2}\ , \tag{145}$$

with μ the physical pion mass. [In the full version of the σ-model, in which the neutral axial-vector current has charged isospin partners, f_π is just the amplitude for weak charged pion decay.] Taking the divergence of Eq. (145), substituting Eq. (144) and using $<\pi(q)|\pi^r|0> = (2q_0)^{-\frac{1}{2}}$, we find the relation

$$-(\mu_1^2/g_0)(Z_3^\pi)^{\frac{1}{2}} = f_\pi/\sqrt{2}\ . \tag{146}$$

So we can eliminate the renormalization constants from Eq. (144) and rewrite the PCAC equation entirely in terms of physical quantities,

$$\partial^{\mu} j^{5}_{\mu} = (f_{\pi}/\sqrt{2})\pi^{r} \qquad (147)$$

So far, we have only discussed the σ-model in the absence of electromagnetism. To include electromagnetism, we simply add to the Lagrangian density the terms

$$-\tfrac{1}{4}F_{\mu\nu}F^{\mu\nu} - e_0\overline{\psi}\gamma_{\mu}\psi A^{\mu} \qquad (148)$$

Because of the triangle diagram, the PCAC equation of Eq. (147) is modified to read

$$\partial^{\mu} j^{5}_{\mu} = (f_{\pi}/\sqrt{2})\ \pi^{r} + \tfrac{1}{2}\,\frac{\alpha_0}{4\pi}\,F^{\xi\sigma}F^{\tau\rho}\epsilon_{\xi\sigma\tau\rho}, \qquad (149)$$

with the factor $\frac{1}{2}$ in the anomaly term in Eq. (149) just a reflection of the factor $\frac{1}{2}$ appearing in the nucleon term in Eq. (141a). By introducing appropriate regulated Feynman rules and carrying out an analog of the argument of the previous section, one can show that Eq. (149) is exact to all orders of perturbation theory in both the electromagnetic and strong interactions. In other words, neither virtual photon nor virtual meson radiative corrections to the triangle diagram change the coefficient of the anomaly term.

All of the above considerations carry over directly to the isospin and full SU_3[18] generalizations of the sigma model. In the full SU_3 case, the proton ψ is replaced by a fermion triplet (ψ_1, ψ_2, ψ_3), the scalar and pseudoscalar mesons are replaced by nonets, and the axial-vector current j_μ^5 becomes the third component $\mathcal{F}_{3\mu}^5$ of the axial-vector current octet. The anomalous PCAC equation now becomes

$$\partial^\mu \mathcal{F}_{3\mu}^5 = (f_\pi/\sqrt{2})\, \pi^{0r} + S \frac{\alpha_0}{4\pi} F^{\xi\sigma} F^{\tau\rho} \epsilon_{\xi\sigma\tau\rho}, \qquad (150)$$
$$S = \sum_j g_j Q_j^2 ,$$

with π^{0r} **the renormalized neutral pion field, with $Q_j e$ the** charge of the **jth fermion and with g_j the fermion coup-**lings appearing in the expression for $\mathcal{F}_{3\mu}^5$ **in terms of** elementary fields,

$$\mathcal{F}_{3\mu}^5 = \sum_j g_j \bar{\psi}_j \gamma_\mu \gamma_5 \psi_j + \text{Meson Terms} . \qquad (151)$$

Again, Eq. (150) is exact to all finite orders of perturbation theory. The interpretation of the expression for S is that the total coefficient of the anomaly is the sum of contributions from triangle graphs involving each of the individual elementary fermions.

Equation (150) generalizes even further, to models in which the naive divergence D_3^5 (i.e., the divergence of $\mathcal{F}_{3\mu}^5$ in the absence of electromagnetism) is not a canonical pion field. Because the argument of Section 4 depended primarily on the multiplicative renormalizability of the naive divergence, we would expect the equation

$$\partial^\mu \mathcal{F}_{3\mu}^5 = D_3^5 + S \frac{\alpha_0}{4\pi} F^{\xi\sigma} F^{\tau\rho} \epsilon_{\xi\sigma\tau\rho} \qquad (152)$$

to be correct in any renormalizable field theory in which the naive axial divergence D_3^5 is multiplicatively renormalizable. Since, by the argument of Eqs. (76) - (79), multiplicative renormalizability of the naive divergence implies finite Z_A/Z_2 in the absence of electromagnetism, we may rephrase the above statement by saying that in any renormalizable field theory with finite axial-vector renormalizations g_A/g_V in the absence of electromagnetism, we expect Eq. (152) to be exact when electromagnetic effects are added. As long as D_3^5 is a smooth interpolating field for the pion, we may effectively make the replacement $D_3^5 \approx (f_\pi/\sqrt{2})\, \pi^{0r}$ in Eq. (152) for small extrapolations away from the pion mass shell. Thus, Eq. (150) is the correct PCAC equation, even in the more general class of models.

5.2 Low Energy Theorem for π^0 Decay

As we saw in Eqs. (125)-(131), the anomalous axial-vector Ward identity equation gives an exact low energy theorem for the vacuum to two photon matrix elements of the naive axial divergence. Since the naive axial divergence in Eq. (150) is the π^0 field, the low energy theorem in this case gives us a statement about the $\pi^0 \to 2\gamma$ decay amplitude, extrapolated off shell to zero pion mass.[19] The standard definition of the $\pi^0 \to 2\gamma$ amplitude $F^\pi(k_1 \cdot k_2)$ is

$$\langle \gamma(k_1, \epsilon_1)\, \gamma(k_2, \epsilon_2) | (\Box^2 + \mu^2)\, \pi^{0r} | 0 \rangle \qquad (153)$$

$$= (4k_{10}k_{20})^{-\frac{1}{2}} k_1^{\xi} k_2^{\tau} \varepsilon_1^{*\sigma} \varepsilon_2^{*\rho} \varepsilon_{\xi\tau\sigma\rho} F^{\pi}(k_1 \cdot k_2) .$$

Comparing with Eqs. (125)-(131), we see that the low energy theorem becomes

$$\widetilde{G}(0) = \mu^{-2} (f_{\pi}/\sqrt{2}) F^{\pi}(0) = S(-2\alpha/\pi), \tag{154a}$$

that is

$$F^{\pi}(0) = (-\alpha/\pi)(2S)\sqrt{2}\,\mu^2/f_{\pi} . \tag{154b}$$

According to Eq. (154b), the off-shell $\pi^0 \to 2\gamma$ amplitude is directly proportional to the anomaly term in Eq. (150). If the anomaly term were omitted [i. e., if Eq. (147) were used to derive the low energy theorem] , one would obtain instead the prediction

$$F^{\pi}(0) = 0 , \tag{155}$$

which states that the $\pi^0 \to 2\gamma$ decay is suppressed.[20] Let us briefly discuss some of the implications of Eq. (154b).

(i) The experimental π^0 decay rate predicted by Eq. (154b) depends on the parameter S, which in turn is determined by the charges and axial couplings of the elementary fermions. In the triplet-model, consisting of an SU_3-triplet of fermions $(\psi_1, \psi_2, \psi_3) \equiv (p, n, \lambda)$ interacting by meson exchange, the axial-vector couplings are $(g_1, g_2, g_3) = (\frac{1}{2}, -\frac{1}{2}, 0)$, and U-spin invariance of the electromagnetic current requires the respective charges of (ψ_1, ψ_2, ψ_3) to be (Q, Q-1, Q-1). We immediately find

$$S = \tfrac{1}{2}Q^2 - \tfrac{1}{2}(Q-1)^2 = Q - \tfrac{1}{2} \equiv Q_{AV} = \tfrac{1}{2}[Q + (Q-1)] \ . \qquad (156)$$

where Q_{AV} denotes the average charge of the fermions participating in the charged β-decay currents. We find, in the fractionally-charged quark model, that $Q = 2/3$ and $S = 1/6$, while in the integrally-charged quark models with $Q = 1$ and $Q = 0$, we find respectively $S = +\frac{1}{2}$ and $S = -\frac{1}{2}$. Using the formula

$$\tau^{-1} = (\mu^3/64\pi)\,|F^\pi(\mu^2)|^2, \qquad (157)$$

taking the experimental value $f_\pi \approx 0.96\,\mu^3$ and approximatimating $F^\pi(\mu^2)$ by its off-shell value $F^\pi(0)$, we find for the π^0 decay rate

$$\tau^{-1} = 0.8 \text{ eV} \quad \text{for } S = \frac{1}{6} \qquad (158a)$$

$$\tau^{-1} = 7.4 \text{ eV} \quad \text{for } S = \pm\frac{1}{2}\,. \qquad (158b)$$

The experimental decay rate quoted by Rosenfeld[21] is

$$\tau^{-1}_{\text{expt}} = (1.12 \pm 0.22)\cdot 10^{16}\ \text{sec}^{-1} \approx (7.37 \pm 1.5)\,\text{eV}, \qquad (159)$$

and may be as large as 11 eV if recent Primakoff effect experiments[22] turn out to be more reliable than earlier counter experiments included in Rosenfeld's average. In any case, we see that the fractionally charged quark model is strongly excluded, while the integrally charged models with triplet charges (1, 0, 0) or (0, -1, -1) are in satisfactory agreement with the experimental rate. Note that the apparent spectacular agreement between Eq. (158b) and Eq. (159)

is somewhat fortuitous, both because of the uncertainty in the experimental rate and because of the expected 10-20 percent extrapolation error involved in PCAC arguments. For example, if instead of using the experimental value of f_π in Eq.(154b) we use the Goldberger-Treiman relation

$$\frac{\sqrt{2}\,\mu^2}{f_\pi} \approx \frac{g_r}{M_N g_A}, \tag{160}$$

M_N = nucleon mass,

g_r = pion-nucleon coupling constant ≈ 13.6,

g_A = nucleon axial-vector coupling constant ≈ 1.22,

the theoretical prediction in the $S = \pm\frac{1}{2}$ cases is increased by 20 percent, to $\tau^{-1} = 9.1$ eV.

(ii) The comparison which we have made with the experimental π^0 decay rate tells us that $|S| \approx 0.5$, but does not determine the sign of S. However, there are a number of different ways of determining the sign of S, all of which, fortunately, seem to agree! The first method is to study $\pi^+ \to e^+\nu\gamma$ decay, the vector part of which is related by isospin rotation to F^π and the axial-vector part of which can be estimated by using hard pion techniques. The analysis,[23] using the experimentally-measured vector to axial-vector ratio for this process, gives a positive value of S. A second method is to make use of forward π^0 photoproduction, where one can observe the interference between the Primakoff amplitude

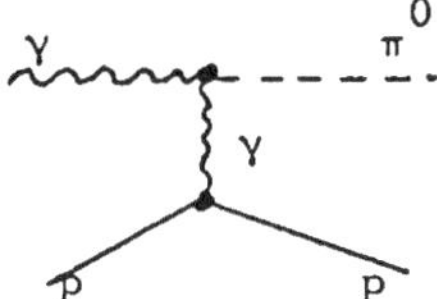

which is proportional to F^π, and the forward strong interaction amplitude. The sign of the latter can be determined by finite energy sum rules from the known sign of the pion photoproduction amplitude in the (3, 3) resonance region; the analysis[24] again indicates S positive. A third method consists of comparing Eq.(154b) with an approximate expression for the $\pi^0 \to 2\gamma$ amplitude derived[25] by applying a pole dominance argument to proton Compton scattering dispersion relations,

$$F^\pi \approx -4\pi\alpha \frac{\kappa_p}{g_r} \frac{1}{M_N}, \tag{161}$$

κ_p = proton anomalous magnetic moment = 1.79.

Eq. (161) gives a $\pi^0 \to 2\gamma$ rate of 2.0 eV, in fair agreement with experiment. The comparison again gives S positive. A fourth method which has been proposed[23] is to use Compton scattering data on protons to try to measure the interference of the pion exchange piece,

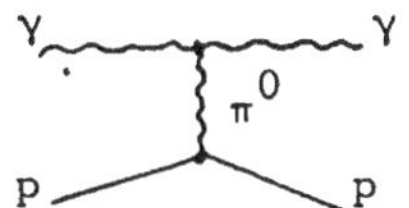

which is proportional to F^{π}, with the nucleon and nucleon isobar exchange pieces. The problem with this proposal[26] is that one does not know whether to take the pion exchange piece in its Born approximation form, $tF^{\pi}/(t-\mu^2)$, or in the polology form $\mu^2 F^{\pi}/(t-\mu^2)$. Since t is negative in the physical region, this uncertainty leads to a sign ambiguity and renders the method dubious. In any case, with fair certainty one learns from the first three methods that S is positive. This means that the triplet model with $Q=1$ and triplet charges (1, 0, 0) is favored.

(iii) Although we have shown that Eq. (154b) is exact to all orders of perturbation theory in an interesting class of theoretical models, we have not dealt with the possibility that Eqs. (150) and (154b) are modified by nonperturbative effects. For example, should the coefficient S receive contributions from triangles involving bound states of the fundamental fields as well as from triangles involving the fundamental fields themselves, or would this be double counting? The answer to this question is not known. Our neglect of possible nonperturbative modifications in the analysis above is pure assumption.

(iv) Let us now backtrack from quantitative predictions to

the more general question of how we know that π^0 decay is not really a suppressed decay, as would be suggested by PCAC with the triangle anomaly omitted. There is in fact an interesting experimental test,[27] which strongly suggests that π^0 decay is not suppressed. To see this, let us return to the suppression. argument in the case when one of the final photons is off mass shell, say $k_1^2 \neq 0$. As we have seen in Eqs. (106) and (107), the vacuum to two photon matrix element of $\partial^\mu \mathcal{F}_{3\mu}^5$ in this case is proportional to $k_1^\xi k_2^\tau \epsilon_1^{*\sigma} \epsilon_2^{*\rho} \epsilon_{\xi\tau\sigma\rho}[\mu^2 + \beta k_1^2]$, with β of order unity. We see that while the on shell part of the amplitude (in the absence of the anomaly) is suppressed by a factor μ^2, the photon off-shell dependence is not suppressed. Since the off-shell amplitude is measured in the reaction $\pi^0 \to e^+e^-\gamma$, the suppression argument predicts that the k_1^2 dependence of this process will have the form $1 + (\beta/\mu^2)k_1^2$, which has a much larger slope than the form $1+(\beta/m_\rho^2)$ (with m_ρ the ρ-meson mass) expected in the absence of suppression of the $\pi^0 \to 2\gamma$ decay. A recent measurement[28] of this slope gives a matrix element $1 + \underline{a}\, k_1^2$, with $\underline{a} = (0.01 \pm 0.11)/\mu^2$. Clearly, this is strong evidence against $\pi^0 \to 2\gamma$ suppression, and therefore some mechanism, like the triangle anomalies which we have discussed in such great detail, is definitely needed to avoid the suppression prediction of Eq. (155).

5.3 Other Ward Identity Anomalies

So far, we have dealt exclusively with the VVA triangle diagram and its Ward identity anomaly. Let us now briefly examine the question of whether there are other diagrams with divergence anomalies. We begin with a study of the Ward identity relating the AAA and the PAA (P = pseudoscalar) fermion triangle diagrams,

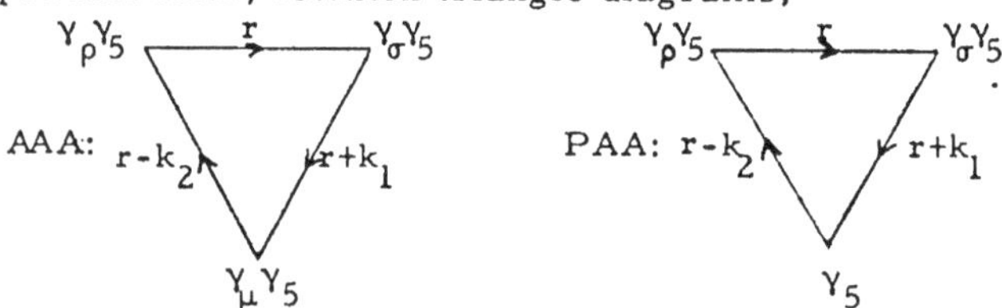

Defining

$$\frac{-iR^5_{\sigma\rho\mu}}{(2\pi)^4} \equiv 2\int \frac{d^4r}{(2\pi)^4}(-1)\ \mathrm{Tr}\{\frac{i}{\not r + \not k_1 - m_0}(-i\gamma_\sigma\gamma_5) \times \frac{i}{\not r - m_0}(-i\gamma_\rho\gamma_5)\frac{i}{\not r - \not k_2 - m_0}\gamma_\mu\gamma_5\} \qquad (162a)$$

$$\frac{-i2m_0R^5_{\sigma\rho}}{(2\pi)^4} \equiv 2\int \frac{d^4r}{(2\pi)^4}(-1)\ \mathrm{Tr}\{\frac{i}{\not r + \not k_1 - m_0}(-i\gamma_\sigma\gamma_5) \times \frac{i}{\not r - m_0}(-i\gamma_\rho\gamma_5)\frac{i}{\not r - \not k_2 - m_0}\,2m_0\gamma_5\}\ , \qquad (162b)$$

we find, by manipulations under the r-integrals which ignore the linear divergence in $R^5_{\sigma\rho\mu}$, the naive Ward identity

$$-(k_1+k_2)^{\mu} R^5_{\sigma\rho\mu} = 2m_0 R^5_{\sigma\rho} \,. \tag{163}$$

To search for possible corrections to Eq. (163), we must first make the definitions of $R^5_{\sigma\rho\mu}$ and $R^5_{\sigma\rho}$ precise. The latter quantity is given by a convergent integral and so is uniquely defined, but the integral in Eq. (162a) for $R^5_{\sigma\rho\mu}$ is linearly divergent, and hence its precise value depends on the choice of origin for symmetric integration. However, $R^5_{\sigma\rho\mu}$ is uniquely specified if we require that it be Bose symmetric under interchange of any pair of vertices. So our question becomes that of finding the extra terms (if any) which appear in Eq. (163) when the Bose-symmetric evaluation of the AAA triangle is used on the left-hand side. One way of doing this would be to explicitly calculate $R^5_{\sigma\rho\mu}$ and its divergence, as we did in our discussion of the VVA case in Eqs. (55)-(63). But this is not actually necessary: we can answer our question by comparison with our result for the VVA triangle. To see this, we note that arguments similar to those of Subsection 2.2 show that any anomaly term in Eq. (163) must be independent of the fermion mass m_0, so that it suffices to consider the case $m_0 = 0$. When $m_0 = 0$, the simple relation

$$(-i\gamma_\sigma\gamma_5)\,\frac{i}{\not{k}}\,(-i\gamma_\rho\gamma_5) = (-i\gamma_\sigma)\,\frac{i}{\not{k}}\,(-i\gamma_\rho) \tag{164}$$

indicates that Eq. (162a) is formally identical to Eq. **(54)** for the VVA triangle $R_{\sigma\rho\mu}$ in the $m_0 = 0$ limit. **Remembering that $R^5_{\sigma\rho\mu}$** is obtained from Eq. **(162a) by symmetrizing with respect to the three vertices, this tells us that in the $m_0 = 0$ limit $R^5_{\sigma\rho\mu}(k_1, k_2)$ is related to $R_{\sigma\rho\mu}(k_1, k_2)$ by**

$$R^5_{\sigma\rho\mu}(k_1, k_2) = (\tfrac{1}{3})[R_{\sigma\rho\mu}(k_1, k_2) + R_{\rho\mu\sigma}(k_2, -(k_1+k_2)) + R_{\mu\sigma\rho}(-(k_1+k_2), k_1)]. \tag{165}$$

Taking the divergence of Eq. (165), and using the fact that $R_{\sigma\rho\mu}$ is divergenceless with respect to the first two indices (the vector indices) and has the third index divergence given by Eq. (63), we find

$$-(k_1+k_2)^\mu R^5_{\sigma\rho\mu}(k_1, k_2) = (\tfrac{1}{3})\, 8\pi^2 k_1^\xi k_2^\tau \epsilon_{\xi\tau\sigma\rho} \tag{166}$$

when $m_0 = 0$. Thus, the AAA diagram has a divergence anomaly which is 1/3 of the anomaly in the VVA diagram. Comparing with Eq. (163), we see that for general m_0 Eq. (166) becomes

$$-(k_1+k_2)^\mu R^5_{\sigma\rho\mu}(k_1, k_2) = 2m_0 R^5_{\sigma\rho} + (8\pi^2/3) k_1^\xi k_2^\tau \epsilon_{\xi\tau\sigma\rho}, \tag{167}$$

which is our final result for the AAA Ward identity. Note that Eq. (166) has the interesting implication that even in the $m_0 = 0$ limit, it is impossible to construct a quantum electrodynamics in which the photon is coupled to the axial vector current, since the AAA triangle diagram cannot be made to simultaneously satisfy the requirements of Bose symmetry and current conservation.

There is yet another way of obtaining Eq. (167) for the AAA diagram [and also our old results of Eq. (63) for the VVA case] without performing a full explicit calculation. This is to introduce a regulator mass M_0 by subtracting from Eqs. (54) and (162a) the corresponding expression with m_0 replaced by M_0:

$$\begin{aligned}
\frac{-ie_0^2}{(2\pi)^4} R^{REG}_{\sigma\rho\mu} &\equiv 2\int \frac{d^4r}{(2\pi)^4} (-1)\, Tr\{I_{VVA}\}, \\
\frac{-i}{(2\pi)^4} R^{5REG}_{\sigma\rho\mu} &\equiv 2\int \frac{d^4r}{(2\pi)^4} (-1)\, Tr\{I_{AAA}\},
\end{aligned} \tag{168}$$

$$\begin{aligned}
I_{VVA} = {} & \frac{i}{\not r + \not k_1 - m_0}(-ie_0\gamma_\sigma)\frac{i}{\not r - m_0}(-ie_0\gamma_\rho)\frac{i}{\not r - \not k_2 - m_0}\gamma_\mu\gamma_5 \\
& - \frac{i}{\not r + \not k_1 - M_0}(-ie_0\gamma_\sigma)\frac{i}{\not r - M_0}(-ie_0\gamma_\rho)\frac{i}{\not r - \not k_2 - M_0}\gamma_\mu\gamma_5 , \\
I_{AAA} = {} & \frac{i}{\not r + \not k_1 - m_0}(-i\gamma_\sigma\gamma_5)\frac{i}{\not r - m_0}(-i\gamma_\rho\gamma_5)\frac{i}{\not r - \not k_2 - m_0}\gamma_\mu\gamma_5 \\
& - \frac{i}{\not r + \not k_1 - M_0}(-i\gamma_\sigma\gamma_5)\frac{i}{\not r - M_0}(-i\gamma_\rho\gamma_5)\frac{i}{\not r - \not k_2 - M_0}\gamma_\mu\gamma_5 .
\end{aligned}$$

Because the subtractions in Eq. (168) remove the linear divergences in both the VVA and AAA cases, $R^{REG}_{\sigma\rho\mu}$ will be automatically divergenceless with respect to the first two indices and $R^{5REG}_{\sigma\rho\mu}$ will be automatically Bose symmetric, with no need for additional subtractions. Thus, the quantities $R_{\sigma\rho\mu}$ and $R^5_{\sigma\rho\mu}$ are simply the limits of the corresponding regulated quantities as $M_0 \to \infty$,

$$R_{\sigma\rho\mu} = \lim_{M_0\to\infty} R^{REG}_{\sigma\rho\mu} \tag{169}$$
$$R^5_{\sigma\rho\mu} = \lim_{M_0\to\infty} R^{5\,REG}_{\sigma\rho\mu}$$

To study the axial vertex divergences, we use the fact that since the regulated triangles have no linear divergences, they have no Ward identity anomalies, but rather satisfy the normal Ward identities

$$-(k_1+k_2)^\mu R^{REG}_{\sigma\rho\mu} = 2m_0 R_{\sigma\rho} - 2M_0 R_{\sigma\rho}|_{m_0\to M_0}, \tag{170}$$
$$-(k_1+k_2)^\mu R^{5\,REG}_{\sigma\rho\mu} = 2m_0 R^5_{\sigma\rho} - 2M_0 R^5_{\sigma\rho}|_{m_0\to M_0}.$$

Taking the limit $M_0 \to \infty$, we find

$$-(k_1+k_2)^\mu R_{\sigma\rho\mu} = 2m_0 R_{\sigma\rho} - \lim_{M_0\to\infty} 2M_0 R_{\sigma\rho}|_{m_0\to M_0},$$
$$-(k_1+k_2)^\mu R^5_{\sigma\rho\mu} = 2m_0 R^5_{\sigma\rho} - \lim_{M_0\to\infty} 2M_0 R^5_{\sigma\rho}|_{m_0\to M_0}. \tag{171}$$

As we saw in Eq. (61), $R_{\sigma\rho}$ is given by

$$R_{\sigma\rho} = k_1^\xi k_2^\tau \epsilon_{\xi\tau\sigma\rho} 8\pi^2 m_0 I_{00}(k_1, k_2), \tag{172a}$$

and a simple calculation shows that

$$R^5_{\sigma\rho} = k_1^\xi k_2^\tau \epsilon_{\xi\tau\sigma\rho} 8\pi^2 m_0 [2(I_{10}+I_{01}) - I_{00}]. \tag{172b}$$

Thus, it is easy to evaluate the limits in Eq. (171), giving

$$-\lim_{M_0\to\infty} 2M_0 R_{\sigma\rho}|_{m_0\to M_0} = -2M_0 k_1^\xi k_2^\tau \epsilon_{\xi\tau\sigma\rho} 8\pi^2 M_0 \int_0^1 dx \int_0^{1-x} dy(-M_0^2)^{-1}$$
$$= 8\pi^2 k_1^\xi k_2^\tau \epsilon_{\xi\tau\sigma\rho}, \tag{173}$$
$$-\lim_{M_0\to\infty} 2M_0 R^5_{\sigma\rho}|_{m_0\to M_0} = -2M_0 k_1^\xi k_2^\tau \epsilon_{\xi\tau\sigma\rho} 8\pi^2 M_0 \int_0^1 dx \int_0^{1-x} dy$$

$$\times [2(x+y)-1](-M_0^2)^{-1} = (8\pi^2/3)k_1^\xi k_2^\tau \epsilon_{\xi\tau\sigma\rho},$$

in agreement with Eqs. (63) and (167). Thus we see that from the regulator point of view, the <u>divergence anomalies result from the failure of the regulator term in the naive divergence to vanish in the limit of infinite regulator mass</u>.

Let us now turn to the question of whether larger loops than triangle diagrams can have Ward identity anomalies. This question has been carefully analyzed, in the case of fermion loops, by both the ϵ-separation method[29] discussed in Subsection 2.4 and by the regulator method[30] discussed immediately above. In addition to considering vector and axial-vector vertices, the analyses allow scalar and pseudoscalar couplings as well, and internal degrees of freedom (such as SU_3 or isotopic spin) are also permitted. The results may be summarized as follows:

(i) No loops involving scalar or pseudoscalar couplings have Ward identity anomalies which cannot be removed by appropriately chosen subtractions. The only loops with true anomalies, which cannot be removed, are ones with <u>only</u> vector and axial-vector vertices, with the number of axial-vector vertices <u>odd</u>. If subtraction terms are chosen so that all vector index Ward identities are normal, the following loops have anomalous axial index Ward identities:

the VVA and AAA triangles, the VVVA and VAAA squares, and the VVVVA, VVAAA and AAAAA pentagons. The triangle anomalies are the ones which we have already discussed, and are the only anomalies when internal degrees of freedom are absent. The squares are anomalous because the naive Ward identity derivations for them involve a translation of integration variable in linearly divergent triangle diagrams. In the case of the pentagons, the naive derivations involve a translation of integration variable in logarithmically divergent square diagrams, which is allowed, but the Ward identities become anomalous as a result of the counter-terms which were added to the square diagrams to satisfy the vector current Ward identities. All diagrams larger than the pentagons have normal Ward identities.

(ii) A compact and explicit description of these anomalies[29] may be obtained by introducing external scalar (S), pseudoscalar (P), vector (V^{μ}) and axial-vector (A^{μ}) fields which couple to the respective currents, and which allow us to write a simple generating functional for all of the fermion loop diagrams. [Note the change in notation from our discussion of QED, where A^{μ} denoted the (vector) photon field.] We start with a Lagrangian density which we write as

$$\mathcal{L}(x) = \mathcal{L}_0(x) + \mathcal{L}_I(x), \tag{174}$$

$$\mathcal{L}_0(x) = \bar{\psi}(x)(i\gamma\cdot\Box - m_0)\psi(x),$$

$$\mathcal{L}_I(x) = \bar{\psi}(x)[S(x) + i\gamma_5 P(x) + \gamma_\mu V^\mu(x) + \gamma_\mu\gamma_5 A^\mu(x) + \Delta m_0]\,\psi(x),$$

with each of the fields $S(x)$, $P(x)$, $V^\mu(x)$, $A^\mu(x)$ a matrix in the internal space,

$$\begin{aligned} S(x) &= \sum_a \lambda_S^a S_a(x), & P(x) &= \sum_a \lambda_P^a P_a(x), \\ V^\mu(x) &= \sum_a \lambda_V^a V_a^\mu(x), & A^\mu(x) &= \sum_a \lambda_A^a A_a^\mu(x). \end{aligned} \tag{175}$$

The fields $S_a(x)$, ... are the external fields and the matrices λ_S^a, ... are their respective coupling matrices. The matrix Δm_0 supplies fermion mass splittings. Going over to the conventional interaction picture, the S-matrix is formally defined by

$$\mathcal{S} = T\,\exp\{i\int d^4x\,\mathcal{L}_I(x)\}, \tag{176}$$

with T the time ordering operator. If we take the fermion vacuum expectation $\langle 0|\mathcal{S}|0\rangle$ and drop disconnected diagrams, we clearly get a generating functional for all of the closed loops. Defining vector and axial vector currents by

$$\mathcal{F}_{a\mu}(x) = -i\frac{\delta}{\delta V_a^\mu(x)}\mathcal{S},\quad \mathcal{F}_{a\mu}^5(x) = -i\frac{\delta}{\delta A_a^\mu(x)}\mathcal{S}, \tag{177}$$

the calculations described above lead to the divergence equations

$$\partial^\mu \mathcal{F}_{a\mu}(x) = D_a(x),$$

$$\partial^\mu \mathcal{F}^5_{a\mu}(x) = D^5_a(x) + \frac{1}{4\pi^2}\epsilon_{\mu\nu\sigma\tau}\mathrm{Tr}_I\{\lambda^a_A[\tfrac{1}{4}F^{\mu\nu}_V(x)F^{\sigma\tau}_V(x)$$
$$+\frac{1}{12}F^{\mu\nu}_A(x)F^{\sigma\tau}_A(x) + \frac{2}{3}i\,A^\mu(x)\,A^\nu(x)\,F^{\sigma\tau}_V(x)$$
$$+\frac{2}{3}iF^{\mu\nu}_V(x)\,A^\sigma(x)\,A^\tau(x) + \frac{2}{3}iA^\mu(x)F^{\nu\sigma}_V(x)\,A^\tau(x) \tag{178}$$
$$-\frac{8}{3}A^\mu(x)\,A^\nu(x)\,A^\sigma(x)\,A^\tau(x)]\},$$

with Tr_I a trace over the internal degrees of freedom, with

$$F^{\mu\nu}_V(x) = \partial^\mu V^\nu(x) - \partial^\nu V^\mu(x) - i[V^\mu(x), V^\nu(x)]$$
$$-i[A^\mu(x), A^\nu(x)], \tag{179}$$
$$F^{\mu\nu}_A(x) = \partial^\mu A^\nu(x) - \partial^\nu A^\mu(x) - i[V^\mu(x), A^\nu(x)]$$
$$-i[A^\mu(x), V^\nu(x)],$$

and with D_a and D^5_a the naive vector and axial-vector divergences. From Eq. (178), all of the fermion loop Ward identities are easily generated by variation with respect to the external fields.

(iii) Finally, we note that a calculation similar to the one described above shows that there are no boson loop Ward identity anomalies.[31]

As a result of the fact that no loops involving scalar or pseudoscalar couplings have Ward identity anomalies, it is easy to see that $\pi^0 \to 2\gamma$, and the SU_3-related processes[32] $\eta \to 2\gamma$ and $X^0 \to 2\gamma$, are the only cases in which the anomalies alter the usual current-algebra-PCAC predictions.

In particular, the anomalies which we have discussed do not alter the predictions of current algebra in the troublesome $\eta \to 3\pi$ decays.

6. CONNECTION BETWEEN WARD IDENTITY ANOMALIES AND COMMUTATOR (BJORKEN-LIMIT) ANOMALIES

As we have seen, the VVA Ward identity anomaly implies other anomalies as well, such as in the renormalization of the axial-vector vertex and in the behavior of γ_5-transformations in massless electrodynamics. In the present section, we will see that the Ward identity anomaly, also implies that anomalous commutators are present: that is, the divergence anomaly requires that certain simple commutators, involving the electromagnetic potential and the currents, have values different from the canonical ones obtained by naive use of canonical commutation relations. The actual, correct expressions for the commutators will be deduced from our formula for the triangle diagram by a technique introduced by Bjorken[33] and Johnson and Low,[34] usually called the "Bjorken limit" method. Commutator anomalies are not a new phenomenon; in the usual QED without axial-vector currents, anomalies in potential-current commutators ("seagulls") and in current-current commutators ("Schwinger terms")[35] have been known for some time. The two anomalies in QED are related, and cancel exactly when the divergence of a covariant matrix element is taken, guaranteeing current conservation. The distinguishing feature of the commutator anomalies associated with the triangle diagram is that when the axial-vector divergence is taken, the seagull and Schwinger term do not cancel.[36] Rather, they combine to give the divergence anomaly found in Section 2, giving an alternative interpretation of the divergence anomaly as the result of non-cancellation of seagull and Schwinger term.

We begin our discussion by reviewing the seagull and Schwinger term, and their cancellation, in QED and also by outlining the Bjorken limit method. Then we will apply the concepts which we have developed to the VVA triangle diagram.

6.1 Schwinger Terms, Seagulls, the Reduction Formula and T and T* Products in QED

Let us consider the equal-time commutator of the time and space components of the ordinary electromagnetic current in QED,

$$[\, j_0(\underset{\sim}{x}, t),\ j_r(\underset{\sim}{y}, t)\,] \,. \tag{180}$$

By naive use of canonical commutation relations, we find for this commutator the value

$$\begin{aligned} &[\psi^\dagger(\underset{\sim}{x}, t)\, \psi(\underset{\sim}{x}, t), \psi^\dagger(\underset{\sim}{y}, t)\gamma_0\gamma_r\, \psi(\underset{\sim}{y}, t)] \\ &= \delta^3(\underset{\sim}{x}-\underset{\sim}{y})\ \psi^\dagger(\underset{\sim}{x}, t)[\,1, \gamma_0\gamma_r\,]\, \psi(\underset{\sim}{x}, t) = 0. \end{aligned} \tag{181}$$

It is easy to see, however, that Eq. (181) is false. To prove this we take the divergence $\partial/\partial y_r$ of Eq. (180) and use current conservation, which tells us that $\partial/\partial y_r\, j_r(\underset{\sim}{y}, t) = -(\partial/\partial t)\, j_0(\underset{\sim}{y}, t)$, giving

$$0=[\, j_0(\underset{\sim}{x}, t),\, \partial^r j_r(\underset{\sim}{y}, t)\,] = -[\, j_0(\underset{\sim}{x}, t), \frac{\partial}{\partial t}\, j_0(\underset{\sim}{y}, t)\,] \tag{182}$$

if Eq. (181) is valid. Taking the vacuum expectation value of Eq. (182), setting $\underset{\sim}{x} = \underset{\sim}{y}$ and inserting a complete set of intermediate states to evaluate the commutator, we find

$$0 = \langle 0 |[\, j_0(\underset{\sim}{x}, t), \frac{\partial}{\partial t}\, j_0(\underset{\sim}{x}, t)\,]\, | 0 \rangle$$

$$= \sum_n \langle 0 | j_0(\underset{\sim}{x}, t) | n\rangle\langle n | \frac{\partial}{\partial t}\, j_0(\underset{\sim}{x}, t) | 0\rangle - \langle 0 | \frac{\partial}{\partial t}\, j_0(\underset{\sim}{x}, t) | n\rangle\langle n | j_0(\underset{\sim}{x}, t) | 0\rangle$$

$$= 2i \sum_n (E_n - E_0)\, |\langle 0 | j_0(\underset{\sim}{x}, t) | n\rangle|^2 \tag{183}$$

All the terms on the right-hand side of Eq. (183) are greater than equal to zero, so Eq. (183) can be satisfied only if $\langle 0 | j_0(\underset{\sim}{x}, t) | n\rangle = 0$ for all intermediate states n, which is manifestly untrue. Thus the commutator of Eq. (180) cannot vanish, as the naive use of canonical commutation relations suggests.

In order to understand this result, let us follow the procedure of Subsection 2.4 and define the space-component[37] of the vector current as the limit of a non-local current, with spacelike separation $\underset{\sim}{\epsilon}$, averaged over the direction of $\underset{\sim}{\epsilon}$,

$$j_r(\underset{\sim}{y}, t) = \lim_{\underset{\sim}{\varepsilon} \to 0} \underset{\underset{\sim}{\varepsilon}}{\mathrm{av}}\, j_r(\underset{\sim}{y}, t; \underset{\sim}{\varepsilon}), \tag{184}$$

$$j_r(\underset{\sim}{y}, t; \underset{\sim}{\varepsilon}) = \overline{\psi}(\underset{\sim}{y}+\tfrac{1}{2}\underset{\sim}{\varepsilon}, t)\gamma_r \psi(\underset{\sim}{y}-\tfrac{1}{2}\underset{\sim}{\varepsilon}, t)\exp[ie_0 \int_{\underset{\sim}{x}-\frac{1}{2}\underset{\sim}{\varepsilon}}^{\underset{\sim}{x}+\frac{1}{2}\underset{\sim}{\varepsilon}} d\underset{\sim}{\ell} \cdot \underset{\sim}{A}(\underset{\sim}{\ell}, t)] .$$

Substituting Eq. (184) into the commutator of Eq. (180) and evaluating by using canonical commutation relations, we find

$$\begin{aligned} [j_0(\underset{\sim}{x}, t), j_r(\underset{\sim}{y}, t; \underset{\sim}{\varepsilon})] &= [\delta^3(\underset{\sim}{x}-\underset{\sim}{y} - \tfrac{1}{2}\underset{\sim}{\varepsilon}) - \delta^3(\underset{\sim}{x}-\underset{\sim}{y}+\tfrac{1}{2}\underset{\sim}{\varepsilon})] \\ &\quad \times \overline{\psi}(\underset{\sim}{y}+\tfrac{1}{2}\underset{\sim}{\varepsilon}, t)\gamma_r\psi(\underset{\sim}{y}-\tfrac{1}{2}\underset{\sim}{\varepsilon}, t)\exp[\ \dots\ ] \\ &= -\underset{\sim}{\varepsilon} \cdot \underset{\sim}{\nabla}_x\, \delta^3(\underset{\sim}{x}-\underset{\sim}{y})\, j_r(\underset{\sim}{y}, t; \underset{\sim}{\varepsilon}) . \end{aligned} \tag{185}$$

Eq. (185) at first glance appears to vanish, because of the factor $\underset{\sim}{\varepsilon}$ in front, but an elementary calculation shows that

$$<0|j_r(\underset{\sim}{y}, t; \underset{\sim}{\varepsilon})|0> = \frac{n\, \varepsilon_r}{(\underset{\sim}{\varepsilon}^2)^2} \tag{186}$$

with n a numerical factor, so that Eqs. (184) and (185) give

$$[j_0(\underset{\sim}{x}, t), j_r(\underset{\sim}{y}, t)] = \partial_r\, \delta^3(\underset{\sim}{x}-\underset{\sim}{y}) \lim_{\underset{\sim}{\varepsilon} \to 0} n/(3\underset{\sim}{\varepsilon}^2) \tag{187}$$
$$+ \text{ possible operator term.}$$

The divergent c-number term on the right-hand side of Eq. (187), called the Schwinger term, eliminates the paradox of Eq. (183). In QED, it appears that no operator term is present, so that the Schwinger term is pure c-number, but this is not true in other field theory models.

Clearly, giving the space-component of the electromagnetic current an explicit dependence on the electromagnetic potential will alter the potential-current commutation relations. Thus, from Eq. (184) we find

$$\begin{aligned}&[\,j_r(\underset{\sim}{y},t),\ \partial A_s(\underset{\sim}{x},t)/\partial t]\\ &=\lim_{\underset{\sim}{\varepsilon}\to 0}\ \underset{\underset{\sim}{\varepsilon}}{\mathrm{av}}\ j_r(\underset{\sim}{y},t;\underset{\sim}{\varepsilon})[\,ie_0\underset{\sim}{\varepsilon}\cdot\underset{\sim}{A}(\underset{\sim}{y},t),\ \partial A_s(\underset{\sim}{x},t)/\partial t]\\ &=e_0\, g_{rs}\ \delta^3(\underset{\sim}{x}-\underset{\sim}{y})\ \lim_{\underset{\sim}{\varepsilon}\to 0} n/(3\underset{\sim}{\varepsilon}^2),\end{aligned} \tag{188}$$

whereas the naive value of this commutator, computed without the $\underset{\sim}{\varepsilon}$-separation, would be zero. We note that the anomalous commutator in Eq. (188) has the same coefficient as the one in Eq. (187); this is not an accident, but rather is necessary to preserve the gauge-invariance of the theory. To understand this, let us study the matrix element for the photon scattering process $\gamma(k_1) + A \to \gamma(k_2) + B$, with A and B arbitrary states of the theory. Applying the LSZ reduction formula[38] to the initial photon, we find that this matrix element is proportional to $\langle B\gamma(k_2)|j_\mu(0)|A\rangle$. Applying the reduction formula a second time, to bring in the final photon, gives

$$\begin{aligned}&\langle B\gamma(k_2)|j_\mu(0)|A\rangle = \varepsilon_\lambda^* M_\mu^\lambda,\\ &M_\mu^\lambda = \frac{-i}{\sqrt{2k_{20}}\sqrt{Z_3}}\int d^4x\, e^{ik_2\cdot x}\,\Box_x^2\,\langle B|\,T(A^\lambda(x)j_\mu(0))\,|A\rangle.\end{aligned} \tag{189}$$

As usual, gauge invariance requires that $\varepsilon_\lambda^* M^\lambda$ be invariant under the gauge transformation $\varepsilon_\lambda \to \varepsilon_\lambda + v k_{2\lambda}$, which implies that

$$k_{2\lambda} M^{\lambda}_{\mu} = 0 \ . \tag{190}$$

Let us now rewrite Eq. (189) by bringing the operator $\Box^2_x$ inside the time-ordered product, so that we can use the equation of motion $\Box^2_x A^{\lambda} = e_0 j^{\lambda}$. Keeping <u>all</u> equal-time commutators which arise from time derivatives of the time ordered product, we get

$$\begin{aligned} M^{\lambda}_{\mu} = \frac{-i}{\sqrt{2k_{20}}\sqrt{Z_3}} \int d^4x \, e^{ik_2 \cdot x} \{ & e_0 \langle B | T(j^{\lambda}(x) j_{\mu}(0)) | A \rangle \\ & + \delta(x_0) \langle B | [\partial A^{\lambda}(x)/\partial x_0, j_{\mu}(0)] | A \rangle \\ & + (\partial/\partial x_0)[\delta(x_0) \langle B | [A^{\lambda}(x), j_{\mu}(0)] | A \rangle] \} . \end{aligned} \tag{191}$$

Even with our $\underset{\sim}{\varepsilon}$-separation present, the third term on the right-hand side of Eq. (191) vanishes. The second term vanishes when either $\mu = 0$ or $\lambda = 0$, but when μ and λ are both spatial components it is just the anomalous potential-current commutator of Eq. (188),

$$\delta(x_0)[\partial A^{\lambda}(x)/\partial x_0, j_{\mu}(0)] = -e_0(g^{\lambda}_{\mu} - g^{\lambda 0} g_{\mu 0}) \delta^4(x) \lim_{\underset{\sim}{\varepsilon} \to 0} n/(3\underset{\sim}{\varepsilon}^2). \tag{192}$$

Eq. (192), which is colloquially called the "seagull" term, describes the coupling of two photons to an electron line at the same space-time point

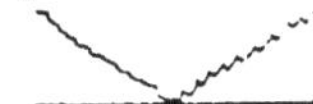

which results from the potential-dependence of the electromagnetic current.

To see that the seagull term plays an essential role in maintaining gauge invariance, let us multiply Eq. (191) by $k_{2\lambda}$. By an integration by parts and by use of Eq. (3) (current-conservation),the contribution of the first term on the right-hand side of Eq. (191) becomes

$$\begin{aligned} &\frac{e_0}{\sqrt{2k_{20}}\sqrt{Z_3}} \int d^4x\, e^{ik_2\cdot x} \frac{\partial}{\partial x^\lambda} <B|\, T(j^\lambda(x) j_\mu(0))\,|A> \\ &= \frac{e_0}{\sqrt{2k_{20}}\sqrt{Z_3}} \int d^4x\, e^{ik_2\cdot x} <B|\delta(x_0)[\, j^0(x), j_\mu(0)]\,|A>. \end{aligned} \tag{193}$$

This expression is just the Schwinger term, and on substitution of Eq. (187) becomes

$$\frac{-ik_{2\lambda} e_0 (g_\mu^\lambda - g^{\lambda 0} g_{\mu 0})}{\sqrt{2k_{20}}\,\sqrt{Z_3}} < B|A > \lim_{\underset{\sim}{\varepsilon}\to 0} n/(3\underset{\sim}{\varepsilon}^2). \tag{194}$$

Thus, because of the presence of the Schwinger term, the divergence of the T-product term in Eq. (191) is not zero. But, combining Eqs. (191) and (192), we see that the divergence of the seagull term is

$$\frac{ik_{2\lambda} e_0 (g_\mu^\lambda - g^{\lambda 0} g_{\mu 0})}{\sqrt{2k_{20}}\,\sqrt{Z_3}} < B|A > \lim_{\underset{\sim}{\varepsilon}\to 0} n/(3\underset{\sim}{\varepsilon}^2), \tag{195}$$

which just cancels away the Schwinger term contributions and gives Eq. (190) for the total matrix element. So we see that gauge invariance is maintained by a cancellation between the divergence of the seagull term and the Schwinger term.

Equation (191) is frequently rewritten in the form

$$M_\mu^\lambda = \frac{-i}{\sqrt{2k_{20}}\sqrt{Z_3}} \int d^4x\, e^{ik_2\cdot x} \langle B|\, T^*(j^\lambda(x) j_\mu(0))\, |A\rangle, \quad (196)$$

with the T*-product defined as the sum of the T-product and the seagull term,

$$T^*(j^\lambda(x) j_\mu(0)) \equiv T(j^\lambda(x)\, j_\mu(0)) \quad (197)$$

$$+\, e_0(g_\mu^\lambda - g^{\lambda 0} g_{\mu 0})\,\delta^4(x) \lim_{\underset{\sim}{\varepsilon}\to 0} \eta/(3\underset{\sim}{\varepsilon}^2).$$

As we have seen, because of the Schwinger term -seagull cancellation, the T* product satisfies the simple current conservation equation

$$\frac{\partial}{\partial x^\lambda} T^*(j^\lambda(x)\, j_\mu(0)) = 0\,. \quad (198)$$

Also, the T*-product transforms as a 2-index Lorentz tensor. (It is covariant because its matrix elements are the covariant Feynman amplitudes.) On the other hand, since the seagull term is not Lorentz covariant, it is clear that the T-product is not Lorentz covariant either. In other words, the properties which are naively attributed to the T-product are actually satisfied by the T*-product, and not by the T-product, when Schwinger terms and seagulls are present.

6.2 The Bjorken-Johnson-Low Method

Although we have found the $\underset{\sim}{\varepsilon}$-separation method to be useful in the above discussion, the method leads to inconsistencies when applied to more general types of commutator anomalies.[39] That this is so should not be too surprising since the averaging procedure of Eq. (184), which excludes timelike separations, is clearly non-covariant. If we include timelike separations to try to get a covariant definition of the current, we are no longer allowed to use canonical equal time commutation relations to evaluate the current-current commutators, but instead must use dynamics (the equations of motion) to follow the time evolution of the fields. Once we have to do this, however, we might just as well abandon our canonical procedure entirely, and instead calculate equal-time commutators as the limit as $t \to 0$ of unequal time commutators, with the latter calculated directly from Feynman diagrams. The Bjorken-Johnson-Low limit gives us a simple way of doing this.

We consider the T-product of two operators $J_{(1)}$ and $J_{(2)}$ and take the matrix element between arbitrary states A and B,

$$< B | \int d^4x \, e^{iq\cdot x} T(J_{(1)}(x) J_{(2)}(0)) | A>. \qquad (199)$$

This matrix element has an analytic continuation into the upper half q_0 plane given by the retarded commutator

$$< B| \int d^4x\, e^{iq\cdot x}\theta(x_0)[\, J_{(1)}(x), J_{(2)}(0)\,]\, |A > \quad (200)$$
$$= \int_0^\infty dt\, e^{iq_0 t}\phi(t),$$
$$\phi(t) \equiv <B|\int d^3x\, e^{-i\underset{\sim}{q}\cdot \underset{\sim}{x}}[\, J_{(1)}(\underset{\sim}{x}, t), J_{(2)}(0)\,]\, |A >.$$

Let us now let q_0 approach infinity in the upper half plane, that is, we set $q_0 = iR$, $R \to \infty$. To find the behavior of Eq. (200), we Taylor expand $\phi(t)$ about $t = 0$,

$$\phi(t) = \phi(t)\,|_{t\to 0} + t\, \phi'(t)\,|_{t\to 0} + \ldots, \quad (201)$$

which on substitution into Eq. (200) gives

$$\int_0^\infty dt\, e^{-Rt}\phi(t) = \frac{1}{R}\phi(t)\,|_{t\to 0} + \frac{1}{R^2}\phi'(t)\,|_{t\to 0} + \cdots \quad (202)$$

Thus we have learned that the matrix elements of the equal time commutators $[\, J_{(1)}, J_{(2)}\,]$, $[\,\partial J_{(1)}/\partial t, J_{(2)}\,]$, ... are just the coefficients of q_0^{-1}, q_0^{-2}, ... as we take $q_0 \to i\infty$ in Eq. (199):

$$\lim_{q_0 \to i\infty} <B| \int d^4x\, e^{iq\cdot x} T(J_{(1)}(x)\, J_{(2)}(0))| A> \quad (203)$$
$$= (-iq_0)^{-1} <B| \int d^3x\, e^{-i\underset{\sim}{q}\cdot \underset{\sim}{x}}[\, J_{(1)}(\underset{\sim}{x}, t), J_{(2)}(0)\,]\, |_{t\to 0}|A>$$
$$+ (-iq_0)^{-2} <B| \int d^3x\, e^{-i\underset{\sim}{q}\cdot \underset{\sim}{x}}[\partial J_{(1)}(\underset{\sim}{x}, t)/\partial t, J_{(2)}(0)\,]\, |_{t\to 0}|A> + \ldots$$

This formula is the recipe of Bjorken and of Johnson and Low. Clearly, the series in Eq. (203) cannot be extended arbitrarily far; a necessary condition for it to be valid out to power q_0^{-n} is that the Taylor coefficients $\phi^j(t)\,|_{t\to 0}$, $0 \le j \le n-1$ must exist. Although Eq. (203) has been formulated in terms of the T-product, it can be applied to the T*-

product as well. To see this, we note that the x-dependence of the seagull term consists of $\delta^4(x)$ and possibly, a finite number of derivatives of $\delta^4(x)$, and therefore the Fourier transform $\int d^4x\, e^{iq\cdot x}$(seagull) is purely a polynomial in q_0. Hence from a Feynman amplitude, or T*-product, we obtain a T-product to which Eq. (203) can be applied by dropping polynomial terms in q_0 which do not vanish as $q_0 \to i\infty$.

The equal-time commutator defined by Eq. (203) has the nice property that, barring pathological oscillatory behavior,[40] it agrees with the usual definition of the commutator as a sum over intermediate states. To show this, we write the Low equation for the left-hand side of Eq. (203),

$$<B| \int d^4x\, e^{iq\cdot x} T(J_{(1)}(x)\, J_{(2)}(0))|A>$$
$$= i\int_{-\infty}^{\infty} dq_0' \,\frac{\rho_{AB}(\underset{\sim}{q}, q_0')}{q_0 - q_0'} \,, \qquad (204)$$

with ρ the spectral function

$$\rho_{AB}(\underset{\sim}{q}, q_0) = (2\pi)^3 \sum_n <B|J_{(1)}|n><n|J_{(2)}|A)\, \delta^4(q+p_B-p_n)$$
$$-(2\pi)^3 \sum_n <B|J_{(2)}|n><n|J_{(1)}|A>\delta^4(q+p_n-p_A). \qquad (205)$$

Provided that the spectral function does not oscillate an infinite number of times [and it cannot in perturbation theory, where we will be applying Eq. (203)] , when the coefficient of q_0^{-1} exists it is equal to the integral

$$\int_{-\infty}^{\infty} dq_0' \, \rho_{AB}(\underset{\sim}{q}, q_0'), \tag{206}$$

which is just the usual sum-over-intermediate-states definition of the commutator.

In conclusion, we note that the results of Eqs. (187) and (188) for the Schwinger term and seagull in QED, which were obtained above by the $\underset{\sim}{\varepsilon}$-separation method, can equally well[33] be obtained by applying the Bjorken-limit technique to the vacuum polarization tensor $\Pi(q)^{\mu\nu}$. Nothing will be lost, then, by abandoning the $\underset{\sim}{\varepsilon}$-separation method of determining commutators in favor of the recipe of Eq. (203).

6.3 Anomalous Commutators Associated with the VVA Triangle Anomaly[41]

Let us now apply what we have learned to the lowest order VVA triangle diagram. We start with the two photon to vacuum matrix element of the axial-vector current, $\langle 0 | j_\mu^5(0) | \gamma(k_1, \varepsilon_1)\gamma(k_2, \varepsilon_2)\rangle$, and apply the reduction formula once to pull in one of the initial photons. This gives

$$\begin{aligned} &\langle 0 | j_\mu^5(0) | \gamma(k_1, \varepsilon_1)\gamma(k_2, \varepsilon_2)\rangle [4k_{10}k_{20}]^{\frac{1}{2}} \\ &= -i\varepsilon_1^\sigma \int d^4x \, e^{-ik_1\cdot x} \Box_x^2 \langle 0 | T(j_\mu^5(0)A_\sigma(x)) | \gamma(k_2, \varepsilon_2)\rangle [2k_{20}]^{\frac{1}{2}} \\ &= -i \, \varepsilon_1^\sigma \varepsilon_2^\rho [e_0^2/(2\pi)^4] \, R_{\sigma\rho\mu}(k_1, k_2), \end{aligned} \tag{207}$$

with $R_{\sigma\rho\mu}$ the explicit expression for the lowest order triangle diagram given in Eqs. (55)-(59). [Since, in Eqs. (207)-

(214), we work to lowest order only, we omit the usual wave-function renormalization factor $Z_3^{\frac{1}{2}}$ from Eq. (207).] Bringing $\Box_x^2$ inside the time-ordered product, as in Eq. (191), we get

$$\int d^4x\, e^{-ik_1\cdot x}\Box_x^2 <0|\, T(j_\mu^5(0)A_\sigma(x))\,|\gamma(k_2,\varepsilon_2)>$$
$$= A_{\mu\sigma}k_{10} + B_{\mu\sigma} + C_{\mu\sigma}(k_{10}), \tag{208}$$

with $A_{\mu\sigma}$ and $B_{\mu\sigma}$ the seagull terms

$$A_{\mu\sigma} = i\int d^4x\, e^{ik_1\cdot x}\delta(x_0)<0|\,[\,A_\sigma(x), j_\mu^5(0)\,]\,|\gamma(k_2,\varepsilon_2)>,$$
$$B_{\mu\sigma} = \int d^4x\, e^{ik_1\cdot x}\delta(x_0)<0|\,[\,\dot{A}_\sigma, j_\mu^5(0)\,]\,|\gamma(k_2,\varepsilon_2)>, \tag{209}$$
$$\dot{A}_\sigma(x) \equiv \partial A_\sigma(x)/\partial x_0,$$

and with $C_{\mu\sigma}(k_{10})$ the T-product

$$C_{\mu\sigma}(k_{10}) = e_0\int d^4x\, e^{-ik_1\cdot x}<0|\, T(j_\mu^5(0)\, j_\sigma(x))\,|\gamma(k_2,\varepsilon_2)> \tag{210}$$
$$= e_0\int d^4x\, e^{i(k_1+k_2)\cdot x}<0|\, T(j_\mu^5(x)j_\sigma(0))\,|\gamma(k_2,\varepsilon_2)>.$$

Let us first show that the assumption that no commutator anomalies are present leads to a contradiction. If all commutators are given by their naive values, $A_{\mu\sigma}$ and $B_{\mu\sigma}$ vanish and no Schwinger term appears when we take the axial-index divergence of $C_{\mu\sigma}$, so that

$$< 0|\partial^\mu j_\mu^5\,|\gamma(k_1\varepsilon_1)\gamma(k_2,\varepsilon_2)>[\,2k_{10}]^{\frac{1}{2}} \qquad \text{(A)}$$
$$= -i(k_1+k_2)^\mu<0|\,j_\mu^5|\,\gamma(k_1,\varepsilon_1)\gamma(k_2,\varepsilon_2)>[\,2k_{10}]^{\frac{1}{2}} \tag{211}$$
$$= -i\,\varepsilon_1^\sigma\,[\,-i(k_1+k_2)^\mu\, C_{\mu\sigma}(k_{10})]$$
$$= -i\,\varepsilon_1^\sigma\, e_0\int d^4x\, e^{i(k_1+k_2)\cdot x}<0|\, T(\partial^\mu j_\mu^5(x)j_\sigma(0))\,|\gamma(k_2,\varepsilon_2)> \qquad \text{(B)}$$

Substituting Eq. (68) into Eq. (211), we get from term (A) the matrix element of $2im_0 j^5$, plus a contribution from the anomaly of order e_0^2. On the other hand, from the term (B) in Eq. (211) we get the reduction formula for the matrix element of $2im_0 j^5$, plus a contribution from the anomaly which is of order e_0^3 at least, but the order e_0^2 contribution of the anomaly is missing. Thus, our assumption that all commutators have their naive values is in contradiction with the anomalous axial-vector divergence equation of Eq. (68).

To determine the required values of the anomalous commutators, we use the $k_{10} \to i\infty$ limit discussed above. The seagull terms in Eq. (208) have polynomial dependence on k_{10}, while Eq. (203) tells us that the large k_{10} limit of $C_{\mu\sigma}$ is

$$C_{\mu\sigma}(k_{10}) = \frac{-ie_0}{k_{10}} \int d^4x \, e^{i\underset{\sim}{k}_1 \cdot \underset{\sim}{x}} \delta(x_0) \langle 0 | [j_\sigma(\underset{\sim}{x}, 0), j_\mu^5(0)] | \gamma(k_2, \varepsilon_2) \rangle$$
$$+ \text{ higher order.} \qquad (212)$$

Thus, the equal-time commutators $[A_\sigma(x), j_\mu^5(0)]$, $[\dot{A}_\sigma(x), j_\mu^5(0)]$ and $[j_\sigma(x), j_\mu^5(0)]$ are to be identified, respectively, with the parts of $R_{\sigma\rho\mu}$ behaving like k_{10}, 1 and k_{10}^{-1} as k_{10} becomes infinite. From Eqs. (55)-(59), we find

$$
\begin{aligned}
&\varepsilon_2^{\rho} R_{\sigma\rho\mu}(k_1, k_2) \\
&= 4\pi^2 (k_2^{\tau}\varepsilon_2^{\rho} - k_2^{\rho}\varepsilon_2^{\tau})\{\varepsilon_{\tau\sigma\rho\mu} + g_{\sigma 0}\varepsilon_{0\tau\rho\mu} - g_{\rho 0}\varepsilon_{0\tau\sigma\mu} \\
&\quad + k_{10}^{-1}[\tfrac{1}{2}(1-g_{\sigma 0})(k_{2\sigma}\varepsilon_{0\tau\rho\mu} + k_{20}\varepsilon_{\sigma\tau\rho\mu}) \\
&\quad + g_{\sigma 0}(1-g_{\eta 0})k_1^{\eta}\varepsilon_{\eta\tau\rho\mu} - g_{\rho 0}(1-g_{\eta 0})k_1^{\eta}\varepsilon_{\eta\tau\sigma\mu} \\
&\quad + (\text{terms which vanish when } \sigma = 0 \text{ or} \\
&\quad \mu = 0)]\} + O(k_{10}^{-2}\ln k_{10})
\end{aligned} \tag{213}
$$

Comparing Eq. (213) with Eqs. (209) and (212), we find the equal-time commutation relations

$$
\begin{aligned}
&[A_{\sigma}(x), j_{\mu}^{5}(y)] = [\dot{A}_{0}(x), j_{\mu}^{5}(y)] = 0, \\
&[\dot{A}_{r}(x), j_{0}^{5}(y)] = (-2i\alpha_0/\pi)\delta^3(\underset{\sim}{x}-\underset{\sim}{y})B^{r}(y), \\
&[\dot{A}_{r}(x), j_{s}^{5}(y)] = (i\alpha_0/\pi)\delta^3(\underset{\sim}{x}-\underset{\sim}{y})\varepsilon^{rst}E^{t}(y), \\
&[j_0(x), j_0^5(y)] = (-ie_0/2\pi^2)\underset{\sim}{B}(y)\cdot\underset{\sim}{\nabla}_x\delta^3(\underset{\sim}{x}-\underset{\sim}{y}), \\
&[j_r(x), j_0^5(y)] = (-ie_0/4\pi^2)[\underset{\sim}{E}(x)\times\underset{\sim}{\nabla}_y\delta^3(\underset{\sim}{x}-\underset{\sim}{y})]^{r}, \\
&[j_0(x), j_s^5(y)] = (ie_0/4\pi^2)[\underset{\sim}{E}(y)\times\underset{\sim}{\nabla}_x\delta^3(\underset{\sim}{x}-\underset{\sim}{y})]^{s},
\end{aligned} \tag{214}
$$

with

$$
\begin{aligned}
&B^{t}(x) = [\underset{\sim}{\nabla}\times\underset{\sim}{A}(x)]^{t} = \varepsilon^{rst}\frac{\partial}{\partial x^{r}}A^{s}(x), \\
&E^{t}(x) = -\dot{A}^{t}(x) - \frac{\partial}{\partial x^{t}}A^{0}(x), \\
&\varepsilon^{123} = 1.
\end{aligned} \tag{215}
$$

We have only listed the current-current commutators containing at least one time component, since these are the only ones which appear when divergences with respect to the vector or axial-vector indices (σ or μ) are brought inside the time-ordered product in Eq. (210). All of the nonvanishing commutators in Eq. (214) are anomalous in the sense that if they are calculated by naive use of canonical commutation relations they vanish.

It is easy to check that the anomalous commutation relations of Eq. (214), together with the reduction formula of Eqs. (208)-(210), correctly reproduce the known divergence properties of the lowest-order triangle diagram. To check gauge invariance for the photon which has been reduced in, we multiply Eq. (208) by k_1^σ and use vector current conservation, giving

$$\begin{aligned} & k_1^\sigma \int d^4x\, e^{-ik_1\cdot x} \Box_x^2 \langle 0| T(j_\mu^5(0)\, A_\sigma(x)) |\gamma(k_2,\epsilon_2)\rangle \\ = {} & k_1^\sigma \int d^4x\, e^{ik_1\cdot x}\delta(x_0) \langle 0| [\dot{A}_\sigma(x), j_\mu^5(0)] |\gamma(k_2,\epsilon_2)\rangle \\ & -ie_0 \int d^4x\, e^{ik_1\cdot x}\delta(x_0) \langle 0| [j_0(x), j_\mu^5(0)] |\gamma(k_2,\epsilon_2)\rangle . \end{aligned} \tag{216}$$

Substituting the commutators of Eq. (214), we easily see that the seagull and Schwinger terms on the right-hand side of Eq. (216) cancel, as expected. To check the axial-vector divergence of the triangle, we multiply by $-i(k_1+k_2)^\mu$, as we did in Eq. (211). When seagulls and Schwinger terms are kept, term (B) of Eq. (211) becomes

$$\begin{aligned} & -i\epsilon_1^\sigma\{ e_0 \int d^4x e^{i(k_1+k_2)\cdot x} \langle 0| T(\partial^\mu j_\mu^5(x)\, j_\sigma(0)) |\gamma(k_2,\epsilon_2)\rangle \\ & +i\int d^4x\, e^{ik_1\cdot x}\delta(x_0)\Big[-(k_1+k_2)^\mu \langle 0| [\dot{A}_\sigma(x), j_\mu^5(0)] |\gamma(k_2,\epsilon_2)\rangle \\ & \qquad +ie_0 \langle 0| [j_\sigma(x), j_0^5(0)] |\gamma(k_2,\epsilon_2)\rangle\Big]\} . \end{aligned} \tag{217}$$

Using Eq.(214) to evaluate the seagull and Schwinger terms in the heavy square brackets, we find

$$\int d^4x\, e^{ik_1\cdot x}\,\delta(x_0)\Big[-(k_1+k_2)^\mu <0|[\dot{A}_\sigma(x),j_\mu^5(0)]\,|\gamma(k_2,\varepsilon_2)>$$
$$+ie_0<0|[j_\sigma(x),j_0^5(0)]\,|\gamma(k_2,\varepsilon_2)>\Big][(2\pi)^3 2k_{20}]^{\frac{1}{2}}$$
$$=-\varepsilon_2^\rho(e_0^2/2\pi^2)\,k_1^\xi k_2^\tau\,\varepsilon_{\xi\sigma\tau\rho}, \qquad (218)$$

which is just the axial-divergence anomaly obtained by substituting Eq. (68) into term (A) of Eq. (211). We see that <u>the anomalous axial divergence arises from a failure of the Schwinger term $[j_\sigma, j_0^5]$ and the seagull $[\dot{A}_\sigma, j_\mu^5]$ to cancel.</u> As a point of consistency, we note that the pseudoscalar-two-photon triangle $R_{\sigma\rho}$ [Eq. (60)] has the asymptotic behavior $R_{\sigma\rho}(k_1,k_2)\to 0$ as $k_{10}\to i\infty$. Thus the naive equal-time commutation relations

$$[A_\sigma(x),\, j^5(y)]=[\dot{A}_\sigma(x),j^5(y)]=0 \qquad (219)$$

remain valid, and no extra seagull terms are picked up when the one photon reduction formula is applied to the matrix element $<0|2im_0 j^5|\gamma(k_1,\varepsilon_1)\gamma(k_2,\varepsilon_2)>$.

We proceed next to check whether the commutation relations of Eqs. (214) and (219) are formally consistent with each other, with the equations of motion, and with the electromagnetic-field canonical commutation relations of Eq. (5). We start with the equation of motion

$$\Box^2 A_\mu = \ddot{A}_\mu - \nabla^2 A_\mu = e_0 j_\mu , \tag{220}$$

and the divergence equations satisfied by the currents $j_\mu(\underset{\sim}{x}, t)$ and $j^5_\mu(\underset{\sim}{x}, t)$,

$$\frac{\partial}{\partial t} j_0 + \underset{\sim}{\nabla} \cdot \underset{\sim}{j} = 0, \tag{221}$$

$$\frac{\partial}{\partial t} j^5_0 + \underset{\sim}{\nabla} \cdot \underset{\sim}{j}^5 = 2im_0 j^5 + (2\alpha_0/\pi) \underset{\sim}{E} \cdot \underset{\sim}{B} .$$

We proceed to combine Eqs. (5), (220) and (221) with Eqs. (214) and (219). All the commutators which we write down are at equal time, with $x_0 = y_0 = t$.

(i) From $[A_\sigma(x), j^5_0(y)] = 0$, we deduce

$$[\dot{A}_\sigma(x), j^5_0(y)] + [A_\sigma(x), (\partial/\partial t) j^5_0(y)] = 0. \tag{222}$$

On substituting Eq. (221) for $(\partial/\partial t) j^5_0(y)$ and using $[A_\sigma(x), j^5(y)] = [A_\sigma(x), \underset{\sim}{j}^5(y)] = 0$, we find

$$[\dot{A}_\sigma(x), j^5_0(y)] = -[A_\sigma(x), 2\alpha_0/\pi) \underset{\sim}{E}(y) \cdot \underset{\sim}{B}(y)] . \tag{223}$$

Using the canonical commutation relations we then get

$$[\dot{A}_0(x), j^5_0(y)] = 0, \tag{224}$$

$$[\dot{A}_r(x), j^5_0(y)] = (-2i\alpha_0/\pi)\, \delta^3(\underset{\sim}{x} - \underset{\sim}{y}) B^r(y) ,$$

in agreement with Eq. (214).

(ii) From $[\dot{A}_0(x), j^5_0(y)] = 0$, we deduce

$$[\ddot{A}_0(x), j^5_0(y)] + [\dot{A}_0(x), (\partial/\partial t) j^5_0(y)] = 0. \tag{225}$$

Substituting Eq. (221) for $(\partial/\partial t) j^5_0(y)$ and Eq. (220) for $\ddot{A}_0(x)$, and using the commutators $[A_0(x), j^5_0(y)] = [\ddot{A}_0(x), j^5(y)] = [\dot{A}_0(x), \underset{\sim}{j}^5(y)] = 0$, we find

$$[e_0 j_0(x), j_0^5(y)] = -[\dot{A}_0(x), (2\alpha_0/\pi)\underset{\sim}{E}(y)\cdot\underset{\sim}{B}(y)]$$
$$= (-2i\alpha_0/\pi)\underset{\sim}{B}(y)\cdot\underset{\sim}{\nabla}_{\underset{\sim}{x}}\delta^3(\underset{\sim}{x}-\underset{\sim}{y}), \qquad (226)$$

that is,

$$[j_0(x), j_0^5(y)] = (-ie_0/2\pi^2)\underset{\sim}{B}(y)\cdot\underset{\sim}{\nabla}_{\underset{\sim}{x}}\delta^3(\underset{\sim}{x}-\underset{\sim}{y}), \qquad (227)$$

in accord with Eq. (214).

(iii) In a similar manner, the relations obtained by time differentiation of $[\dot{A}_r(x), j_0^5(y)] = -(2i\alpha_0/\pi)\delta^3(\underset{\sim}{x}-\underset{\sim}{y})B^r(y)$ and $[j_0(x), j_0^5(y)] = -(ie_0/2\pi^2)\underset{\sim}{B}(y)\cdot\underset{\sim}{\nabla}_{\underset{\sim}{x}}\delta^3(\underset{\sim}{x}-\underset{\sim}{y})$ are found to be consistent with Eqs. (214), (219), (220) and (221).

(iv) Finally, to check the consistency of quantization in the Feynman gauge, we must verify that

$$L \equiv \dot{A}_0 + \underset{\sim}{\nabla}\cdot\underset{\sim}{A} \qquad (228)$$

and $\dot{L}$ remain dynamically independent of the axial-vector current. That is, we must verify that

$$[L(x), j_\mu^5(y)] = 0 \qquad (229)$$

and that

$$[\dot{L}(x), j_\mu^5(y)] = 0. \qquad (230)$$

Equation (229) follows immediately from the first line of Eq. (214). To check Eq. (230), we substitute Eq. (220) for $\ddot{A}_0$ and use $[A_0(x), j_\mu^5(y)] = 0$, giving

$$[\dot{L}(x), j_\mu^5(y)] = [e_0 j_0(x), j_\mu^5(y)] + [\underset{\sim}{\nabla}_{\underset{\sim}{x}}\cdot\dot{\underset{\sim}{A}}(x), j_\mu^5(y)]. \qquad (231)$$

Substituting commutators from Eq. (214) then shows that right-hand side of Eq. (231) vanishes.

We conclude that the commutation relations of Eq. (214), which were obtained from the triangle graph in lowest-order perturbation theory, are consistent with the equations of motion and canonical commutation relations. Moreover, the fact that Eq. (224) for $[\dot{A}_r(x), j_0^5(y)]$ and Eq. (227) for $[j_0(x), j_0^5(y)]$ were deduced from simpler, exact commutators and equations of motion suggests that Eqs.(224) and (227) are themselves exact to all orders of perturbation theory. The values given in Eq. (214) for $[\dot{A}_r(x), j_s^5(y)]$, $[j_r(x), j_0^5(y)]$ and $[j_0(x), j_s^5(y)]$ cannot, on the other hand, be deduced from the consistency argument. To see this, we note that the consistency checks of items (iii) and (iv) above are unchanged if we modify these commutators to read

$$[\dot{A}_r(x), j_s^5(y)] = \frac{i\alpha_0}{\pi}\delta^3(\underset{\sim}{x}-\underset{\sim}{y})\epsilon^{rst}E^t(y) - ie_0\delta^3(\underset{\sim}{x}-\underset{\sim}{y})S^{rs}(y),$$

$$[j_r(x), j_0^5(y)] = \frac{-ie_0}{4\pi^2}[\underset{\sim}{E}(x)\times\underset{\sim}{\nabla}_{\underset{\sim}{y}}\delta^3(\underset{\sim}{x}-\underset{\sim}{y})]^r + i\frac{\partial}{\partial y^s}[\delta^3(\underset{\sim}{x}-\underset{\sim}{y})S^{rs}(y)], \quad (232)$$

$$[j_0(x), j_s^5(y)] = \frac{ie_0}{4\pi^2}[\underset{\sim}{E}(y)\times\underset{\sim}{\nabla}_{\underset{\sim}{x}}\delta^3(\underset{\sim}{x}-\underset{\sim}{y})]^s - i\frac{\partial}{\partial x^r}[\delta^3(\underset{\sim}{x}-\underset{\sim}{y})S^{rs}(y)],$$

with $S^{rs}(y)$ a pseudotensor operator. In other words, the consistency check does not rule out the possibility that higher orders of perturbation theory may modify Eq. (214) by adding Schwinger terms and seagulls of the usual type, which cancel against each other when vector or axial-vector divergences are taken. It is expected[42] on general grounds that the seagull commutator $[\dot{A}_r(x), j_s^5(y)]$ does not involve derivatives of the δ function and the Schwinger term commutators $[j_r(x), j_0^5(y)]$ and $[j_0(x), j_s^5(y)]$ do not involve derivatives of the delta function higher than the first. [This structure for seagulls and Schwinger terms was found in the pure QED example discussed above in Subsection 6.1.] Under this assumption, Eq. (232) represents the most general form for these commutators consistent with Eqs. (220) and (221).

Using Eq. (224), we can easily complete the argument, sketched in Subsection 3.3, that the operator

$$\bar{Q}^5 = \int d^3x [\, j_0^5(x) + (\alpha_0/\pi)\, \underset{\sim}{A}(x) \cdot \underset{\sim}{\nabla}_x \times \underset{\sim}{A}(x)] \tag{233}$$

is the conserved generator of the γ_5 transformations in massless electrodynamics. We have already shown that $\bar{Q}^5$ is conserved and that it satisfies the correct commutation relations with the fermion fields. We now show that $\bar{Q}^5$ commutes with the photon field variables. From the first line of Eq. (214) we find

$$[\bar{Q}^5, A_\sigma(y)] = [\bar{Q}^5, \dot{A}_0(y)] = 0, \tag{234a}$$

while from Eq. (224) we find

$$[\bar{Q}^5, \dot{A}_r(y)] = \left[\int d^3x\, j_0^5(x), \dot{A}_r(y)\right] + \left[\int d^3x (\frac{\alpha_0}{\pi}) \underset{\sim}{A}(x) \cdot \underset{\sim}{\nabla}_x \times \underset{\sim}{A}(x), \dot{A}_r(y)\right]$$

$$= \frac{2i\alpha_0}{\pi} B^r(y) - \frac{2i\alpha_0}{\pi} B^r(y) = 0, \qquad (234b)$$

as required.

7. APPLICATIONS OF THE BJORKEN LIMIT

The Bjorken limit formula of Eq. (203) has been extensively applied over the past several years to the study of radiative corrections to the hadronic β decay[43] and to the derivation of asymptotic sum rules[44] and asymptotic cross section relations[45] for high energy inelastic electron and neutrino scattering. In all of these applications, it is assumed that the equal-time commutators appearing on the right-hand side of Eq. (203) are the same as the "naive" commutators obtained by straightforward use of canonical commutation relations and equations of motion. As we have seen in the previous section, in the case of vacuum polarization and triangle diagrams in QED, this assumption is not borne out, and we find cases in which the Bjorken-limit and the naive commutator do not agree. Because of special features of the diagrams which lead to these counter-examples, they do not directly invalidate the applications of Eq. (203) mentioned above. However, when detailed perturbation theory calculations are made on a wider class of diagrams,[46] one does find anomalous commutators which invalidate all of the above-mentioned applications. In the present section, we will briefly derive various consequences of the Bjorken-limit formula in the case when anomalous behavior is neglected. Then, in Section 8, we will discuss the changes that result from the presence of perturbation theory anomalies.

7.1 Radiative Corrections to Hadronic β Decay

We begin by considering the theory of second order radiative corrections to hadronic β decay in the local current-current theory of weak interactions. For definiteness, we will discuss only the <u>vector amplitude</u> for the specific process of neutron decay, $n \rightarrow p + e^- + \bar{\nu}_e$. The lowest order matrix element M for this process is represented by the diagram

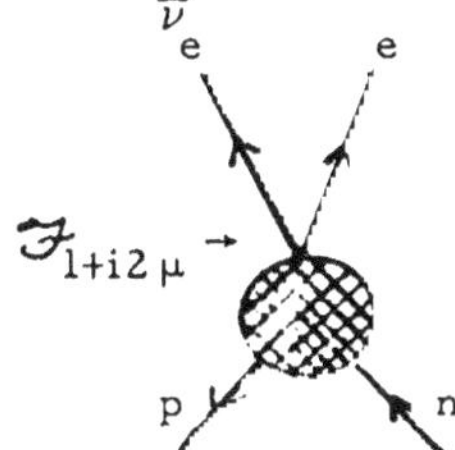

with $\mathcal{F}_{1+i2\mu}$ the component of the hadronic current to which the leptons couple. The radiative corrections to this process come from the following four diagrams (As before, the wavy line denotes a virtual photon)

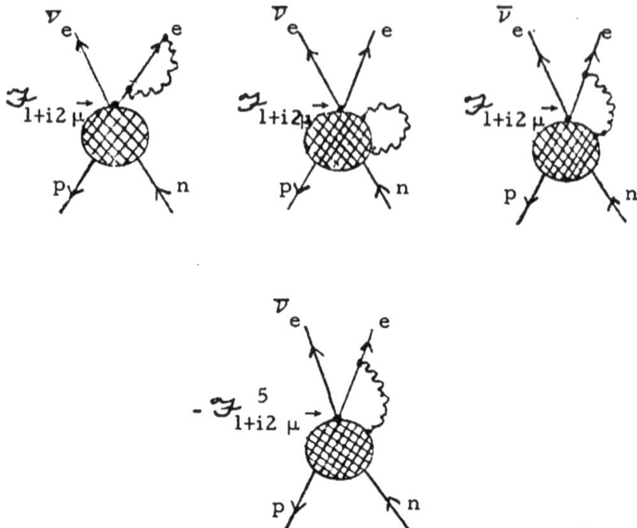

Although the fourth diagram involves the axial-vector current $\mathcal{F}^5_{1+i2\,\mu}$, it has a piece which contains the pseudotensor $\epsilon_{\mu\nu\lambda\sigma}$ and therefore transforms as a vector coupling and contributes to the radiative-corrected vector amplitude.

The three radiative correction diagrams on the first line can be analyzed by the standard methods of <u>time component</u> current algebra, without the use of Bjorken limits. In the approximation of zero momentum transfer to the leptons (an eminently reasonable approximation, since the ratio of the leptonic momentum transfer to the nucleon mass is of the same order as higher order electromagnetic corrections) one finds[47] the remarkable result that these three diagrams sum to a universal, structure-independent, divergent correction to the vector amplitude,

$$\delta M^{\text{first line}} = \frac{3\alpha}{8\pi}(\ell n\, \Lambda^2)M,$$
$$M = (\tilde{G}/\sqrt{2})\, \bar{u}_e\, \gamma^\mu (1-\gamma_5)\, v_\nu < p|\mathcal{F}_{1+i2\,\mu}(0)|n> . \tag{235}$$

Here $\tilde{G} = G\cos\theta_C$ is the effective Fermi constant, and Λ is the mass of a regulator photon which has been introduced to permit evaluation of the integrals. The fourth diagram cannot be treated by using solely the techniques of time-component current algebra, but it can be evaluated by use of the Bjorken limit. The contribution of this diagram may be written as

$$\delta M^{\text{second line}} = \frac{ie^2}{(2\pi)^4}\int \frac{d^4k}{k^2}\frac{\tilde{G}}{\sqrt{2}}\, \bar{u}_e \gamma^\sigma \frac{1}{\not{\ell}-\not{k}-m_e}\gamma^\mu(1-\gamma_5)v_\nu$$
$$\times\, [\, T_{\sigma\mu}(k) + \text{seagull}] \tag{236}$$

with

$$T_{\sigma\mu}(k) = -i\int d^4x\, e^{ik\cdot x} < p|\, T(j_\sigma^{EM}(x)\, \mathcal{F}^5_{1+i2\mu}(0))\,|n> \tag{237}$$

and with ℓ the electron four-momentum, m_e the electron mass and j_σ^{EM} the hadronic electromagnetic current. Making our approximation of zero momentum transfer to the leptons, we drop ℓ and m_e. Since the factor multiplying the seagull is odd in k and since we expect the seagull to be proportional to $\int d^4x\, e^{ik\cdot x}\,\delta^4(x)$, which is k-independent, the seagull drops out and we get

$$\delta M^{\text{second line}} = \frac{-ie^2}{(2\pi)^4}\int \frac{d^4k}{k^2}\frac{\tilde{G}}{\sqrt{2}}\bar{u}_e\gamma^\sigma \frac{\not{k}}{k^2}\gamma^\mu(1-\gamma_5)v_\nu T_{\sigma\mu}(k). \quad (238)$$

In order to isolate the divergent part of Eq. (238), we need only calculate the large-k behavior of $T_{\sigma\mu}(k)$. According to the Bjorken limit formula of Eq. (203), the large-k_0 behavior of $T_{\sigma\mu}(k)$ is given by

$$T_{\sigma\mu}(k)_{k_0 \to i\infty} k_0^{-1}\int d^4x\, e^{-ik\cdot x}\delta(x_0) < p|[j_\sigma^{EM}(x), \mathcal{F}^5_{1+i2\mu}(0)]|n> \quad (239)$$

In order to evaluate Eq. (239), we will adopt a specific model of the strong interactions, in which the basic fields are a fermion SU_3-triplet $\psi \equiv (\psi_1, \psi_2, \psi_3)$ with electric charges (Q, Q-1, Q-1), bound by the exchange of SU_3-singlet vector, scalar and pseudoscalar bosons. The electromagnetic and weak axial-vector currents in this model are given by the simple expressions

$$j_\sigma^{EM}(x) = \bar{\psi}(x)\lambda_Q\gamma_\sigma\psi(x), \quad \mathcal{F}^5_{1+i2\mu}(x) = \bar{\psi}(x)\lambda_+\gamma_\mu\gamma_5\psi(x),$$
$$\lambda_Q = \begin{bmatrix} Q & 0 & 0 \\ 0 & Q-1 & 0 \\ 0 & 0 & Q-1 \end{bmatrix}, \quad \lambda_+ = \begin{bmatrix} 0 & 1 & 0 \\ 0 & 0 & 0 \\ 0 & 0 & 0 \end{bmatrix}. \quad (240)$$

Let us further make the assumption that the equal-time commutator appearing in Eq. (239) is the same as the naive canonical equal time commutator, giving

$$\delta(x_0)[j_\sigma^{EM}(x), \mathcal{F}^5_{1+i2\mu}(0)] = \delta^4(x)\bar{\psi}(x)\gamma_0[\lambda_Q\gamma_0\gamma_\sigma, \lambda_+\gamma_0\gamma_\mu\gamma_5]\psi(x)$$

$$= \delta^4(x)\, \bar{\psi}(x)\gamma_0\big(\tfrac{1}{2}\{\lambda_Q, \lambda_+\}[\gamma_0\gamma_\sigma, \gamma_0\gamma_\mu\gamma_5] + \tfrac{1}{2}[\lambda_Q, \lambda_+]\{\gamma_0\gamma_\sigma, \gamma_0\gamma_\mu\gamma_5\}\big)\psi(x) \quad . \tag{241}$$

The second term on the right-hand side of Eq. (241) is pure axial-vector in character, and so can be dropped. Using the elementary relations

$$\begin{aligned} \tfrac{1}{2}\{\lambda_Q, \lambda_+\} &= \tfrac{1}{2}(Q+Q-1)\,\lambda_+ = Q_{AV}\,\lambda_+\,, \\ \gamma_\sigma\gamma_0\gamma_\mu - \gamma_\mu\gamma_0\gamma_\sigma &= -2i\,\epsilon_{\sigma 0 \mu\eta}\gamma^\eta\gamma_5, \end{aligned} \tag{242}$$

the first term can be rewritten as

$$-2i\delta^4(x)\, Q_{AV}\,\epsilon_{\sigma 0\mu\eta}\,\mathcal{F}^\eta_{1+i2}(x), \tag{243a}$$

which, on substitution into Eq. (239) gives

$$T_{\sigma\mu}(k)\underset{k_0\to i\infty}{\to}\frac{-2iQ_{AV}}{k_0}\epsilon_{\sigma 0\mu\eta}\langle p|\,\mathcal{F}^\eta_{1+i2}|n\rangle. \tag{243b}$$

Let us now define[48] the auxiliary tensor $t_{\sigma\mu}(k)$ by

$$t_{\sigma\mu}(k) \equiv \frac{-2iQ_{AV}}{k^2}\,\epsilon_{\sigma\lambda\mu\eta}k^\lambda\langle p|\,\mathcal{F}^\eta_{1+i2}|n\rangle, \tag{244}$$

so that, by construction, $\Delta_{\sigma\mu}(k) \equiv T_{\sigma\mu}(k) - t_{\sigma\mu}(k)$ approaches zero faster than k^{-1} as $k_0 \to i\infty$. Arguments based on dispersion theory[49] can then be used to show that $\Delta_{\sigma\mu}(k)$ decreases faster than k^{-1} as k approaches infinity in an arbitrary direction, which means that to extract the ultraviolet divergent part of Eq. (238), we need only evaluate the k-integrals with $T_{\sigma\mu}(k)$ replaced by $t_{\sigma\mu}(k)$. This is a completely straightforward calculation, which gives

$$\delta M^{\text{second line}} = \frac{3\alpha}{8\pi}\,(\ln \Lambda^2)\, 2Q_{AV}\, M. \tag{245}$$

Adding Eqs. (245) and (235), we get for the total second order radiative correction to the vector part

$$\delta M = \frac{3\alpha}{8\pi} (1 + 2Q_{AV}) \ln \Lambda^2 M . \tag{246}$$

Thus, if $Q_{AV} = -\frac{1}{2}$, which corresponds to the choice of triplet charges (0, -1, -1), the radiative corrections are finite; for other choices of Q_{AV}, such as the value $+\frac{1}{2}$ favored by our analysis of $\pi^0 \to 2\gamma$ decay, the radiative corrections diverge. A similar calculation can be done for the axial-vector part of the amplitude and for more general β decay processes, with again the conclusion that the radiative corrections are finite only for $Q_{AV} = -\frac{1}{2}$. *If* the analysis leading to Eq. (246) were correct, one could adopt one of two points of view (just as in our discussion of purely leptonic processes in Subsection 3.2): (i) The radiative corrections should be finite in the local current-current theory, requiring the choice $Q_{AV} = -\frac{1}{2}$; (ii) The radiative corrections need not be finite, since the divergence in Eq. (246) is a weak one which becomes a significant correction to the weak decay amplitude only for large values of Λ, where the local current-current theory must fail in any case. In actual fact, we will see that when interactions of the fermion triplet with the mesons are taken into account, the assumption of identity of the Bjorken-limit and naive commutators used to derive Eq. (246) breaks down. As a result, even the choice $Q_{AV} = -\frac{1}{2}$ leaves the radiative corrections infinite, and so the second point of view seems to be the more reasonable one.

We note, in conclusion, that the analysis given above also applies to the second order radiative corrections to μ-meson decay. Since the average charge of the μ-meson and the μ-neutrino is $-\frac{1}{2}$, Eq. (246) predicts that the radiative corrections in this case are finite, as is indeed found both by explicit calculation and by the Ward identity arguments of Subsection 3.1.

7.2 Asymptotic Sum Rules and Asymptotic Cross Section Relations

We consider next the use of the Bjorken limit formula to derive asymptotic sum rules and asymptotic cross section relations for lepton-nucleon scattering. In inelastic electron-nucleon scattering, one observes the reaction

$$e_i(k_i) + N(p) \rightarrow e_f(k_f) + \Gamma(p_\Gamma), \tag{247}$$

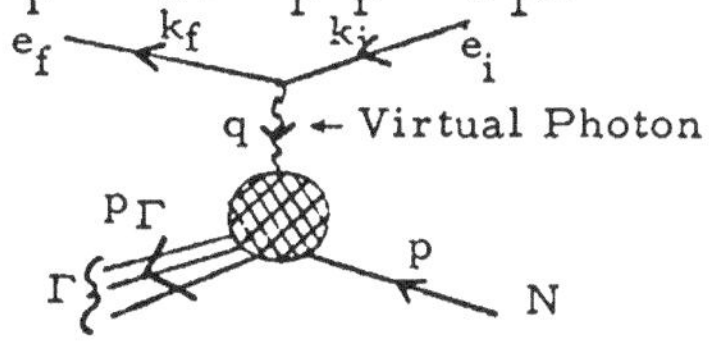

with e_i and e_f the initial and final electrons, with N the nucleon (typically, a free proton or a neutron bound in a deuteron) and with Γ any inelastic hadron final state. Four-momenta of the particles are indicated in parentheses. In the experiments done at SLAC and other laboratories,[50] one measures the incident and final electron energies E_i and E_f and the laboratory scattering angle θ between the electron directions, but obtains no detailed information about the conposition of the state Γ. Thus, the experimentally measured differential cross section is $d^2\sigma/d\Omega_f\, dE_f$, with Ω_f the final electron solid angle. Since the matrix element for Eq. (247) is proportional to $\langle \Gamma(p_\Gamma)|j_\lambda^{EM}|N(p)\rangle$, after squaring, averaging over nucleon spin and summing over final states Γ we are clearly measuring the quantity

$$\frac{1}{2}\sum_{\text{spin}(N),\,\Gamma} \langle N(p)|j_\mu^{EM}|\Gamma(p_\Gamma)\rangle\langle\Gamma(p_\Gamma)|j_\lambda^{EM}|N(p)\rangle(2\pi)^3\delta^4(p_\Gamma - p - q), \quad (248)$$

$$q = k_i - k_f,$$

which is essentially the imaginary part of the amplitude for forward Compton scattering of virtual photons of $(\text{mass})^2 = q^2$ on the target nucleon. Using Lorentz invariance and gauge invariance, Eq. (248) can be rewritten in the form

$$W_1(\nu, q^2)\left(-g_{\mu\lambda} + \frac{q_\mu q_\lambda}{q^2}\right) + M_N^{-2} W_2(\nu, q^2)\left(p_\mu - \frac{\nu}{q^2}q_\mu\right)\left(p_\lambda - \frac{\nu}{q^2}q_\lambda\right), \quad (249)$$

$$\nu = q\cdot p, \quad M_N = \text{nucleon mass},$$

and in terms of W_1 and W_2 the experimentally measured differential cross section is

$$\frac{d^2\sigma}{d\Omega_f dE_f} = \frac{\alpha}{\pi}\frac{E_f^2}{(q^2)^2}[2W_1(\nu, q^2)\sin^2(\frac{\theta}{2})+W_2(\nu,q^2)\cos^2(\frac{\theta}{2})] . \quad (250)$$

Roughly speaking, at small scattering angles one measures W_2 and at large scattering angles one measures W_1. In the neutrino scattering reaction

$$\nu_\mu(k_i) + N(p) \rightarrow \mu(k_f) + \Gamma(p_\Gamma), \quad (251)$$

the doubly differential cross section is given by a formula similar in form to Eq. (250), but containing a third term, proportional to $(E_i+E_f)\sin^2(\frac{\theta}{2})$, arising from vector-axial-vector interference.[51]

Rather than considering the physically realistic electron-nucleon and neutrino nucleon processes themselves, we will illustrate the application of Bjorken-limit techniques to these reactions by studying the cross sections for the absorption of fictitious charged, isovector virtual photons by nucleons,

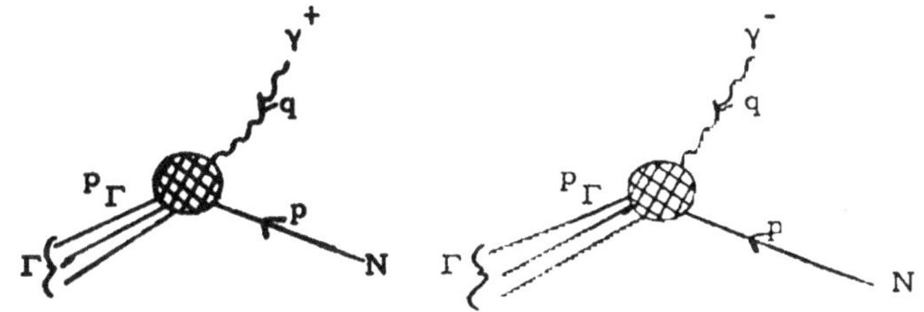

These cross sections differ only in isospin structure from the corresponding virtual photon cross sections appearing in the electron scattering case, and differ from the cross

sections appearing in νN and $\bar{\nu}$N scattering in that the axial-vector current terms have been omitted. The results which we will obtain for the simpler, fictitious reactions are readily extended to the realistic cases.

To proceed, we first develop some properties of the nucleon-spin averaged amplitude for the forward scattering of isovector photons, from initial isotopic state b to final isotopic state a, on a nucleon target. This is given by

$$(M_N/p_0)\, T^*_{a\mu\, b\lambda}(p, q) = \tfrac{1}{2}\sum_{\mathrm{spin}(N)} -i\int d^4x\, e^{iq\cdot x}$$
$$\times \langle N(p)|\, T^*(\mathcal{F}_{a\mu}(x)\mathcal{F}_{b\lambda}(0))\,|N(p)\rangle \qquad (252)$$
$$= \text{seagull} + \tfrac{1}{2}\sum_{\mathrm{spin}(N)} -i\int d^4x\, e^{iq\cdot x}\langle N(p)|\, T(\mathcal{F}_{a\mu}(x)\mathcal{F}_{b\lambda}(0))\,|N(p)\rangle,$$
$$a, b = 1, 2, 3,$$

where, as before, the seagull term is a polynomial in q_0. Since we have seen that amplitudes involving only vector currents have normal Ward identities, by using both isospin current conservation and the ordinary time component current algebra[52] we find that the divergence of Eq. (252) is given by

$$q^\mu T^*_{a\mu\, b\lambda} = (p_0/M_N)\, \tfrac{1}{2}\sum_{\mathrm{spin}(N)} \int d^4x\, e^{iq\cdot x}$$
$$\times \langle N(p)|\, \delta(x_0)[\mathcal{F}_{a0}(x), \mathcal{F}_{b\lambda}(0)]\,|N(p)\rangle$$
$$= (p_0/M_N)\, \tfrac{1}{2}\sum_{\mathrm{spin}(N)} \langle N(p)|\mathcal{F}_{[a,b]\lambda}|N(p)\rangle \qquad (253)$$
$$= \tfrac{1}{2}\,\mathrm{tr}\left\{\left(\frac{\not{p}+M_N}{2M_N}\right)\gamma_\lambda\left[\tfrac{1}{2}\lambda a, \tfrac{1}{2}\lambda b\right]\right\}$$

[For notational convenience, we have omitted the initial and final nucleon isospinors, and so Eqs. (252) and (253) are really matrix equations in isospin space.] If we examine the Born approximation to $T^{*}_{a\mu\, b\lambda}(p,q)$,

$$T^{*\mathrm{BORN}}_{a\mu\, b\lambda}(p,q)=\tfrac{1}{2}\mathrm{tr}\{(\frac{\not{p}+M_N}{2M_N})\gamma_\mu\tfrac{1}{2}\lambda_a\frac{1}{\not{p}+\not{q}-M_N}\gamma_\lambda\tfrac{1}{2}\lambda_b+\gamma_\lambda\tfrac{1}{2}\lambda_b \times\frac{1}{\not{p}-\not{q}-M_N}\gamma_\mu\tfrac{1}{2}\lambda_a)\} \tag{254}$$

we easily see that $q^\mu T^{*\mathrm{BORN}}_{a\mu\, b\lambda}(p,q)$ agrees with Eq. (253), and that $q^\lambda T^{*\mathrm{BORN}}_{a\mu\, b\lambda}(p,q)$ agrees with the λ-index analog of Eq. (253). This means that the non-Born part of Eq. (252) is divergenceless, and therefore Eq. (252) has the general structure

$$T^{*}_{a\mu\, b\lambda}(p,q)=T^{*\mathrm{BORN}}_{a\mu\, b\lambda}(p,q)+T_{1\,ab}(q^2,\omega)(-g_{\mu\lambda}+\frac{q_\mu q_\lambda}{q^2}) + M_N^{-2}T_{2ab}(q^2,\omega)(p_\mu-\frac{\nu}{q^2}q_\mu)(p_\lambda-\frac{\nu}{q^2}q_\lambda). \tag{255}$$

In writing Eq. (255) we have eliminated ν in terms of the dimensionless variable

$$\omega=-q^2/\nu\,. \tag{256}$$

To proceed further, we separate off the isospin dependence of the non-Born amplitudes,

$$T_{1,2\,ab}(q^2,\omega)=T^{(+)}_{1,2}(q^2,\omega)\,\{\tfrac{1}{2}\lambda_a,\tfrac{1}{2}\lambda_b\}+T^{(-)}_{1,2}(q^2,\omega)\,[\tfrac{1}{2}\lambda_a,\tfrac{1}{2}\lambda_b]\,; \tag{257}$$

the crossing symmetry relation $T^{*}_{a\mu\, b\lambda}(p,q)=T^{*}_{b\lambda\, a\mu}(p,-q)$ implies that the (+) amplitudes are even and the (-) amplitudes are odd functions of ω. A standard forward dispersion relations analysis, very similar to that for the familiar case[53] of pion-nucleon scattering, shows that the amplitudes $T_{1,2}^{(\pm)}$ satisfy the following dispersion relations,

$$\begin{aligned}
T_1^{(+)}(q^2,\omega) &= T_1^{(+)}(q^2,0) - \int_0^2 d\omega'[W_1^-(q^2,\omega')+W_1^+(q^2,\omega')]\left(\frac{1}{\omega'-\omega}+\frac{1}{\omega'+\omega}\right), \\
T_1^{(-)}(q^2,\omega) &= -\int_0^2 d\omega'\,[W_1^-(q^2,\omega')-W_1^+(q^2,\omega')]\left(\frac{1}{\omega'-\omega}-\frac{1}{\omega'+\omega}\right), \\
T_2^{(+)}(q^2,\omega) &= -\omega\int_0^2 \frac{d\omega'}{\omega'}\,[W_2^-(q^2,\omega')+W_2^+(q^2,\omega')]\left(\frac{1}{\omega'-\omega}-\frac{1}{\omega'+\omega}\right), \\
T_2^{(-)}(q^2,\omega) &= -\omega\int_0^2 \frac{d\omega'}{\omega'}\,[W_2^-(q^2,\omega')-W_2^+(q^2,\omega')]\left(\frac{1}{\omega'-\omega}+\frac{1}{\omega'+\omega}\right),
\end{aligned} \tag{258}$$

with absorptive parts given by

$$\begin{aligned}
&-(2\pi)^3(p_0/M_N)\tfrac{1}{2}\sum_{\text{spin}(N)}\sum_{\Gamma}\langle N(p)|2^{-\frac{1}{2}}(\mathcal{F}_{1\mu}\mp i\mathcal{F}_{2\mu})|\Gamma(p_\Gamma)\rangle \\
&\quad\times\langle\Gamma(p_\Gamma)|2^{-\frac{1}{2}}(\mathcal{F}_{1\lambda}\pm i\,\mathcal{F}_{2\lambda})|N(p)\rangle\,\delta^4(p_\Gamma-p-q). \qquad (259)\\
&= W_1^{\pm}(q^2\omega)\left(-g_{\mu\lambda}+\frac{q_\mu q_\lambda}{q^2}\right)+M_N^{-2}W_2^{\pm}(q^2,\omega)\left(p_\mu-\frac{\nu}{q^2}q_\mu\right)\left(p_\lambda-\frac{\nu}{q^2}q_\lambda\right).
\end{aligned}$$

The structure functions $W_{1,2}^{\pm}$ appearing here are the charged-photon analogs of the electron scattering structure functions defined in Eqs. (248) - (249). In writing Eq. (258), we have assumed one subtraction each for $T_1^{(\pm)}$ [the subtraction constant $T_1^{(-)}(q^2,0)$ vanishes by crossing symmetry] and no subtraction for $T_2^{(\pm)}$, as is suggested both by perturbation theory calculations and by a simple Regge model[54] for the high energy (large ν, small ω) behavior of the amplitudes.

Having finished our preliminaries, we can now derive a so-called asymptotic sum rule satisfied by the structure functions $W_1^{\pm}(q^2, \omega)$ in the asymptotic limit as $q^2 \to -\infty$. Referring back to Eq. (255), we set $\underset{\sim}{q} = \underset{\sim}{p} = 0$, $\mu = \lambda = 1$ and take the Bjorken limit $q_0 \to i\infty$. Using Eqs. (254) - (258), we find that

$$\begin{aligned} T^*_{a1b1}(p,q)_{q_0 \to i\infty} &\to q_0^{-1} [\tfrac{1}{2}\lambda_a, \tfrac{1}{2}\lambda_b]\{1- \lim_{q^2 \to -\infty} 2M_N \int_0^2 d\omega' [W_1^-(q^2, \omega') \\ &- W_1^+(q^2, \omega')]\} + \text{symmetric in } a, b + O(q_0^{-2} \ln q_0), \end{aligned} \tag{260}$$

while applying the Bjorken-Johnson-Low recipe of Eq. (203) to Eq. (252) gives the alternative evaluation

$$\begin{aligned} T^*_{a1b1}(p,q)_{q_0 \to i\infty} &\to \text{polynomial} + q_0^{-1}(p_0/M_N)\tfrac{1}{2}\sum_{\text{spin}(N)} \int d^4x\, e^{iq\cdot x} \\ &\times \delta(x_0)\langle N(p)| [\mathcal{F}_{a1}(\underset{\sim}{x}, 0), \mathcal{F}_{b1}(0)]\, |N(p)\rangle + O(q_0^{-2} \ln q_0). \end{aligned} \tag{261}$$

To calculate the equal-time commutator appearing in Eq. (261), we will again adopt the SU_3-triplet model of the strong interactions described in the preceding subsection, and again we assume the identity of the Bjorken-limit and the naive canonical commutator. We thus get

$$\begin{aligned} &\delta(x_0)[\mathcal{F}_{a1}(\underset{\sim}{x}, 0), \mathcal{F}_{b1}(0)] \\ &= \delta(x_0)[\overline{\psi}(\underset{\sim}{x}, 0)\tfrac{1}{2}\lambda_a\gamma_1\psi(\underset{\sim}{x}, 0), \overline{\psi}(0)\tfrac{1}{2}\lambda_b\gamma_1\psi(0)] \\ &= \delta^4(x)\, \overline{\psi}\, \gamma_0[\tfrac{1}{2}\lambda_a, \tfrac{1}{2}\lambda_b]\,\psi\,, \end{aligned} \tag{262}$$

and the right-hand side of Eq. (261) becomes

$$\text{polynomial} + q_0^{-1}[\tfrac{1}{2}\lambda_a, \tfrac{1}{2}\lambda_b] + O(q_0^{-2} \ln q_0). \tag{263}$$

Comparing with Eq. (260), we thus get the asymptotic sum

rule

$$0 = \lim_{q^2 \to -\infty} 2M_N \int_0^2 d\omega' \, [W_1^-(q^2,\omega') - W_1^+(q^2,\omega')] \; . \qquad (264)$$

The value 0 appearing on the left-hand side of Eq. (264) is peculiar to the SU_3-triplet model; other field theoretic models, which also satisfy the Gell-Mann time component current algebra, have different values for the space-space commutator of Eq. (262) and therefore lead to modified sum rules. For example, in the algebra of fields model,[55] the commutator in Eq. (262) vanishes and the left-hand side of Eq. (264) is 1. Eq. (264) is readily generalized to the physically realistic cases, where it yields a similar sum rule[49] for the W_1 structure functions in ν and $\bar{\nu}$-nucleon scattering, and an inequality for the W_1 structure functions in electron-nucleon scattering. We note finally that in the usual form in which Eq. (264) appears in the literature, the Born terms are not separated out of W_1; if the Born terms are included in W_1, the left-hand side of Eq. (264) becomes, respectively, -1 in the quark-model and 0 in the field algebra cases.

A useful alternative form of Eq. (264) is obtained by recalling that the usual fixed-q^2 sum rule, following from the local time component algebra alone,[56] is in our present notation

$$0=\int_0^2 \frac{d\omega'}{\omega'^2}[W_2^-(q^2,\omega')-W_2^+(q^2,\omega')] \tag{265}$$

Multiplying Eq. (265) by $2q^2/M_N$ and adding to Eq. (264), we get the modified sum rule[57]

$$0=\lim_{q^2\to-\infty} 2\int_0^2 d\omega'[L^-(q^2,\omega')-L^+(q^2,\omega')] \tag{266}$$

with

$$L^{\mp}(q^2,\omega)=M_N[W_1^{\mp}(q^2,\omega)+\frac{q^2}{\omega^2}\frac{1}{M_N^2}W_2^{\mp}(q^2,\omega)] \tag{267}$$

the total longitudinal cross section[58] for inelastic charged-photon-nucleon scattering. Eq. (266) has a very interesting analog in the case of electron scattering, obtained by Callan and Gross[45] by a more complicated derivation involving the commutator $\delta(x_0)[\partial j_r^{EM}/\partial x_0, j_s^{EM}]$, which appears as the coefficient of the q_0^{-2} term when the recipe of Eq. (203) is applied to the T-product of two electromagnetic currents. In the quark model, this analog reads

$$0=\lim_{q^2\to-\infty}\int_0^2 d\omega\,\omega\, L^{EM}(q^2,\omega)\ , \tag{268}$$

and since L^{EM} is positive definite Eq.(268) has the very strong implication that

$$\lim_{q^2\to-\infty} L^{EM}(q^2,\omega)=0 \tag{269}$$

for all ω in the range $0\le\omega\le 2$. Just as in our previous examples, the derivation of Eq. (269) involves in a crucial way the assumption that the Bjorken-limit commutator and the naive canonical commutator are the same.

8. BREAKDOWN OF THE BJORKEN LIMIT IN PERTURBATION THEORY

As we have greatly emphasized, all of the applications of the Bjorken limit method discussed in the preceding section involve the assumption that the commutators appearing in the Bjorken-Johnson-Low recipe can be evaluated by naive application of canonical commutation relations. We have also seen, in Section 6, that this assumption is not generally true in perturbation theory, since it fails, for example, in the simple case of the triangle diagram. Consequently, it is natural to ask whether the assumption is valid for the Compton-like perturbation theory diagrams involved in the applications of Section 7; we will find, upon detailed examination, that it is not, and that all of the applications fail in perturbation theory. For definiteness, we will consider the SU_3-triplet model described above, consisting of a fermion triplet bound by the exchange of SU_3-singlet boson "gluons" which can be spatially scalar, pseudoscalar or vector in character. It will not be possible to test the applications of the Bjorken limit in precisely the formulations given in Section 7, where we always assumed involvement of a nucleon, which appears (conjecturally) in the gluon model only as a complicated multiparticle bound state. However, it is easy to see that the derivations of Section 7 are equally valid if the nucleons p and n are replaced by the first two fermion triplets ψ_1 and ψ_2, with the nucleon mass M_N replaced in the kinematic formulas by the triplet mass m. Eq. (246) then becomes a statement about radiative corrections to the vector amplitude for the triplet β-decay $\psi_2 \rightarrow \psi_1 + e + \bar{\nu}_e$, while Eqs. (264), (266) and (269) become asymptotic sum rules and cross section relations for charged photon-triplet and electron-triplet scattering. It is the validity of these relations which we will directly test in perturbation theory.

8.1 Computational Results

We begin by summarizing the results of some perturbation theory calculations in the triplet-gluon model. In order to treat simultaneously commutators involving scalar, pseudoscalar and tensor currents as well as the usual vector and axial-vector currents, we introduce the abbreviated notation

$$J_{(1)} = \bar{\psi}\gamma_{(1)}\psi , \quad J_{(2)} = \bar{\psi}\gamma_{(2)}\psi ,$$
$$\gamma_{(1)} = \gamma_\mu \tfrac{1}{2}\lambda_a (\gamma_\mu\gamma_5\tfrac{1}{2}\lambda_a, \tfrac{1}{2}\lambda_a, \ldots) \tag{270}$$
$$\gamma_{(2)} = \gamma_\lambda \tfrac{1}{2}\lambda_b (\gamma_\lambda\gamma_5\tfrac{1}{2}\lambda_b, \tfrac{1}{2}\lambda_b, \ldots)$$

according to whether the first or second current is a vector (axial-vector, scalar, ...) current. The naive equal time commutator of the two currents is

$$\delta(x_0)[J_{(1)}(x), J_{(2)}(0)] = \delta^4(x)\,\overline{\psi}(x)\,C\,\psi(x), \qquad (271)$$

$$C = \gamma_0[\gamma_0\gamma_{(1)}', \gamma_0\gamma_{(2)}] = \gamma_{(1)}\gamma_0\,\gamma_{(2)} - \gamma_{(2)}\gamma_0\gamma_{(1)}.$$

We wish to compare the Bjorken-limit commutator with the naive commutator in the special case in which Eqs. (270) and (271) are sandwiched between triplet states. To do this, we calculate the renormalized current-fermion scattering amplitude $\widetilde{T}^*_{(1)(2)}(p,p',q)$ in the limit $q_0 \to i\infty$, and compare the coefficient of the q_0^{-1} term with the renormalized vertex $\widetilde{\Gamma}(C;p,p')$ of the naive commutator. Identity of the Bjorken-limit and the naive commutators would mean that

$$\lim_{\substack{q_0\to i\infty \\ \underset{\sim}{q},\,p,\,p'\ \text{fixed}}} \widetilde{T}^*_{(1)(2)}(p,p',q) = \underset{\text{in } q_0}{\text{polynomial}} + q_0^{-1}\widetilde{\Gamma}(C;p,p') + O(q_0^{-2}\ln q_0), \qquad (272)$$

with the polynomial, as usual, coming from the seagull term. In the calculations which follow, we test the validity of Eq. (272) in perturbation theory.

(i) Second order. To second order in the gluon-fermion coupling constant g_r, there are two classes of diagrams which contribute to $\widetilde{T}^*_{(1)(2)}$. The diagrams of the first class

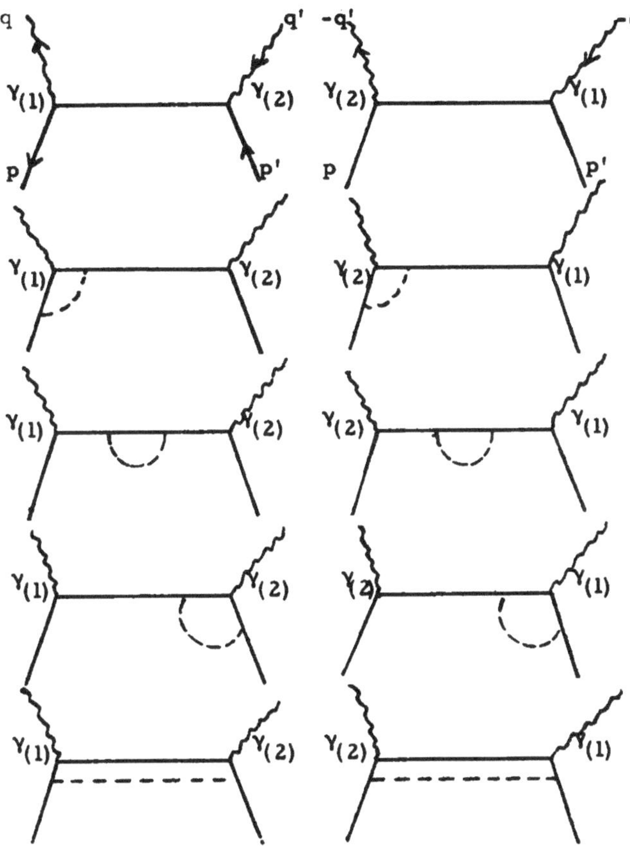

consist of the lowest order current-fermion diagrams and the second order diagrams obtained from the lowest order ones by insertion of a single virtual gluon (denoted by the dashed line). The diagrams of the second class

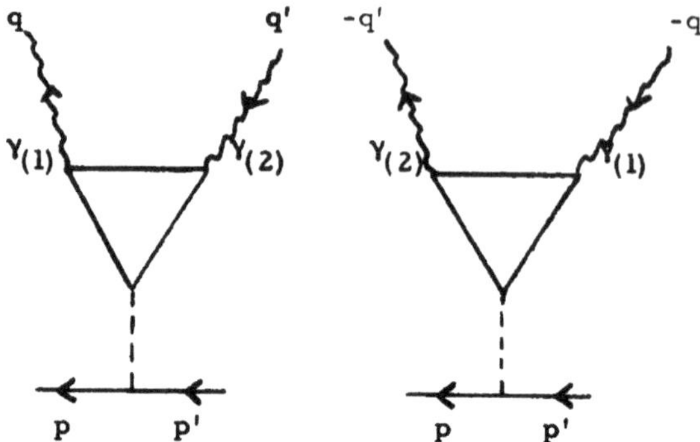

involve a fermion triangle graph. We denote the contributions of these two classes to $\tilde{T}^{*}_{(1)(2)}$ by $\tilde{T}^{*\ \text{Compt}}_{(1)(2)}$ and $\tilde{T}^{*\ \text{Triang}}_{(1)(2)}$, respectively.

The first-class diagrams are evaluated by the standard technique of regulating the gluon propagator with a regulator of mass Λ, which defines an unrenormalized amplitude $T^{*\ \text{Compt}}_{(1)(2)}$. To get the renormalized amplitude we multiply by the fermion wave function renormalization constant Z_2 (the Feynman rules supply us with a factor $\sqrt{Z_2}$ for each of the two external fermion legs) and take the limit $\Lambda \to \infty$,

$$\tilde{T}^{*\ \text{Compt}}_{(1)(2)} = \lim_{\Lambda \to \infty} Z_2\, T^{*\ \text{Compt}}_{(1)(2)} . \tag{273}$$

In certain cases, as discussed below, this limit diverges logarithmically; in these cases we take Λ to be finite but very large, dropping terms which vanish as $\Lambda \to \infty$ but retaining all terms which are proportional to $\ln \Lambda^2$. The renormalized vertex $\tilde{\Gamma}(C;p,p') = \lim(\Lambda\to\infty)\ Z_2\Gamma(C;p,p')$ is calculated by the same techniques from the diagram

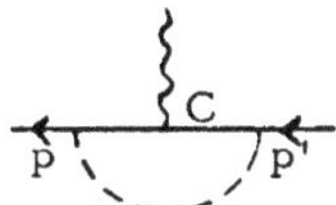

Finally, we take the limit $q_0 \to i\infty$ in our expression for $\tilde{T}^{*\,\mathrm{Compt}}_{(1)(2)}$ and compare with $\tilde{\Gamma}(C;p,p')$, giving the results[59]

$$\lim_{\substack{q_0\to i\infty \\ \underset{\sim}{q},\,p,\,p'\ \mathrm{fixed}}} \tilde{T}^{*\,\mathrm{Compt}}_{(1)(2)}(p,p',q) = q_0^{-1}[\tilde{\Gamma}(C;p,p') + \Delta^{\mathrm{Compt}}] + O(q_0^{-2}\ln q_0), \tag{274a}$$

$$\begin{aligned}
\Delta^{\mathrm{Compt}} = \frac{g_r^2}{32\pi^2}\Big\{ & \ln\Big(\frac{\Lambda^2}{|q_0|^2}\Big)\Big[-\gamma_{(1)}\gamma_0\underset{\sim}{\gamma}\gamma_0\underset{\sim}{\gamma}\gamma_0\gamma_{(2)} + \tfrac{1}{2}\underset{\sim}{\gamma}\gamma_\tau\gamma_{(1)}\gamma_0\gamma_{(2)}\gamma^\tau\underset{\sim}{\gamma} \\
& -\tfrac{1}{2}\gamma_{(1)}\gamma_0\underset{\sim}{\gamma}\gamma_\tau\gamma_{(2)}\gamma^\tau\underset{\sim}{\gamma} - \tfrac{1}{2}\underset{\sim}{\gamma}\gamma_\tau\gamma_{(1)}\gamma^\tau\underset{\sim}{\gamma}\gamma_0\gamma_{(2)}\Big] \qquad (274b) \\
& -\frac{3}{2}\gamma_{(1)}\gamma_0\underset{\sim}{\gamma}\gamma_0\underset{\sim}{\gamma}\gamma_0\gamma_{(2)} - \tfrac{1}{2}\underset{\sim}{\gamma}\gamma_0\gamma_{(1)}\gamma_0\gamma_{(2)}\gamma_0\underset{\sim}{\gamma} \\
& + \gamma_{(1)}\gamma_0\underset{\sim}{\gamma}\gamma_0\gamma_{(2)}\gamma_0\underset{\sim}{\gamma} + \underset{\sim}{\gamma}\gamma_0\gamma_{(1)}\gamma_0\underset{\sim}{\gamma}\gamma_0\gamma_{(2)} \\
& -\tfrac{1}{4}\gamma_{(1)}\gamma_0\underset{\sim}{\gamma}\gamma_\tau\gamma_{(2)}\gamma^\tau\underset{\sim}{\gamma} - \tfrac{1}{4}\underset{\sim}{\gamma}\gamma_\tau\gamma_{(1)}\gamma^\tau\underset{\sim}{\gamma}\gamma_0\gamma_{(2)} \\
& +\tfrac{1}{4}\underset{\sim}{\gamma}[\gamma_\tau\gamma_{(1)}\gamma^\tau\gamma_{(2)}\gamma_0 + \gamma_0\gamma_{(1)}\gamma_\tau\gamma_{(2)}\gamma^\tau]\underset{\sim}{\gamma} - (1)\longleftrightarrow(2)\Big\}.
\end{aligned}$$

In Eq. (274), the notation $\underset{\sim}{\gamma}\ldots\underset{\sim}{\gamma}$ is a shorthand for $1\ldots1$ in the scalar gluon case, $i\gamma_5\ldots i\gamma_5$ in the pseudoscalar gluon case and $(-\gamma_\rho)\ldots\gamma^\rho$ in the vector gluon case. If more than one type of gluon is present in the theory, the quantity Δ^{Compt} appearing in Eq. (274a) is simply the sum of contributions as in Eq. (274b) for each gluon.

Because our model contains only SU_3-singlet gluons, the second class diagrams contribute only to the SU_3-singlet part of the commutator. Taking the Bjorken limit, and comparing with the bubble diagram contribution to $\widetilde{\Gamma}(C;p,p')$

one finds[34]

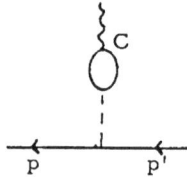

$$\lim_{\substack{q_0 \to i\infty \\ \underset{\sim}{q},\, \underset{\sim}{p},\, \underset{\sim}{p}' \text{ fixed}}} \widetilde{T}^{*\,\mathrm{Triang}}_{(1)(2)}(p,p',q) = \text{constant} + q_0^{-1}[\,\widetilde{\Gamma}(C;p,p')^{\mathrm{Bubble}} + \Delta^{\mathrm{Triang}}] + O(q_0^{-2}\,\ell n q_0). \qquad (275a)$$

We will not exhibit the detailed form of Δ^{Triang}, but only remark that in all cases Δ^{Triang} vanishes when the three-momenta $\underset{\sim}{q}$ and $\underset{\sim}{q}' = \underset{\sim}{q} + \underset{\sim}{p} - \underset{\sim}{p}'$ associated with the currents $J_{(1)}$ and $J_{(2)}$ vanish,

$$\Delta^{\mathrm{Triang}}\Big|_{\underset{\sim}{q} = \underset{\sim}{q}' = 0} = 0 \,. \qquad (275b)$$

[Eq. (275) holds when the triplet of fermions are degenerate in mass. The effect of mass splittings is discussed in Ref. 34.] Thus, for the physically-interesting case of the commutator of spatially integrated currents, the entire answer is given by Eq. (274). No cancellation between the SU_3-singlet part of Δ^{Compt} and Δ^{Triang} is possible, and we conclude that the Bjorken-limit and the naive commutator in our models differ in second order perturbation theory.

For future reference it will be useful to write out in detail some special cases of Eq. (274). We consider first the commutator of two vector currents, with $\gamma_{(1)}=\gamma_\mu \frac{1}{2}\lambda_a$, $\gamma_{(2)}=\gamma_\lambda \frac{1}{2}\lambda_b$ and with naive commutator $\widetilde{\Gamma}(C;p,p')$,

$$C=\tfrac{1}{2}\{\tfrac{1}{2}\lambda_a,\tfrac{1}{2}\lambda_b\}(\gamma_\mu\gamma_0\gamma_\lambda-\gamma_\lambda\gamma_0\gamma_\mu)+\tfrac{1}{2}[\tfrac{1}{2}\lambda_a,\tfrac{1}{2}\lambda_b](\gamma_\mu\gamma_0\gamma_\lambda+\gamma_\lambda\gamma_0\gamma_\mu). \tag{276}$$

In the vector gluon case we find

$$\Delta^{\text{Compt}}=\frac{g_r^2}{16\pi^2}\{2(g_{\mu\lambda}-g_{\mu 0}g_{\lambda 0})\gamma_0[\tfrac{1}{2}\lambda_a,\tfrac{1}{2}\lambda_b] \tag{277}$$
$$+\tfrac{3}{2}(\gamma_\lambda\gamma_0\gamma_\mu-\gamma_\mu\gamma_0\gamma_\lambda)\{\tfrac{1}{2}\lambda_a,\tfrac{1}{2}\lambda_b\}\},$$

while in the scalar and pseudoscalar gluon cases we find

$$\Delta^{\text{Compt}}=\frac{g_r^2}{16\pi^2}\{(g_{\mu\lambda}-g_{\mu 0}g_{\lambda 0})\gamma_0[\tfrac{1}{2}\lambda_a,\tfrac{1}{2}\lambda_b] \tag{278}$$
$$-\tfrac{1}{2}(\gamma_\lambda\gamma_0\gamma_\mu-\gamma_\mu\gamma_0\gamma_\lambda)\{\tfrac{1}{2}\lambda_a,\tfrac{1}{2}\lambda_b\}[\ln(\frac{\Lambda^2}{|q_0|^2})-1]\}.$$

When one or both of the currents in the vector gluon case is an axial-vector current, Δ^{Compt} is obtained from the expression in Eq. (277) by the following simple substitutions:

Current $J_{(1)}$	Current $J_{(2)}$	Change in Eq. (277)
V	V	none
A	V	$\Delta^{\text{Compt}} \to -\gamma_5\, \Delta^{\text{Compt}}$
V	A	$\Delta^{\text{Compt}} \to \Delta^{\text{Compt}}\gamma_5$
A	A	$\Delta^{\text{Compt}} \to -\gamma_5 \Delta^{\text{Compt}}\gamma_5$

The results for axial-vector currents in the scalar and pseudoscalar gluon cases are not so simple.

(ii) Fourth order.

To fourth order in g_r, the number of diagrams contributing to $\tilde{T}^*_{(1)(2)}$ is so large that a direct calculation of the Bjorken limit, in analogy with our treatment of the second order case, is prohibitively complicated. However, as we have seen in Eqs. (260), (265) and (266), dispersion relations and unitarity provide a connection between the Bjorken limit for two vector currents and an integral over the longitudinal current-fermion inelastic cross section,

$$\lim_{q_0 \to i\infty} T^*_{a1\,b1}(p,q) = \lim_{q_0 \to i\infty} \tfrac{1}{2}\mathrm{tr}\{(\tfrac{\not{p}+m}{2m})\tilde{T}_{(1)(2)}(p,p,q)\}\Big|_{\substack{\gamma_{(1)}=\gamma_1\frac{1}{2}\lambda_a \\ \gamma_{(2)}=\gamma_1\frac{1}{2}\lambda_b}}$$

$$= q_0^{-1}[\tfrac{1}{2}\lambda^a, \tfrac{1}{2}\lambda^b]\{1 - \lim_{q^2 \to -\infty} 2\int_0^2 d\omega' [L^-(q^2,\omega') - L^+(q^2,\omega')]\}$$

$$+ \text{symmetric in } a, b + O(q_0^{-2} \ln q_0). \tag{279}$$

From this equation we can calculate the part of Δ which is proportional to $[\tfrac{1}{2}\lambda_a, \tfrac{1}{2}\lambda_b]$ if the longitudinal cross sections are known. The longitudinal cross sections themselves are, in general, extremely complicated. But multiplying Eq. (259) by $p^\lambda p^\mu$ and comparing with Eq. (267) (with M_N replaced by m) shows that, in the limit as the triplet mass m approaches zero, the longitudinal cross sections are given by the simple expression (280)

$$L^{\mp} = \lim_{m\to 0} \frac{-p_0\omega^2}{q^2}(2\pi)^3 \sum_{\text{spin}(\psi)} \sum_{\Gamma} |\langle\psi(p)|\tfrac{1}{2}p^\mu(\mathcal{F}_{1\mu} \pm \mathcal{F}_{2\mu})|\Gamma(p_\Gamma)\rangle|^2 \times \delta^4(p+q-p_\Gamma).$$

The factor p_0 in front of this equation is of purely kinematic origin, and cancels against a factor p_0^{-1} arising from our choice of normalization. The important point is that the factor p^μ in the matrix element in Eq. (280) leads to a considerable simplification in the calculation of $L^{\mp}$ in the zero triplet mass limit. Eqs. (279) and (280) have been applied,[59] in the scalar and pseudoscalar gluon cases, to the calculation of the part of the vector-vector commutator which is proportional to $[\frac{1}{2}\lambda_a, \frac{1}{2}\lambda_b]$, independent of m, $\underset{\sim}{q}$, $\underset{\sim}{q}'$, $\underset{\sim}{p}$ and $\underset{\sim}{p}'$ and which is logarithmically divergent as $|q_0|^2$ becomes infinite, with the result that

$$\Delta = (g_{\mu\lambda} - g_{\mu 0}g_{\lambda 0})\gamma_0[\tfrac{1}{2}\lambda_a, \tfrac{1}{2}\lambda_b]\left[\frac{g_r^2}{16\pi^2} + 7\left(\frac{g_r^2}{16\pi^2}\right)^2 \ln|q_0|^2 + g_r^4 \times \text{constant}\right] \tag{281}$$

+ symmetric in a, b + terms proportional to m, $\underset{\sim}{q}$, $\underset{\sim}{q}'$, $\underset{\sim}{p}$, $\underset{\sim}{p}'$.

Even the calculation of this one special case is very

complicated, and involves the consideration of three-body final states containing either a triplet plus two gluons or a triplet plus a triplet-antitriplet pair.

8.2 Discussion

We proceed next to discuss a number of features of the results of Eqs. (274), (276-279) and (281), and in particular, to indicate their effect on the applications of the Bjorken limit developed above.

(i) We begin by noting that to second order in g_r^2, Δ^{Compt} contains terms $\ell n(\Lambda^2/|q_0|^2)$ which diverge logarithmically both in the Bjorken limit $q_0 \to i\infty$ and in the infinite cutoff limit $\Lambda \to \infty$. It is easy to see that the $\ell n\Lambda^2$ divergences result from a mismatch between the multiplicative factors needed to make $T^{*\,\text{Compt}}_{(1)(2)}(p, p', q)$ and $\Gamma(C;p, p')$ finite (i.e., $\ell n\,\Lambda^2$-independent) as $\Lambda \to \infty$. As we recall, the <u>renormalized</u> quantities $\tilde{T}^{*\,\text{Compt}}_{(1)(2)}(p, p', q)$ and $\tilde{\Gamma}(C;p, p')$ are obtained from $T^{*\,\text{Compt}}_{(1)(2)}(p, p', q)$ and $\Gamma(C;p, p')$ by multiplying by the wave function renormalization Z_2 and taking the limit $\Lambda \to \infty$, keeping any residual $\ell n\Lambda^2$ dependence. On the other hand, the <u>finite</u> quantities $T^{*\,\text{Compt}}_{(1)(2)}(p, p', q)^{\text{finite}}$ and $\Gamma(C;p, p')^{\text{finite}}$ are obtained by multiplying by appropriate vertex and propagator renormalization factors which completely remove the $\ln \Lambda^2$ dependence,

$$\Gamma(C;p,p')^{\text{finite}} = Z(C)\Gamma(C;p,p'), \tag{282}$$

$$T^{*\,\text{Compt}}_{(1)(2)}(p,p',q)^{\text{finite}} = Z(\gamma_{(1)})Z(\gamma_{(2)})Z_2^{-1}T^{*\,\text{Compt}}_{(1)(2)}(p,p',q).$$

In general, the vertex renormalizations $Z(C)$, $Z(\gamma_{(1)})$ and $Z(\gamma_{(2)})$ are not equal to each other, or to Z_2. For example, in the case of the vector gluon model, we have seen in Subsection 3.1 that the pseudoscalar vertex renormalization factor is $Z(\gamma_5) = Z_2 m_0$, with m_0 the divergent fermion bare mass. If we write

$$Z(C) = 1 + \Lambda(C), \tag{283}$$

$$Z_2 = 1 + \Lambda(\gamma_\mu) = 1 + \Lambda_2,$$

then we find to second order that

$$\tilde{T}^{*\,\text{Compt}}_{(1)(2)}(p,p',q) = T^{*\,\text{Compt}}_{(1)(2)}(p,p',q)^{\text{finite}} \tag{284}$$

$$+[2\Lambda_2 - \Lambda(\gamma_{(1)}) - \Lambda(\gamma_{(2)})][\gamma_{(1)} \frac{1}{\not p + \not q - m} \gamma_{(2)} + \gamma_{(2)} \frac{1}{\not p - \not q' - m} \gamma_{(1)}],$$

$$\tilde{\Gamma}(C;p,p') = \Gamma(C;p,p')^{\text{finite}} + [\Lambda_2 - \Lambda(C)]C.$$

Since finite quantities on the left and right hand sides of Eq. (284) must match up, we see that

$$\Delta^{\text{Compt}} = [\Lambda_2 + \Lambda(C) - \Lambda(\gamma_{(1)}) - \Lambda(\gamma_{(2)})]C + \text{finite}, \tag{285}$$

confirming that the $\ln \Lambda^2$ dependence in Δ^{Compt} results from a mismatch between the renormalization factors on the left and right hand sides of Eq. (274a). A simple calculation shows that

$$\Lambda(C)C = \frac{g_r^2}{32\pi^2}\, \tfrac{1}{2}\, \underset{\sim}{\gamma}\, \gamma_\tau\, C\, \gamma^\tau\, \underset{\sim}{\gamma}\, \ln \Lambda^2, \tag{286}$$

which on substitution into Eq. (285) does indeed give the $\ln \Lambda^2$ terms in Eq. (274).

The presence of terms which diverge as $\ln|q_0|^2$ in Eq. (274) indicates that, <u>in the general case, the Bjorken limit does not exist in perturbation theory.</u> The fact that the $\ln|q_0|^2$ and $\ln \Lambda^2$ terms occur in the combination $\ln(\Lambda^2/|q_0|^2)$ means that, <u>to second order, the existence of the Bjorken limit is directly connected with the matching of renormalization factors on the left and right hand sides of Eq. (274a)</u>: When the renormalization factors match, the Bjorken limit exists; when the factors do not match, the Bjorken limit diverges.[60] Unfortunately, we shall see that this simple result does not hold in higher orders in perturbation theory.

(ii) There are a number of interesting cases in which the renormalization factors do match, and hence the Bjorken limit exists in second order. In the vector gluon model, Eq. (277) and the table following Eq. (278) show that this is true for all commutators involving vector and axial-vector currents. In the scalar and pseudoscalar gluon models, it is true for the vector piece of the V-V commutator [Eq. (278)]

and for the axial-vector piece of the V-A and A-V commutators. The remarkable result that emerges from these examples is that, <u>even when the Bjorken limit exists in second order, it does not agree with the naive commutator.</u> According to the discussion of Eqs. (204)-(206) above, this means that the Bjorken limit agrees with the spectral function integral of Eq. (206), but the naive commutator does not. So it is really somewhat of a misnomer to talk about Bjorken limit breakdown; it is the naive commutator, and not the Bjorken limit, which breaks down.

Armed with the explicit formulas of Eqs. (277) and (278), we can now go back to see what happens to the various Bjorken limit applications developed above. First we consider the discussion of radiative corrections to β-decay given in Subsection 7.1.[61] As we have seen, the term proportional to Q_{AV} in Eq. (246) comes from the spatially vector, isospin symmetric part of a V-A commutator [the $\{\lambda_Q, \lambda_+\}$ $[\gamma_0\gamma_\sigma, \gamma_0\gamma_\mu\gamma_5]$ term in Eq. (241)]. In the vector gluon model, this corresponds to the $\{\frac{1}{2}\lambda_a, \frac{1}{2}\lambda_b\}$ term in the V-A version of Eq. (277). Comparing Eq. (277) with Eq. (276), we see that to second order the coefficient of $\frac{1}{2}\{\frac{1}{2}\lambda_a, \frac{1}{2}\lambda_b\}(\gamma_\mu\gamma_0\gamma_\lambda - \gamma_\lambda\gamma_0\gamma_\mu)$ is changed from 1 to $1 - 3g_r^2/(16\pi^2)$, and hence Eq. (246) is modified in the vector gluon model to read

$$\delta M = \frac{3\alpha}{8\pi} \left[1 + \left(1 - \frac{3g_r^2}{16\pi^2}\right) 2Q_{AV} \right] \ln \Lambda^2 M, \qquad (287)$$

which diverges even in the special case $Q_{AV} = -\frac{1}{2}$. In the scalar and pseudoscalar gluon models, the situation is even worse, since the vector part of the V-A commutator is divergent in these models, as a result of mismatch of renormalization factors, and consequently the coefficient of $\ln \Lambda^2$ in Eq. (287) is itself logarithmically divergent. In fact, detailed field theoretic analyses[11] show that the vector gluon model with $Q_{AV} = -\frac{1}{2}$ is the only renormalizable, SU_3-symmetric model of the strong interactions which has the possibility of having finite radiative corrections to β-decay, so our result of Eq. (287) shows that there are in fact <u>no</u> renormalizable, SU_3-symmetric models with this property.

It is important to note that Eq. (287) does not contradict the result, mentioned above, that the radiative corrections to μ-meson decay are finite to all orders in QED. The point is that while the vector gluon is an SU_3-singlet, and hence couples symmetrically to all of the triplet fermions, the photon couples to the muon and electron but <u>not</u> to the neutrino. As a result, certain diagrams which are present in the triplet decay process are absent in muon decay, e.g.

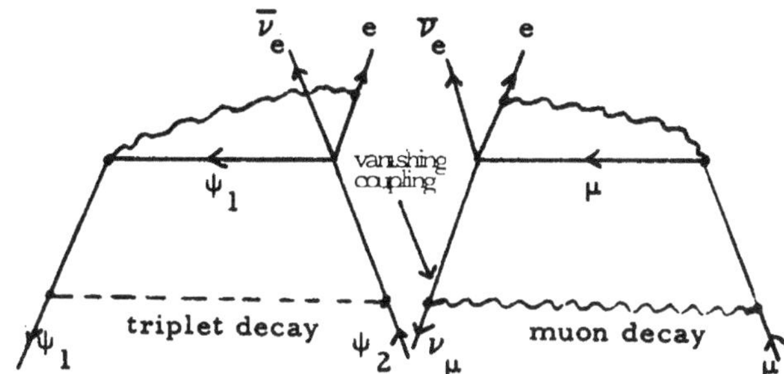

It turns out to be precisely these missing diagrams which cause the disagreement between the Bjorken limit and naive commutators.

Next, we consider the asymptotic sum rules and cross section relations of Eqs. (260) and (266). We have actually already seen what happens in these cases: according to Eq. (279), the integrals no longer vanish, but instead are proportional to the coefficient of the $[\frac{1}{2}\lambda_a, \frac{1}{2}\lambda_b]$ term in Eq. (277) or Eq. (278),

$$\begin{aligned}
&\lim_{q^2 \to -\infty} 2M_N \int_0^2 d\omega' \,[\, W_1^-(q^2, \omega') - W_1^+(q^2, \omega')\,] \\
&= \lim_{q^2 \to -\infty} 2 \int_0^2 d\omega' \,[\, L^-(q^2, \omega') - L^+(q^2, \omega')\,] \qquad (288) \\
&= \begin{cases} g_r^2/(8\pi^2) & \text{vector gluon} \\ g_r^2/(16\pi^2) & \text{scalar or pseudoscalar gluon} \end{cases}
\end{aligned}$$

In other words, the backward neutrino sum rule and the backward electron-scattering inequality depend on the dynamics of the triplet-gluon interaction, not just on the kinematic structure of the weak and electromagnetic currents. If, for definiteness, we take the fermion state ψ in Eq. (280) to be ψ_1, we find that the entire contribution to the longitudinal cross section to order g_r^2 comes from the intermediate state $\Gamma = \psi_2$ + gluon. The resulting two-body phase space integral is easily evaluated in the center of mass frame, giving

$$L^+ = 0 \tag{289}$$

$$\lim_{q^2 \to -\infty} L^-(q^2,\omega) = \begin{cases} g_r^2\omega/32\pi^2 & \text{vector gluon} \\ g_r^2\omega/64\pi^2 & \text{scalar or pseudoscalar gluon,} \end{cases}$$

in agreement with Eq. (288). The corresponding electromagnetic longitudinal cross section is obtained by replacing $\mathcal{F}_{1\mu} - i\,\mathcal{F}_{2\mu}$ in Eq. (280) by j_μ^{EM}, and receives its only contribution from the intermediate state $\Gamma = \psi_1$ + gluon. Clearly, the ratio $L^{EM}(q^2,\omega)/L^-(q^2,\omega)$ is just the squared charge Q^2 of ψ_1, indicating that the Callan-Gross relation of Eq. (269) also fails in our perturbation theory models.[61,62] We conclude that none of the principal applications of the Bjorken limit method are valid in perturbation theory.

(iii) From an inspection of Eq. (277) we see that in the vector gluon case, for all commutators involving vector and axial-vector currents, Δ^{Compt} vanishes when either $\mu=0$ or $\nu=0$. In other words, only the space-component-space-component commutators are anomalous. When $J_{(1)}$ and $J_{(2)}$ are both vector currents, this result can be deduced directly from the Ward identity of Eq. (253), which in our present notation reads, on the mass shell,

$$\tilde{T}^{*\,\text{Compt}}_{(1)(2)}(p,p',q)\Big|_{\substack{\gamma_{(1)}=q^{\mu}\gamma_{\mu}\frac{1}{2}\lambda_a \\ \gamma_{(2)}=\gamma_{\lambda}\frac{1}{2}\lambda_b}} = \tilde{\Gamma}([\tfrac{1}{2}\lambda_a,\tfrac{1}{2}\lambda_b]\,\gamma_{\lambda};p,p'). \quad (290)$$

Multiplying by q_0^{-1} and taking the limit $q_0 \to i\infty$ gives immediately

$$\lim_{q_0\to i\infty} \tilde{T}^{*\,\text{Compt}}_{(1)(2)}(p,p',q)\Big|_{\substack{\gamma_{(1)}=\gamma_0\frac{1}{2}\lambda_a \\ \gamma_{(2)}=\gamma_{\lambda}\frac{1}{2}\lambda_b}} = q_0^{-1}\tilde{\Gamma}([\tfrac{1}{2}\lambda_a,\tfrac{1}{2}\lambda_b]\,\gamma_{\lambda};p,p') + O(q_0^{-2}\ln q_0), \quad (291)$$

confirming our explicit calculation. A similar derivation holds in the cases involving axial-vector currents, provided that the divergence of the axial-vector current is "soft",[63] as it is in the vector gluon model. We thus see that the breakdown of the Bjorken limit which we have found in Compton-like graphs is consistent with the constraints imposed by Ward identities, just as we found a similar consistency in the triangle graph case in Subsection 6.3. This means that except in cases such as $\pi^0 \to 2\gamma$, where Ward identity anomalies occur, the standard results of the Gell-Mann time component algebra, which are derived directly from Ward identities, remain valid. From the point of view of trying to distinguish between different models of the hadronic current, it is unfortunate that the Bjorken-limit fails precisely in the case of space-space commutators, where the usual current algebra "infinite momentum limit" and "low energy theorem" methods also do not work.

(iv) Basically, the origin of Bjorken limit breakdown is the very singular nature of perturbation expansions in field theory.[64] To see this, we note that if we take the Bjorken limit $q_0 \to i\infty$ <u>before</u> letting the regulator mass Λ approach infinity, so that we are dealing with a convergent cutoff field theory, the Bjorken limits and naive commutators agree.[59;65] The order of limits is thus of crucial importance here, unlike the situation in the low energy theorem discussion of Section 4. The reason is that when Λ is held finite while the Bjorken limit is taken, the spectral function integral of Eq. (206) contains contributions from the regulator particle, which cancel away the anomalous terms which we have found. On the other hand, if the limit $\Lambda \to \infty$ is taken first, so that we are dealing with renormalized perturbation theory, and then the Bjorken limit is taken, these regulator contributions are absent. Although the regulator theory has no Bjorken limit anomalies, it is not a very satisfactory physical model, since cross sections for regulator particle production can be negative. For instance, the Callan-Gross limit of Eq. (269) is satisfied in the regulator theory, but only by virtue of a cancellation between the cross section for fermion + gluon production, which is positive, and a negative cross section for fermion + regulator particle production.

(v) We turn next to the order g_r^4 result of Eq. (281), which gives the V-V→V commutator in the scalar and pseudoscalar gluon models. We see that <u>even though the renormalization factors match, the Bjorken limit in this case diverges in fourth order.</u> We note, however, that the divergence behaves as $g_r^4 \ell n|q_0|^2$, whereas in fourth order terms behaving like $g_r^4(\ell n|q_0|^2)^2$ could in principle be present. On the basis of this behavior and our second order results, the following <u>conjecture</u> seems reasonable: When the renormalization factors needed to make $T^{*\mathrm{Compt}}_{(1)(2)}(p, p', q)$ and $\Gamma(C;p,p')$ finite are the same, the Bjorken limit in order 2n of perturbation theory contains no terms $g_r^{2n}(\log|q_0|^2)^n$, but begins in general with terms

$g_r^{2n}(\log|q_0|^2)^{n-1}$.

(vi) Finally, we must emphasize that all of our results have been obtained in perturbation theory,[66] whereas strong interactions are notoriously non-perturbative in behavior. Thus, one is always free to postulate that non-perturbative effects somehow conspire to "damp out" the anomalous terms when all orders of perturbation theory are summed, although the need for this assumption would mean that asymptotic sum rules would not give a test of the space-space current algebra alone, but would involve deep dynamical considerations as well. There is an alternative point of view which has been analyzed in detail recently.[67] This is that Bjorken limits and naive commutators may well have little relation to each other, but still Bjorken limits may be interesting because, via asymptotic sum rules and Eq. (206), they furnish an experimental means for measuring equal-time commutators and other singular behavior of time-ordered products. In order for this point of view to bear fruit, it will be necessary to find new ways, not involving naive commutators, of correlating this singular behavior with the underlying structure of the theory or of relating to one another the singular behavior measured in different types of experiments.

References

1. The reader is referred to several of the excellent texts available, in particular, J. D. Bjorken and S. D. Drell, "Relativistic Quantum Fields" (McGraw-Hill, New York, 1965); S. S. Schweber, "An Introduction to Relativistic Quantum Field Theory" (Row, Peterson, Evanston, 1961); F. Mandl, "Introduction to Quantum Field Theory" (Interscience, New York, 1959); J. Schwinger, ed., "Quantum Electrodynamics" (Dover, New York, 1958).

2. Our notation will always be that of Bjorken and Drell, Ref. 1, pp. 377-390. In particular, $g_{00} = -g_{11} = -g_{22} = -g_{33} = 1$, and $\epsilon_{0123} = -\epsilon^{0123} = 1$. Unless specifically indicated by sub- or superscripts, field quantities ψ, A^{μ}, etc. will always denote the unrenormalized fields.

3. For a full discussion of the renormalization of QED, the reader is referred to the final chapter of Bjorken and Drell, Ref. 1.

4. For a discussion of the Fierz transformation, see the lectures by J. D. Jackson in "Elementary Particle Physics and Field Theory," VI of the 1962 Brandeis lectures, p. 280.

5. See J. D. Jackson, Ref. 4, p. 409.

6. J. M. Jauch and F. Rohrlich, "The Theory of Photons and Electrons" (Addison-Wesley, Cambridge, Mass. 1955), pp. 458-461.

7. L. Rosenberg, Phys. Rev. 129, 2786 (1963).

8. S. Weinberg, Phys. Rev. 118, 838 (1960). For a simplified exposition of Weinberg's results, see J. D. Bjorken and S. D. Drell, Ref. 1, pp. 317-330 and pp. 364-368. Weinberg's theorem applies for an arbitrary spacelike four-vector q.

9. S. L. Adler, Phys. Rev. 177, 2426 (1969).

10. J. Schwinger, Phys. Rev. 82, 664 (1951), Sec. V; C. R. Hagen, Phys. Rev. 177, 2622 (1969); R. Jackiw and K. Johnson, Phys. Rev. 182, 1457 (1969); B. Zumino, "Proceedings of the Topical Conference on Weak Interactions" (CERN, Geneva, 1969), pp. 361-370.

11. G. Preparata and W. I. Weisberger, Phys. Rev. 175, 1965 (1968), Appendix C.

12. For recent discussions of the sicknesses of the local current-current theory and their possible remedies, see N. Christ, Phys. Rev. 176, 2086 (1968); and M. Gell-Mann, M. L. Goldberger, N. M. Kroll and F. E. Low, Phys. Rev. 179, 1518 (1969).

13. D. C. Cundy et. al., Phys. Letters 31B, 478 (1970).

14. For details, see S. L. Adler and R. F. Dashen, "Current Algebras" (W. A. Benjamin, New York, 1968), pp. 15-18.

15. Our development in this section follows S. L. Adler and W.A. Bardeen, Phys. Rev. 182, 1517 (1969).

16. For a discussion of PCAC, see Ref. 14, pp. 40-48.

17. For further details about the σ-model and its renormalization, see Ref. 15 and also B. W. Lee, Nucl. Phys. B9, 649 (1969).

18. See Ref. 17, pp. 24-27.

19. The connection between anomalous divergences and π^0 decay was first pointed out by J. S. Bell and R. Jackiw, Nuovo Cimento 60A, 47 (1969).

20. D. G. Sutherland, Nucl. Phys. B2, 433 (1967).

21. Particle Data Group, Revs. Modern Phys. 42, 87 (1970).

22. B. Richter, rapporteurs talk in "Proceedings of the International Conference on High Energy Physics" (CERN, Geneva, 1968), p. 13.

23. S. Okubo, Phys. Rev. 179, 1629 (1969).

24. F. J. Gilman, Phys. Rev. 184, 1964 (1969).

25. M. L. Goldberger and S. B. Treiman, Nuovo Cimento 9, 451 (1958); H. Pagels, Phys. Rev. 158, 1566 (1967).

26. A. Hearn, unpublished.

27. R. F. Dashen, unpublished.

28. S. Devons et. al., Phys. Rev. 184, 1356 (1969).

29. W. A. Bardeen, Phys. Rev. 184, 1848 (1969).

30. R. W. Brown, C-C. Shih and B.-L. Young, Phys. Rev. 186, 1491 (1969).

31. W. A. Bardeen, talk at the Trieste Conference on Renormalization Theory, 1969.

32. S. L. Glashow, R. Jackiw and S. S. Shei, Phys. Rev. 187, 1916 (1969).

33. J. D. Bjorken, Phys. Rev. 148, 1467 (1966).

34. K. Johnson and F. E. Low, Progr. Theoret. Phys. (Kyoto) Suppl. Nos. 37-38, 74 (1966).

35. J. Schwinger, Phys. Rev. Letters 3, 296 (1959).

36. R. Jackiw and K. Johnson, Ref. 10.

37. Putting the $\underset{\sim}{\varepsilon}$ separation only into the spatial components of j can be justified by a more careful treatment.

38. See. J. D. Bjorken and S. D. Drell, Ref. 1, Ch. 16.

39. C. R. Hagen, Phys. Rev. 188, 2416 (1969).

40. R. Brandt and J. Sucher, Phys. Rev. Letters 20, 1131 (1968).

41. The treatment here follows S. L. Adler and D. G. Boulware, Phys. Rev. 184, 1740 (1969). See also R. Jackiw and K. Johnson, Ref. 10.

42. S. L. Adler and R. F. Dashen, Ref. 14, Ch. 3; D. G. Boulware and L. S. Brown, Phys. Rev. 156, 1724 (1967).

43. For references, see G. Preparata and W. I. Weisberger, Ref. 11.

44. For a survey, see lectures by J. D. Bjorken in "Selected Topics in Particle Physics, Proceedings of the International School of Physics 'Enrico Fermi' Course XLI," edited by J. Steinberger (Academic Press, New York, 1968).

45. C. G. Callan and D. J. Gross, Phys. Rev. Letters 22, 156 (1969).

46. S. L. Adler and W.-K. Tung, Phys. Rev. Letters 22, 978 (1969); R. Jackiw and G. Preparata, Phys. Rev. Letters 22, 975 (1969).

47. E. S. Abers, R. E. Norton and D. A. Dicus, Phys. Rev. Letters 18, 676 (1967); E. S. Abers, D. A. Dicus, R. E. Norton, and H. Quinn, Phys. Rev. 167, 1461 (1968).

48. We follow here C. G. Callan, Phys. Rev. 169, 1175 (1968).

49. See J. D. Bjorken, Ref. 44.

50. For a recent review, see F. J. Gilman in the Proceedings of the 1969 International Symposium on Electron and Photon Interactions at High Energies, Liverpool, 1969.

51. See J. D. Bjorken, Ref. 44.

52. See S. L. Adler and R. F. Dashen, Ref. 14, pp. 27-34.

53. G. F. Chew, M. L. Goldberger, F. E. Low and Y. Nambu, Phys. Rev. 106, 1337 (1957).

54. See S. L. Adler and R. F. Dashen, Ref. 14, p. 261 and pp. 354-357.

55. T. D. Lee, S. Weinberg and B. Zumino, Phys. Rev. Letters 18, 1029 (1967).

56. S. L. Adler and R. F. Dashen, Ref. 14, pp. 239-246.

57. F. J. Gilman, Phys. Rev. 167, 1365 (1968).

58. L. N. Hand, Phys. Rev. 129, 1834 (1963).

59. S. L. Adler and W.-K. Tung, Phys. Rev. (to be published).

60. This was first noted by A. I. Vainshtein and B. L. Ioffe {Zh. Eksperim-i Teor. Fiz. --Pis'na Redakt 6, 917 (1967) [translation: Soviet Phys--JETP Letters 6, 341 (1967)]}.

61. S. L. Adler and W.-K. Tung, Phys. Rev. Letters 22, 978 (1969).

62. R. Jackiw and G. Preparata, Phys. Rev. Letters 22, 975 (1969).

63. See S. L. Adler and R. F. Dashen, Ref. 14, pp. 257-260.

64. For a detailed discussion, see W.-K. Tung, Phys.

Rev. 188, 2404 (1969).

65. C. R. Hagen, Phys. Rev. 188, 2416 (1969).

66. Further results on the Bjorken limit in perturbation theory have been given recently by J. C. Polkinghorne (to be published).

67. R. Jackiw, R. Van Royen and G. B. West (to be published); L. S. Brown and D. G. Boulware, as described in lectures by L. S. Brown at the Summer Institute for Theoretical Physics, University of Colorado, 1969 (to be published); H. Leutwyler and J. Stern (to be published); R. A. Brandt (to be published).

The assistance of H. Pendleton, D. Sidhu, and K. Trabert in checking the typescript is gratefully acknowledged.

Dynamical Applications of the Veneziano Formula

Stanley Mandelstam
Department of Physics
University of California
Berkeley, California

CONTENTS

1. Considerations Motivating the Veneziano Formula

The Veneziano formula provides us with a model in which Regge trajectories rise indefinitely instead of turning over as they do in potential theory. As we shall see below, such a behavior is indicated, though not proved conclusively, by the experimental results, and it fits with the further fact that low-energy resonances are observed to be narrow. A model for an amplitude with such properties might be useful for phenomenological analysis of experimental data, and it might also form the basis for a more fundamental approach to the problem of strong interactions. In these lectures we shall stress this latter aspect, even though, at the present time, there exist defects in the model which remain to be overcome.

I shall begin by outlining the experimental evidence in favor of indefinitely rising Regge trajectories. I shall then indicate how one might attempt to construct a dynamical model which incorporates such trajectories; such a model would not contain a separate class of "elementary particles". I shall indicate the approximation scheme which might be adopted, and the consistency conditions which must be satisfied. Dynamical schemes based on such principles had been proposed before the discovery of the Veneziano formula, and they are quite different in nature from previous schemes such as the N/D method or the strip approximation. I shall then come to the main topic of the lectures and shall discuss how the Veneziano formula and its generalizations provide us with a model which incorporates the consistency conditions referred to above.

One further feature of such dynamical models may be mentioned. It has always been a mystery why the observed meson and baryon spectra agree with those predicted by the quark model, even though quarks have not been seen. Within the framework of the Veneziano model one can understand how this may occur. One can also understand $SU(6)_W$ symmetry, which had previously been difficult to interpret within the framework of dispersion relations. These last remarks must be qualified by mentioning that the main defect of the Veneziano model, namely the existence of "ghost" resonances with negative decay probability, is particularly severe in models with fermions or with quarks. Nevertheless, the fact that so many things fit together leads us to hope that the defects of the model may be able to be overcome.

Let us now summarize the experimental evidence in favor of indefinitely rising Regge trajectories. Such a trajectory has been shown

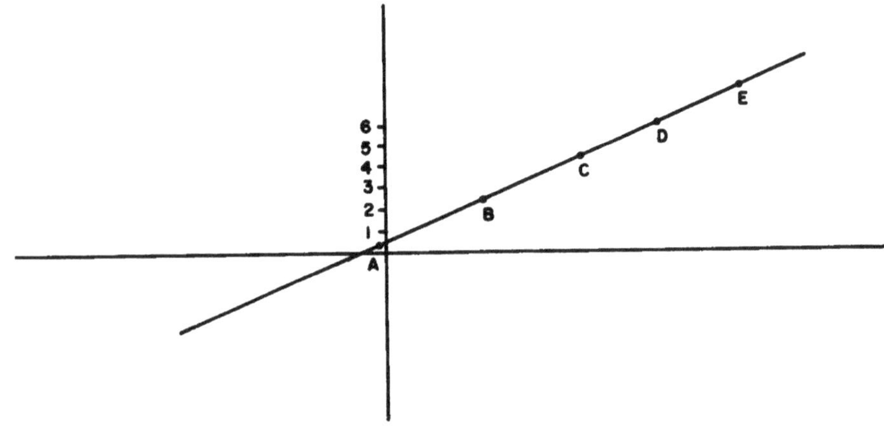

Figure 1. A Rising Regge Trajectory

diagramatically in Figure 1. Ideally, one would wish to identify resonances at all points where the trajectory passes through even or odd integers or half-integers, and to measure their spins. In actual fact the spins have usually been measured only for the two or three lowest particles. We thus have the points A, B, C on the trajectory in Figure 1. If a straight line drawn through these points is extrapolated to higher energies, resonances are observed at points D, E, etc., where the trajectory passes through a succession of higher even or odd (half) integral values of j, reaching as far as $\frac{19}{2}$ in one case. The spins of these higher particles have not been measured, though the combination of long lifetime and high Q-value indicates that they are high. It is tempting to identify the particles with the higher members of the Regge sequence.

The fact that the Regge trajectories are almost linear is correlated with the observed narrowness of the resonances. This may be seen at once from the dispersion relation for the trajectory function $\alpha(t)$. Thus, with two subtractions,

$$\alpha(t) = at + b + \frac{1}{\pi}\int dt' \ \frac{\text{Im}\ \alpha(t')}{t'-t} \tag{1.1}$$

if $\text{Im}\ \alpha(t) \to 0$ as $t \to \infty$. Since the width of the resonance is proportional to Im α, we notice that the trajectory is linear if and only if the resonances are narrow. Furthermore, in any model where all resonances are narrow the Regge trajectories must rise indefinitely.[1]

The experimental facts which we have summarized suggest that we attempt to construct a model scattering amplitude where all trajectories are exactly linear and all resonances infinitely narrow. Such a model can at best provide only an approximation to nature, but it may well serve as a starting-point in a new approach to strong interactions. Finite-width effects may hopefully be included at subsequent stages of an approximation scheme. In other words, one may regard the narrow-resonance approximation as a new kind of weak-coupling approximation. However, it is a weak-coupling approximation without elementary particles; all resonances lie on Regge trajectories. The old type of weak-coupling approximation, if it has any meaning at all in strong coupling physics, is certainly a very bad approximation. The new narrow-resonance approximation, though not very accurate, may nevertheless possess many of the qualitative features of the complete system, particularly with regard to the spectrum of particles.[2]

One way in which the present scheme differs from previous schemes is that Levinson's theorem is not satisfied. If the Regge trajectory rises and falls again, the phase shift will also rise and fall, and the energies at which the Regge trajectory and the phase shift turn over will be of the same order of magnitude. If the Regge trajectories continue to rise, the phase shifts would be expected to do so as well, and the quantity $\delta(\infty) - \delta(0)$ would be infinite. The "bootstrap" requirement that there be no elementary particles is now introduced, not by means of Levinson's theorem, but by demanding that all particles lie on Regge trajectories.

In the conventional weak-coupling approximation, there are no self-consistency conditions to apply. One can construct Born terms where all masses and coupling constants are arbitrary. By demanding that all particles lie on Regge trajectories, as we do in the present narrow-resonance approximation, we impose consistency conditions which are decidedly non-trivial.

Papers claiming to prove the impossibility of constructing such models, even for scattering of scalar particles, began to appear shortly before Veneziano constructed one explicitly. In the remainder of this section we shall indicate the consistency conditions which are necessary.

The first and most fundamental consistency condition is that the amplitude have Regge asymptotic behavior in all three channels simultaneously. If for simplicity we consider an amplitude with resonances in two out of the three channels only (s and t, say), the amplitude must have Regge asymptotic behavior in both channels simultaneously. It may well be questioned whether we should assign such a fundamental role to Regge asymptotic behavior, and we may alternatively require that the resonances in both channels lie on Regge trajectories or, in other words, that they be members of rotational sequences. We may then attempt to weaken the asymptotic assumption by simply requiring that there be no unphysical features such as exponential increase in the physical region. The only narrow-resonance models so far known which satisfy these requirements do have Regge asymptotic behavior. In any case, the crucial aspect of the assumptions is that they apply to both channels simultaneously.

We may formulate this self-consistency condition in another way. Let us write a dispersion relation for A at fixed t:

$$A(s,t) = \frac{1}{\pi}\int ds' \, \frac{\operatorname{Im} A(s',t)}{s'-s} \, . \tag{1.3}$$

Since we are assuming that the amplitude has no u-channel resonances, the second term in the dispersion relation will be absent. Let us also assume that the dispersion relation is true without substractions for some range of values of t. This will certainly be the case if we have Regge asymptotic behavior with a linear trajectory function $\alpha(t)$. (Figure 1). In the narrow-resonance approximation, the function Im A(s,t) consists entirely of delta-functions:

$$\mathrm{Im}\, A(s,t) = \sum_{\ell, s_\ell} \delta(s-s_\ell)\, P_\ell \left\{1+ \frac{2t}{s_\ell - 4\mu^2}\right\}, \qquad (1.4a)$$

so that

$$A(s,t) = \sum_{\ell, s_\ell} \frac{1}{s-s_\ell}\, P_\ell \left\{1+ \frac{2t}{s_\ell - 4\mu^2}\right\}. \qquad (1.4b)$$

The amplitude can thus be written as a sum of contributions from s-channel resonances if t is sufficiently small. In the same way we can show that the amplitude may be written as a sum of contributions from t-channel resonances if s is sufficiently small. We emphasize that we do not sum over s-channel and t-channel resonances, as we would if the resonances were elementary. The entire amplitude can be represented either as a sum over s-channel resonances or as a sum over t-channel resonances. This is one definition of a principle emphasized by Dolen, Horn and Schmid and known as duality.[3]

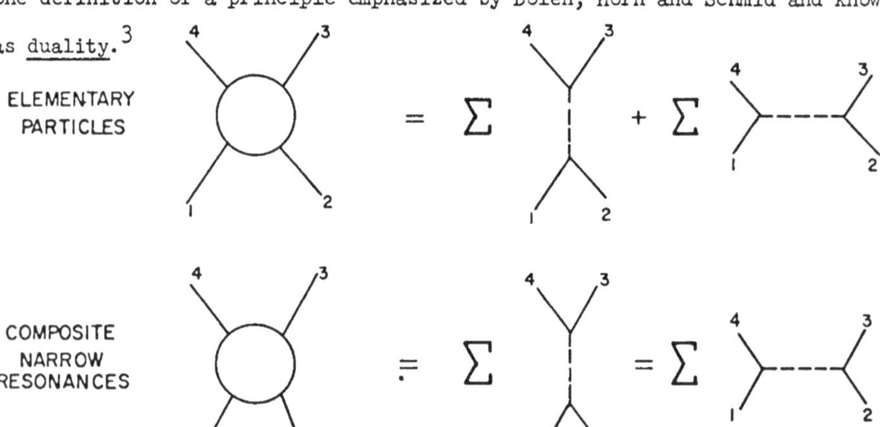

Figure 2. The difference between amplitudes with elementary particles and dual-resonance amplitudes.

So far we have only treated the scattering of scalar particles. However, each resonance on a Regge trajectory itself corresponds to a stable or unstable particle in the system, and one must be able to construct amplitudes for the scattering of such particles. Whenever we have a system of several particles, the residues at the poles in the scattering amplitude must satisfy the factorization condition. Let us examine a general amplitude A+B → C+D, and let us isolate a pole due to a single-particle intermediate state E [Figure 3(a)]. The residue at such a pole

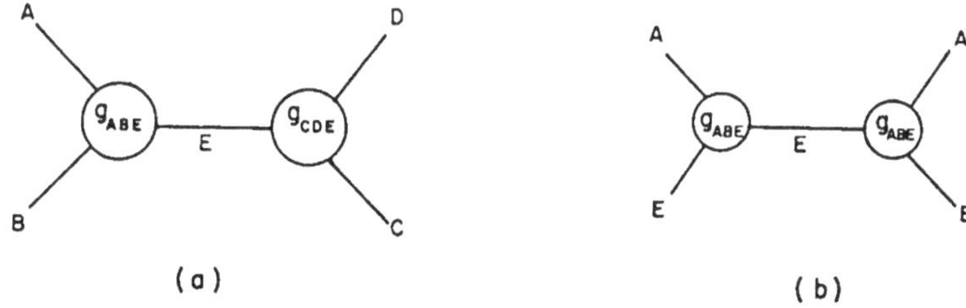

Figure 3. Factorization of Scattering Amplitudes.

must factorize into two "coupling constants" g_{ABE} and g_{CDE}. If we apply the factorization restriction to the amplitudes A+B → A+B, A+B → C+D and C+D → C+D, we notice that the three residues can be expressed in terms of two coupling constants g_{ABE}, g_{CDE}. In systems with more than two communicating channels we obtain correspondingly more restrictions.

Factorization has further consequences when we consider an amplitude A+E → A+E and isolate the pole due to an intermediate state B. The coupling constant g_{ABE} which occurs in this reaction is the same as that of the previous reaction. This requirement is sometimes known as vertex symmetry.

Our general requirements are thus that the amplitude for the process

A+B → C+D should consist of narrow resonances in the s and t channels, that it should have Regge asymptotic behavior in both channels, that the factorization requirement should be satisfied in both channels, and that the coupling constants should have vertex symmetry. Corresponding restrictions exist for production amplitudes; in fact, the four-point and n-point problems are so intimately linked that they must be treated simultaneously. The requirements, taken together, are very restrictive, and it is remarkable that functions can be constructed which satisfy them exactly. Such functions appear to be defined with a fair degree of uniqueness.

As we have already mentioned, narrow-resonance models may in a sense be regarded as a new type of Born approximation, and effects due to finite widths must subsequently be included in a complete theory. A very appealing extension of the Veneziano model has been proposed for including such effects in a perturbation series. This approach works in terms of Feynman-like diagrams, the original Veneziano model corresponding to the Born (or "tree") diagrams. It is doubtful whether one can treat the strong-interaction problem by straight perturbation methods, and the higher-order terms of the Feynman-like series do appear to possess a divergence associated with the large phase space available to high-energy intermediate states. However, it may well be possible to use semi-perturbation methods based on the series of Feynman-like diagrams, and we shall discuss the higher-order terms in subsequent lectures of this series.

2. The Veneziano Formula for the Four-Point Scalar Amplitude

In the present section we shall discuss the original Veneziano[4] model for the elastic scattering of a pair of equal-mass scalar particles. We require an amplitude with narrow resonances and Regge asymptotic behavior in the s and the t channels; we are still supposing for simplicity that there are no resonances in the u-channel. Since we are only considering one kind of external particle, questions of factorization do not arise yet.

The Veneziano formula is as follows:

$$A(s,t) = \frac{\Gamma\{-\alpha(s)\}\Gamma\{-\alpha(t)\}}{\Gamma\{-\alpha(s)-\alpha(t)\}} \equiv B\{-\alpha(s),-\alpha(t)\} = \int_0^1 dx\, x^{-\alpha(s)-1}(1-x)^{-\alpha(t)-1}. \tag{2.1a}$$

As in the last section, the Regge trajectory function $\alpha(s)$ is assumed linear:

$$\alpha(s) = as+b. \tag{2.1b}$$

The combination of gamma functions is (2.1a) is known as a beta-function (B-function), and its properties are outlined in most books on special functions. It has the integral representation given in (2.1).

In studying the characteristics of our amplitude, we may use either the representation in terms of gamma functions or the integral representation. Veneziano, in his original paper, worked mainly in terms of gamma functions. He was led to his formula by a study of such functions, since they occur in many models with Regge asymptotic behavior, including those of Veneziano and his co-workers quoted in Reference 2. Most generalizations of the Veneziano model start from the integral representation for the beta function and, since the generalizations greatly extend the power of the model, we shall derive the results form the integral representation.

Let us first examine the singularities of A as a function of s and t. We wish to show that A is analytic except for poles in s when $\alpha(s)$ is a positive integer, or poles in t when $\alpha(t)$ is a positive integer. We also wish to show that the residues at the poles in s are polynomials in z_s or, in other words, in t, and to examine the order of the polynomials. Similarly, we must show that the residues at the poles in t are polynomials in s. Since the integral in (2.1a) is obviously symmetric in s and t, we need only examine the singularities in one of the two variables.

To examine the singularities of (2.1a) as a function of s, we first notice that the integral converges uniformly when $\alpha(s)$ and $\alpha(t)$ are negative. Thus, since the integrand is analytic in s and t, the integral is also analytic. However, as $\alpha(s)$ approaches zero, the integral in (2.1a) becomes divergent at x = 0. We may write the non-divergent factor $(1-x)^{-\alpha(s)-1}$ as

$$(1-x)^{-\alpha(s)-1} = 1 + \{(1-x)^{-\alpha(s)-1} - 1\} .$$

The second term, when inserted into (2.1a), gives an integral which is convergent near x=1 when $\alpha(s) = 0$. Thus

$$\int_0^1 dx\, x^{-\alpha(s)-1} (1-x)^{-\alpha(t)-1} = \int_0^1 dx\, x^{-\alpha(s)-1} + \text{terms finite at } \alpha(s)=0$$

$$= -\frac{1}{\alpha(s)} + \text{terms finite at } \alpha(s)=0. \qquad (2.2)$$

The function A(s,t) therefore has a pole in s at $\alpha(s) = 0$, and the residue at the pole is independent of t. This is the property we require, since the amplitude is expected to possess an S-wave resonance when $\alpha(s)=0$.

The integral on the right of (2.1a) is not defined when $\alpha(s) > 0$. We therefore define it by analytic continuation. To do so, we first re-express the integral in such a way that the expression is unchanged for $\alpha(s) < 0$, but has a meaning for $\alpha(s) > 0$. This may be achieved by integration by parts:

$$\int_0^1 dx\, x^{-\alpha(s)-1}(1-x)^{-\alpha(t)-1} = \frac{\alpha(t)+1}{\alpha(s)}\int_0^1 dx\, x^{-\alpha(s)}(1-x)^{-\alpha(t)-2} \tag{2.3}$$

The right-hand side of (2.3) is defined if $\alpha(s) < 1$. As $\alpha(s)$ approaches 1, the integral begins to diverge at x=0. In the same way as before, we can easily see that

$$A(s,t) = -\frac{\alpha(t)+1}{\alpha(s)-1} + \text{terms finite at } \alpha(s) = 1. \tag{2.4}$$

We notice that the amplitude has a pole at $\alpha(s)=1$, but now the residue is a linear function of $\alpha(t)$, and therefore of t. The amplitude thus possesses S-wave and P-wave resonances at s=1.

By repeated integration by parts we can extend the definition of the integral in (2.1) to an arbitrarily large value of s. The amplitude has poles at any integer, and, near $\alpha(s) = n$:

$$A(s,t) = -\frac{\{\alpha(t)+1\}\{\alpha(t)+2\}\ldots\{\alpha(t)+n\}}{n!}\;\frac{1}{\alpha(s)-n} + \text{finite terms.} \tag{2.5}$$

The residue is thus a polynomial of the nth degree in t, and the amplitude possess resonances with angular momentum ranging from 0 to n.

The resonances of our amplitude lie on the Regge trajectories shown in Figure 4. When $\alpha(s)=0$ there is an S-wave resonance, when $\alpha(s)=2$ an

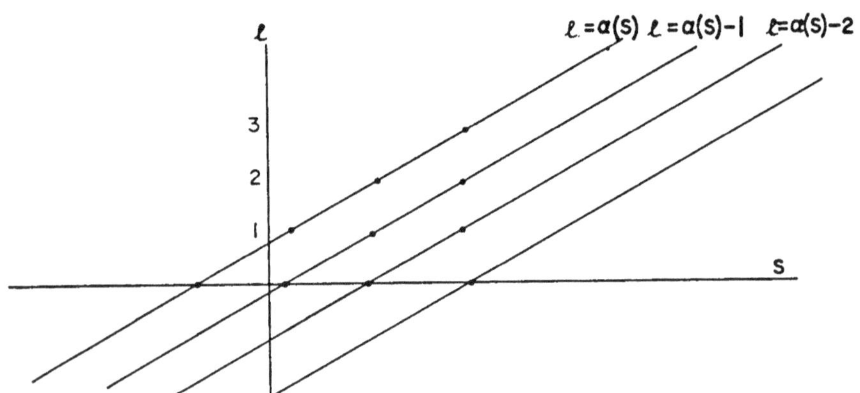

Figure 4. Regge trajectories in the Veneziano Model.

S-wave and a P-wave resonance, when $\alpha(s)=2$ an S-wave, a P-wave and a D-wave resonance, and so on. The resonances have been denoted by dots in Figure 4. Resonances at such points would occur if there were an infinite number of Regge trajectories satisfying the equations $\ell=\alpha(s)-n$, $n=0, 1,2,\ldots$ We cannot obtain a model with a finite number of Regge trajectories which has Regge behavior in both channels. Since our present amplitude is symmetric in s and t, the resonances in the t-channel also lie on the Regge trajectories shown in Figure 4.

We next verify that the amplitude defined by (2.1a) does have Regge asymptotic behavior. It is necessary to establish such behavior when s approaches infinity in any direction in the complex plane, and we begin with the simplest case, which is the limit $s \to -\infty$ along the real axis. The integral (2.1a) is then dominated by the region $x \approx 1$. To obtain an asymptotic series in s, we make the substitution

$$x = e^{-w} \tag{2.6}$$

In terms of this variable, the integral becomes

$$A(s,t) = \int_0^\infty dw\; e^{w\,\alpha(s)}\; (1-e^{-w})^{-\alpha(t)-1} \tag{2.7}$$

The dominant region is w=0, where we may make the approximation:

$$1-e^{-w} \approx w. \tag{2.8}$$

Substituting (2.8) in (2.7), we can perform the integral, to obtain the result:

$$A(s,t) \approx \Gamma\{-\alpha(t)\}\; \{-\alpha(s)\}^{\alpha(t)}$$

$$\approx \Gamma\{-\alpha(t)\}\; (-a\; s)^{\alpha(t)} . \tag{2.9}$$

The expression (2.9) does have the required behavior as s approaches $-\infty$. We also confirm the presence of poles when $\alpha(t)$ is a positive integer.

It is not very difficult to apply this method to obtain an asymptotic expansion for A(s,t) as a function of s. The last factor of (2.7) can be expanded in a power series in w as follows:

$$(1-e^{-w})^{-\alpha(t)-1} = w^{-\alpha(t)-1} \sum_{r=0}^{\infty} a_r w^r \tag{2.10}$$

On substituting (2.10) in (2.7) and evaluating the integral term by term, we find:

$$A(s,t) \approx \sum_{r=0}^{\infty} f_r(t)\ \{-\alpha(s)\}^{\alpha(t)-r}$$

$$\approx \sum_{r=0}^{\infty} g_r(t)(-s)^{\alpha(t)-r}. \tag{2.11}$$

We thus obtain an asymptotic behavior corresponding to a series of Regge trajectories in the t-channel at $\alpha(s)-r$. Our amplitude therefore has the required Regge asympotic behavior in the s and t channels when the appropriate variable approaches $-\infty$.

The reasoning we have just given applies as long as s (or t) approaches ∞ along a ray directed into the left half-plane. The factor $x^{-\alpha(s)-1}$ in (2.1a) is then very small unless x is near 1, and the integral is dominated by this limit. When s approaches infinity along a ray directed into the right half-plane, the factor $x^{-\alpha(s)-1}$ becomes very large unless x is near 1. We therefore have to deform the path of integration. The manner of doing so is shown in Figure 5, and we find that:

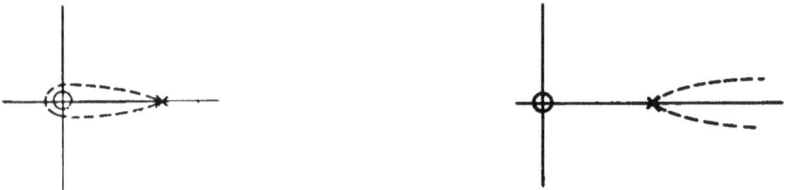

Figure 5. Deformation of the contour of integration in the Veneziano formula.

$$A(s,t) = \frac{\sin\pi\{\alpha(s)+\alpha(t)+1\}}{\sin\pi\{\alpha(s)+1\}} \int_1^\infty dx\, x^{-\alpha(s)-1} (x-1)^{\alpha(t)-1} . \tag{2.12}$$

By repeating the reasoning which led to (2.9), we obtain the asympotic approximation:

$$A(s,t) \approx \frac{\sin\pi\{\alpha(s)+\alpha(t)+1\}}{\sin\pi\{\alpha(s)+1\}} \Gamma\{-\alpha(t)\} (a\ s)^{\alpha(t)} . \tag{2.13}$$

We can also obtain an asymptotic expansion as in (2.11).

If s approaches infinity along a ray directed into the right half-plane in a complex direction, the first factor of (2.13) becomes $e^{-i\pi\alpha(t)}$, and we again obtain the usual Regge asymptotic formula. If s approaches infinity along the real axis, we can obviously not obtain the Regge formula without going beyond the narrow-resonance approximation. The Regge asymptotic formula $B(t)\ \{-s\}^{\alpha(t)}/\ \sin\pi\ \alpha(t)$ does not allow for poles in s, whereas such poles are clearly present in the narrow-resonance approximation and in (2.13). When we improve on the narrow-resonance approximation these poles will be smoothed out and we may expect Regge asymptotic behavior along the real axis. Within the framework of the narrow-resonance approximation, we can only demand Regge asymptotic behavior along the negative real axis and along a ray directed into the complex plane.

One crucial feature of the integral (2.1) is that it must be taken over the range $0 < x < 1$. If for instance we were to remove a segment within this range, we would not be able to carry out the contour deformations just described. The amplitude would then increase exponentially as s approached infinity in the right half-plane or along the positive real axis, and it would be physically unacceptable.

The two functions $\alpha(s)$ and $\alpha(t)$ in (2.1) need not be the same. We could have an amplitude of the form

$$A(s,t) = B\{-\alpha_s(s), \alpha_t(t)\} , \tag{2.14a}$$

where

$$\alpha_s(s) = as+b_s \quad , \quad \alpha_t(t) = at+b_t \; . \tag{2.14b}$$

Though the two intercepts b_s and b_t may be different, the two slopes a must be the same, or we would have increasing exponential behavior when s or t approaches infinity at fixed angle for a certain range of angles within one of the physical regions. If the slopes are the same, we have a decreasing exponential behavior when s or t approaches infinity at fixed angle (other than 0 or π) in any of the physical regions. When s or t approaches infinity at fixed u we have Regge asymptotic behavior, just as we did at fixed t or fixed s. These results can all be proved by suitably deforming the contour of integration in (2.1).

In order to generalize (2.1) to an amplitude with resonances in all three channels, we simply take a sum of three terms:

$$A(s,t) = C_1\, B\{-\alpha_s(s),-\alpha_t(t)\} + C_2\, B\{-\alpha_s(s),-\alpha_u(u)\} + C_3\, B\{-\alpha_t(t),-\alpha_u(u)\}. \tag{2.15}$$

The individual terms in (2.15) are obviously analogous to the three terms of the double-dispersion representation, even though the double spectral functions are zero in this approximation. The third term gives rise to Gribov-Pomeranchuk poles in the s-channel at the wrong-signature nonsense points, just like the third term of the double-dispersion representation.

Equation (2.1a) certainly does not define the only possible amplitude with Regge behavior in both channels, even if we require that there be no trajectories other than those shown in Figure 4. The function

$$B\{-\alpha(s) + n, -\alpha(t) + m\} , \qquad (2.16)$$

where m and n are positive integers, has trajectories which coincide with those of Figure 4, except that the first n trajectories in the s-channel and the first m in the t-channel are missing. If we add terms of the form (2.14), with arbitrary coefficients, to (2.1), we obtain an amplitude whose trajectories coincide with those of (2.1). We may further generalize the amplitude by adding terms of the form

$$p_{mn}(s,t)\, B\{-\alpha(s) + n, -\alpha(t) + m\} , \qquad (2.15)$$

where p_{mn} is a polynomial of the mth degree in s and the nth degree in t. The requirements which we have so far proposed thus leave us with a wide choice of amplitudes; we shall see later that the choice will be greatly restricted by the factorization requirement.

There exists another type of generalization of the Veneziano formula which applies to amplitudes with resonances in all three channels, but where the amplitude cannot be written as a sum of terms such as (2.15), each of which has resonances in only two channels.[5] We shall not discuss this generalization here. From our present standpoint the generalized amplitude appears to be as satisfactory as the original Veneziano amplitude, except perhaps for the fact that it does not allow one to construct models with exchange degeneracy. From the point of view of factorization properties, which we shall study in a later section, the original amplitude does appear to be preferable to the generalized amplitude.

We shall conclude this section by remarking that one can eliminate every alternate trajectory of Figure 4 if we restrict the constants a and b by the condition

$$4a\,\mu^2 + 3b + 1 = 0, \tag{2.16}$$

where μ is the mass of the external particle. (For simplicity we are assuming that $b_s=b_t$). This relation was pointed out by Veneziano in his original paper, where he showed that it appeared to be satisfied experimentally in certain amplitudes. We have no compelling reason for removing alternate trajectories, but later we shall find the condition (2.16) reappearing in a completely different connection.

3. The Veneziano Formula for the n-point Scalar Amplitude

Requirements of the Model

We have already pointed out that the n-point Veneziano formula is not only of importance in its own right, but that it also allows us to construct a Veneziano formula for an arbitrary elastic or inelastic process with spinning external particles, and that the resulting formula satisfies the factorization condition.

The four-point formula consists of a sum of three terms, each of which may be associated with a particular ordering of the four external particles. A single term has resonances only in channels which consist of adjacent particles. For instance, the term corresponding to Figure 2 has no resonance in the u-channel, which consists of particles 1 and 3. The n-point formula will consist of a sum of $\frac{1}{2}(n-1)!$ terms, each associated with a particular ordering of the external particles. In the remainder of this section we shall select one ordering and study the appropriate term. By associating each term with a particular ordering of the external particles we do make a restriction, but we shall not examine more general formulas at the moment.

The general requirements of our formula are an extension of those for the four-point function. Each channel must have resonances which lie on linearly rising Regge trajectories. Now, however, certain sets of channels can have simultaneous resonances, i.e., the amplitudes can have terms of the form

$$\frac{f(s_{i+1}\ldots s_j)}{(s_1-m_1^2)\ldots\ldots(s_i-m_i^2)} \tag{3.1}$$

In the four-point function, the residues at the poles in s were polynomials in t or u, and there were no simultaneous resonances.

In Figure 6a) a pair of channels such as A and B is known as a non-overlapping pair, a pair such as A and C is an overlapping pair. In general, a pair of channels is non-overlapping either if they have no particles in

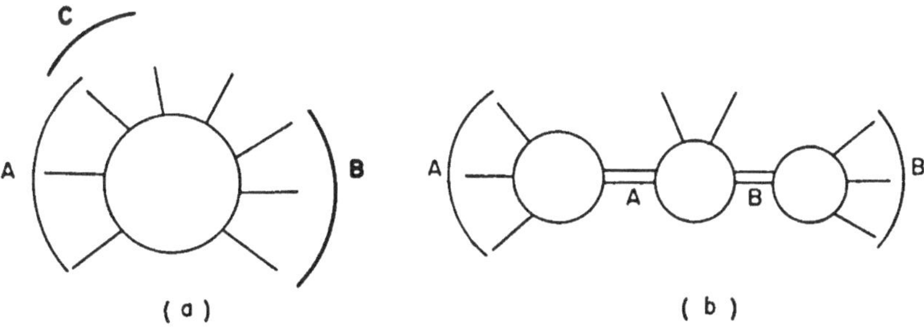

Figure 6. Overlapping and non-overlapping Channels.

common or if one is completely contained within the other. From Figure 6(b), we notice that two non-overlapping channels can have simultaneous resonances. Two overlapping channels cannot, since there exists no diagram with single-particle intermediate states in both channels. In fact, if s is the energy of a particular channel, the residue at the pole when $\alpha(s)=\ell$ must be a polynomial of the ℓth degree in the over-lapping-channel variables. This is a consequence of the requirement that the resonance at this energy should have a maximum angular momentum of ℓ.

The n-point function can have at most n-3 mutually non-overlapping channels. Any particular set of n-3 non-overlapping channels can have simultaneous resonances, but the residue at the (n-3)-fold pole must be a polynomial in the other channel variables. The entire amplitude must be expressible as a sum of contributions from the (n-3)-fold poles, provided that the other channel variables lie within a certain range. The result must be independent of the choice of the n-3 non-overlapping channels.

The discovery of a formula for the five-point function which fulfils these requirements is due independently to Bardakci and Ruegg and to Virasoro.[6] It was then independently generalized to the n-point function by Bardakci and Ruegg, Chan and Tsou, Goebel and Sakita, and Koba and Nielsen.[7] In the work of all these authors, a particular set of n-3 channels was chosen, and each of these channels was associated with an integration variable analogous to the integration variable x in (2.1). By making a change of variables it was shown that the integral was independent of the choice of the n-3 channels. Koba and Nielsen[8] then proposed a formalism which was manifestly symmetric in all r-particle channels ("cyclic symmetry"). As this symmetry is a fundamental requirement of our formula, we shall depart from the historical order and shall introduce the Koba-Nielsen formalism from the start. We feel that it is worth while to do so in spite of the slightly more formal mathematics involved.

Projective Transformations

The Koba-Nielsen formalism is based on the invariance of functions under a type of transformation known as a projective transformation (also as a Möbius transformation or a homographic transformation). We shall therefore summarize the more important features of such transformations.

A projective transformation is defined by the formula:

$$z_i' = \frac{Az_i+B}{Cz_i+D}\,, \tag{3.2}$$

where A, B, C and D are arbitrary. Multiplication of all four parameters A, B, C and D by the same constant leaves the transformation unchanged, so that a general projective transformation depends on three independent parameters. Transformations similar to (3.2), where z_i is a vector, play a fundamental role in projective geometry, but we are only interested in the analytic properties of projective transformations. An infinitesimal projective transormation is given by the formula:

$$z_i' = z_i + \epsilon(\alpha+\beta z_i + \gamma z_i^2). \tag{3.3}$$

We shall quote the properties of projective transformations without giving the straightforward proofs. The first fundamental property is that the cross-ratio (or anharmonic ratio) of four points z_1, z_2, z_3, z_4, defined by the expression

$$\frac{(z_1-z_3)(z_2-z_4)}{(z_1-z_4)(z_2-z_3)} \tag{3.4}$$

is invariant when each of the z's undergoes the same transformation (3.2) or (3.3). We shall see that cross-ratios play an important part in the Koba-Nielsen formalism.

We may generalize the invariance property of the cross-ratio as follows: the product

$$\prod_{i,j=1}^{N} (z_i-z_j)^{a_{ij}} \tag{3.5a}$$

is invariant under a projective transformation provided that

$$\sum_{j=1}^{N} a_{ij}+ a_{ji} = 0 \tag{3.5b}$$

for all i.

In the Koba-Nielsen formalism, the z's are integration variables, and we therefore require an invariant form involving differentials. The expression

$$\frac{dz_1 \ldots dz_n}{(z_2-z_1)(z_3-z_2)\ldots(z_n-z_{n-1})(z_1-z_n)} \tag{3.6}$$

is invariant under a projective transformation. If we regard the variables $z_1 \ldots z_n$ as being cyclically ordered, the denominator consists of the differences between adjacent points. Another possible invariant is

$$\frac{dz_1 \ldots dz_n}{(z_3-z_1)(z_4-z_2)\ldots(z_n-z_{n-2})(z_1-z_{n-1})(z_2-z_n)} \cdot \tag{3.7}$$

The differentials which we shall require are not precisely those given in (3.6) and (3.7). Our integrand will be a function of n variables z_i and it will remain invariant under an arbitrary projective transformation. Since the projective transformation has three parameters, the integrand is really a function of n-3 variables. In fact, by applying a suitable projective transformation, we can transform any three of the z's to preassigned values. We therefore integrate over the remaining n-3 z's. The differential

$$\frac{dz_1 \ldots [dz_a] \ldots [dz_b] \ldots [dz_c] \ldots dz_n (z_b-z_a)(z_c-z_b)(z_a-z_c)}{(z_2-z_1)(z_3-z_2)\ldots(z_n-z_{n-1})(z_1-z_n)} , \tag{3.8}$$

and the differential

$$\frac{dz_1 \ldots [dz_a] \ldots [dz_b] \ldots [da_c] \ldots dz_n (z_b-z_a)(z_c-z_b)(z_a-z_c)}{(z_3-z_1)(z_4-z_2)\ldots(z_n-z_{n-2})(z_1-z_{n-1})(z_2-z_n)} \tag{3.9}$$

are invariant under projective transformations. The square brackets indicate that the differentials within them are to be omitted. If a projectively invariant function is multiplied by the differential (3.8) or

(3.9) and integrated, the result is independent of the choice of a, b and c.

Next we require to know how the relative ordering of the n points $z_1 \ldots z_n$ may be changed by a projective transformation. It is necessary to emphasize that the point at infinity must not be regarded as different from any other point. If $z_i = \infty$, then the transformed value z_i' (Equation 3.2) is not necessarily infinite, and conversely z_i' may be infinite if z_i is finite. When the z's are restricted to be real, we regard the points ∞ and $-\infty$ on the real line as identical, and we consider only the cyclic ordering of points on the real line. A projective transformation either leaves the cyclic ordering of the points unchanged, or it reverses it. If the z's lie on a given contour in the complex plane (e.g., the unit circle), and the projective transformation is such that it takes a point on the contour to another point on the contour, the transformation will again either leave the cyclic ordering of the points unchanged or will reverse it.

We end this sub-section by quoting certain properties of projective transformations which will not be required in discussing the Koba-Nielsen formalism, but which will be useful for later work:

i) The product of two transformations $P(z_i)=(Az_i+B)/(Cz_i+D)$ and $P'(z_i)=(A'z_i+B')/(C'z_i+D')$ is equal to $(A''z_i+B'')/(C''z_i+D'')$, where the matrix

$$\begin{pmatrix} A'' & B'' \\ C'' & D'' \end{pmatrix}$$

is simply the product of the matrices

$$\begin{pmatrix} A & B \\ C & D \end{pmatrix} \text{ and } \begin{pmatrix} A' & B' \\ C' & D' \end{pmatrix}$$

ii) Let us denote the projective transformation (2.1) by $P(z_i)$. If the variables z_i and z_i of (2.1) are subjected to a projective transformation

$$\tilde{z}_i = Q(z_i), \qquad \tilde{z}_i' = Q(z_i'), \tag{3.10}$$

then the variables $\tilde{z}_i$ and $\tilde{z}_i$ will be related by a new projective transformation

$$\tilde{z}_i' = \tilde{P}(\tilde{z}_i).$$

Let the transformation P and Q be associated respectively with the matrices

$$\begin{pmatrix} A & B \\ C & D \end{pmatrix} \text{ and } \begin{pmatrix} D & E \\ F & G \end{pmatrix}.$$

Then $\tilde{P}$ is associated with the matrix

$$\begin{pmatrix} \tilde{A} & \tilde{B} \\ \tilde{C} & \tilde{D} \end{pmatrix} = \begin{pmatrix} D & E \\ F & G \end{pmatrix}^{-1} \begin{pmatrix} A & B \\ C & D \end{pmatrix} \begin{pmatrix} D & E \\ F & G \end{pmatrix}. \tag{3.11}$$

iii) Any projective transformation leaves two points, which we shall call x_1 and x_2, invariant (i.e., $x_1'=x_1$, $x_2'=x_2$). We may express the parameters A, B, C and D in terms of x_1, x_2, and a third parameter which we shall call w. The projective transformation then takes the form

$$P(z_i) = \frac{(x_2-x_1w)z_i-x_1x_2(1-w)}{(1-w)z_i-(x_1-x_2w)} \tag{3.12}$$

If the z's are subjected to a projective transformation (3.10), the new "transformed projective transformation" $\tilde{P}$ is obtained by subjecting x_1 and x_2 to the projective transformation (3.10), and leaving w unchanged. The parameter w is thus an invariant associated with the projective transformation P.

iv) Given any projective transformation P, it is possible to find a projective transformation Q such that, if the z's are transformed according to (3.10), the transformed projective transformation will be simply

$$\tilde{P}(z_i) = w^{-1}z_i \; . \tag{3.13}$$

The parameter w in (3.13) is the same as the invariant w associated with the transformation P or $\tilde{P}$. When the projective transformation has the form (3.13), the invariant points are $x_1 = 0$, $x_2 = \infty$.

The Koba-Nielsen Formalism

We shall now quote the Koba-Nielsen formula for the n-point Veneziano amplitude. We number the particles cyclically from 1 to n. In Figure 7

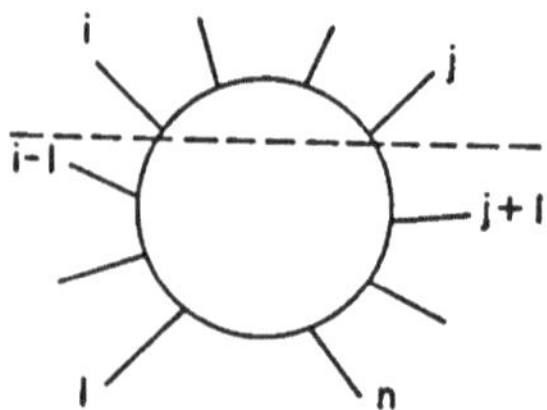

Figure 7. The n-point diagram.

we may associate one channel with each pair of particles i and j satisfying the inequalities:

$$2 \leq i < j \leq n \qquad j \neq n \text{ if } i = 2. \tag{3.14}$$

The square of the centre-of-mass energy of the ij channel is denoted by s_{ij}. We associate one integration variable z_i with each particle i. As we have explained above, there are really only n-3 integration variables. The Koba-Nielsen formula is then as follows:

$$A = -\int \frac{dz_1 \ldots [dz_a] \ldots [dz_b] \ldots [dz_c] \ldots dz_n (z_b - z_a)(z_c - z_b)(z_a - z_c)}{(z_3 - z_1)(z_4 - z_2) \ldots (z_n - z_{n-2})(z_1 - z_{n-1})(z_2 - z_n)}$$

$$\times \prod \left\{ \frac{(z_j - z_i)(z_{j+1} - z_{i-1})}{(z_j - z_{i-1})(z_{j+1} - z_i)} \right\}^{\alpha(s_{ij}) - 1} \tag{3.14}$$

The range of integration is restricted only by the condition that the variables z be cyclically ordered, either along the real line or along a contour in the complex plane. The product $\prod$ is over all channels. If i=1, i-1 is taken to be n and, similarly, if j=n, j+1 is taken to be 1.

We notice that the factor involving the differentials is precisely the invariant (3.9). The remainder of the integrand depends only on cross-ratios and is therefore projectively invariant. It follows that the integral on the right of (3.14) is independent of the values we assign to the three fixed z's z_a, z_b and z_c, and it is also independent of the choice of a, b and c.

Since our definition of the channel in terms of the particle indices was of necessity unsymmetrical, it may be worth while to state in words how the cross-ratio in (3.14) was constructed. The first factor z_i-z_j is the difference between the two z's on one side of the dotted line, the second factor $z_{i-1}-z_{j+1}$ is the difference between the two z's on the other side. The factors z_i-z_{j+1} and $z_{i-1}-z_j$ in the denominator are the differences between one z on one side of the dotted line and another, non-adjacent z on the other side.

Let us first verify that our n-point formula reduces to the four-point formula of Section 2 when n=4. If we take a, b, c = 1, 3, 4, the formula becomes:

$$-\int \frac{dz_2(z_3-z_1)(z_4-z_3)(z_1-z_4)}{(z_3-z_1)(z_4-z_2)(z_1-z_3)(z_2-z_4)}\left\{\frac{(z_2-z_1)(z_3-z_4)}{(z_2-z_4)(z_3-z_1)}\right\}^{-\alpha(s)-1}\left\{\frac{(z_4-z_1)(z_3-z_2)}{(z_4-z_2)(z_3-z_1)}\right\}^{-\alpha(t)-1} \tag{3.15}$$

We have emphasized that we may choose the three quantities z_1,z_3,z_4 arbitrarily, and that the result will be independent of the choice. Let us therefore choose $z_1=0$, $z_3=1$, $z_4=\infty$ - , a choice that we shall find convenient for the three arbitrary z's in the sequel. The expression (3.15) is then:

$$\int dz_2\, z_2^{-\alpha(s)-1}\,(1-z_2)^{-\alpha(t)-1},$$

which is identical to (2.1). The Koba-Nielsen formula is of course unnecessarily elaborate for the four-point amplitude, but we do see how the simple case emerges from the general case. Alternatively, without making any particular choice of the variables, we may define

$$x = \frac{(z_2-z_1)(z_3-z_4)}{(z_2-z_4)(z_3-z_1)}$$

$$1-x = \frac{(z_4-z_1)(z_3-z_2)}{(z_4-z_2)(z_3-z_1)}$$

$$d\,x = -\frac{dz_2(z_4-z_3)(z_1-z_4)}{(z_4-z_2)(z_1-z_3)(z_2-z_4)} \tag{3.16}$$

On substituting (3.16) into (3.15) we recover (2.1), and we again verify that the four-point formula is a particular case of the n-point formula.

We next show that the Koba-Nielsen formula does satisfy the requirements stated at the beginning of the section. The cross-ratio associated with a particular channel in (3.14) is analogous to the variable x or 1-x in (2.1), and the integrand will be an analytic function of the variables s_{ij} except where the cross-ratio vanishes. It is therefore important to study how one cross-ratio behaves when another cross-ratio vanishes. We mention the following three theorems:

i) The cross-ratio in (3.14) must lie between zero and unity.

Since we know that the cross-ratio is invariant under a projective transformation, we can simplify the proof of this fact by fixing three of the z's. We take $z_i=0$, $z_{j+1}=1$, $z_{i-1}=\infty$. The condition of cyclic ordering then restricts z_j to be between 0 and 1, and the theorem follows immediately.

ii) If a particular cross-ratio is equal to zero, the cross-ratio associated with any overlapping channel is unity.

Without loss of generality, we can suppose that the channel whose cross-ratio vanishes is the 1j channel. Again we take $z_1=0$, $z_{j+1}=1$ $z_n=\infty$ (we remind the reader that, if i=1, i-1=n). The vanishing of the cross-ratio then implies that $z_i=0$.

Let us examine an overlapping channel. The most general overlapping channel i'j' has $1 < i' \leq j'$, $j < j' < n$. In the cross-ratio $\{(z_{j'}-z_{i'})(z_{j'+1}-z_{i'-1})\}/\{(z_{j'}-z_{i'-1})(z_{j'+1}-z_{i'})\}$, the indices i' and k'-1 lie between j and 1, so that $z_{i'}$ and $z_{i'-1}$ lie between z_1 and z_i and are therefore zero. The indices j' and j'+1 lie between j and n, so that the variables $z_{j'}$ and $z_{j'+1}$ lie between z_{i+1} and z_n and are not zero. The cross-ratio is therefore equal to unity.

iii) Any two cross-ratios associated with non-overlapping channels can be zero simultaneously.

The proof is similar to that of the last theorem, and we leave it to the reader.

We can now easily prove that our amplitude has the required properties. The reasoning is similar to the corresponding analysis in Section 2. We examine the singularities in the variable s_{ij}, and again we set $z_i=0$, $z_{j+1}=1$, $z_{i-1}=\infty$, so that the cross-ratio in (3.14) is simply equal to z_j. The factor $z_j^{-\alpha(s_{ij})-1}$, integrated over z_j, will provide the expected poles when $\alpha(s_{ij})$ is equal to a positive integer. In order to investigate the other factors, we must take into account the fact that the variables z_k, $i < k < j$, tend to zero with z_j, since they are retricted to lie between z_i and z_j. We therefore redefine

$$y_k=z_k/z_j, \quad i < k < j. \tag{3.17}$$

The integrand is regarded as a function of $y_k(i < k < j)$, z_j and z_ℓ ($\ell < i-1$ or $\ell > j+1$), and we can let z_j approach zero while keeping the other variables constant. All cross-ratios other than that associated with the channel ij approach constant values in this limit, the cross-ratios associated with the overlapping channels approaching the value unity. Both the numerator and the denominator of the differential factor in (3.14) contain a factor z_j^{j-i-1}, so that this factor, too, approaches a constant. We may therefore repeat the reasoning of Section 2 to show that the z_j-integration leads to a pole at $\alpha(s_{ij})=0$. The residue at the pole will depend on the variables s_{ij} for non-overlapping channels, but not on the variables for the overlapping channels, since the cross-ratios associated with the latter channels are unity when $z_j=0$.

To evaluate the integral when $\alpha(s_{ij}) > 1$, we again integrate by parts as before. The cross-ratios associated with the overlapping channels do depend on z_j before $z_j \to 0$; any cross-ratio which involves $z_k(i < k < j)$ will depend on z_j through (3.17). An n-fold integration by parts will therefore bring down a polynomial of the nth degree in the α's for the overlapping channels, and the residue at the pole where $\alpha(s_{ij})=n$ will contain such a polynomial.

We have mentioned that the residue at the pole may depend on the non-overlapping channel variables and, since the cross-ratios associated with any two non-overlapping channels can vanish simultaneously, we can have simultaneous poles in the variables s associated with the two channels. All the required results have therefore been proved.

It is worth while mentioning a further property which is actually a particular case of the factorization condition to be studied in the following section. We shall identify the external particle with the internal particle which occurs at $\alpha(s)=0$. The residue at the lowest pole of the n-point function should therefore be equal to the product of an (r+1)-point function and an (n-r+1)-point function. (Figure 8) We leave it to the reader to verify that this is so.

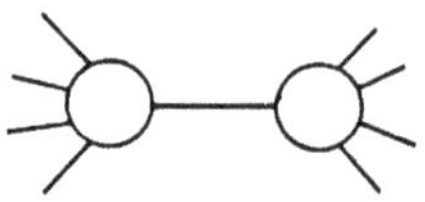

Figure 8. Residue at the lowest pole of the n-point function.

We shall not examine the question of asymptotic behavior here. We refer the interested reader to Reference 6.

Alternative Expression for the Koba-Nielsen Formula

The formula (3.14), in which a cross-ratio of the integration variables is associated with each channel, may be rewritten in a form which involves scalar products of the external momenta and which is often more convenient. To do so we examine a particular factor (z_j-z_i), and we find the sum of the exponents associated with this factor. The factor occurs in four cross: ratios:

Channel ij	Exponent	$-\alpha(s_{ij})-1$
Channel i+1, j-1	"	$-\alpha(s_{i+1,j-1})$
Channel i+1, j	"	$+\alpha(s_{i+1,j})+1$
Channel i, j-1	"	$+\alpha(s_{i,j-1})+1.$

As long as the particles i and j are separated by at least two particles on either side, all four channels enumerated above exist. Using the fact that

$$\alpha(s_{ij}) = a(\sum_{k=i}^{j} p_i)^2 + b, \tag{3.18}$$

we immediately find that the total exponent is simply $-2ap_ip_j$.

If the particles i and j are separated by one particle, only three of the above channels exist. If for instance $j=i+2$, the channel $i+1$, $j-1$ consists of only one particle and does not exist. In evaluating the total exponent, we now have to include an extra term -1 which arises from the factor $z_{i+2}-z_i$ in the denominator of the first factor of (3.14). The exponent is thus equal to

$$-a(p_i+p_{i+1}+p_{i+2})^2-b-1+a(p_{i+1}+p_{i+2})^2+b+1+a(p_i+p_{i+1})^2+b+1-1$$

$$=-2ap_ip_{i+2}+ap_{i+1}^2+b. \tag{3.19}$$

The exponent will again be equal to $-2ap_ip_{i+2}$ provided that

$$a\mu^2+b=0, \tag{3.19}$$

where μ is the mass of the external particle. In other words, the mass of the external particle must be equal to the mass of the lightest internal particle. The external and internal particles of our model must ultimately be the same, and it is therefore reasonable to impose the condition (3.19). When we do so, the exponent is equal to $-2ap_ip_j$ for this case as well.

We do obtain an extra term in the exponent associated with the factor $z_{i+1}-z_i$. The channel i, i+1 is the only channel of the four enumerated above which exists, and the exponent is

$$\begin{aligned} -\alpha(s_{i,i+1})-1 &= -a(p_i+p_{i+1})^2-b-1 \\ &= -2p_ip_{i+1}-2a\,\mu^2-b-1 \\ &= -2a\,p_ip_{i+1}-a\,\mu^2-1, \end{aligned} \qquad (3.20)$$

from (3.19).

The Koba-Nielsen formula may therefore be written as follows:

$$A=\int \frac{dz_1\ldots[dz_a]\ldots[dz_b]\ldots[dz_c]\ldots dz_n(z_b-z_a)(z_c-z_b)(z_a-z_c)}{(z_2-z_1)(z_3-z_2)\ldots(z_n-z_{n-1})(z_1-z_n)}$$

$$\times \prod_{j>i}(z_j-z_i)^{-2a\,p_ip_j}\prod(z_{i+1}-z_i)^{-a\,\mu^2} \qquad (3.21)$$

The product $\prod(z_{i+1}-z_i)$ includes a factor z_1-z_n. We have separated the factors $(z_{i+1}-z_i)^{-a\,\mu^2-1}$ into two in writing (3.21). The differential factor is then just the invariant quantity (3.8), while the other factor is a particular case of (3.5); the exponents are easily seen to satisfy (3.5b) as a cosequence of the equations $\sum p_i=0$, $p_i^2=\mu^2$. The whole integrand is thus projectively invariant.

It is worth while pointing out that if one of the three arbitrary z's is chosen to be ∞, the factors involving that variable simply disappear from (3.21), because the sum of their exponents is zero.

We can arrange the positions of the trajectories so that the extra factors $(z_{i+1}-z_i)^{-a\mu^2-1}$ are unity, and all factors (z_i-z_j) have the exponent $-2a\, p_i p_j$. We require

$$a\,\mu^2 = -1, \tag{3.20a}$$

or, from (3.19),

$$b = 1. \tag{3.20b}$$

Equation (3.20) is precisely the Veneziano supplementary condition (2.16), together with the condition (3.19). The requirement is that the leading trajectory be in the position of the Pomeron! However, the residue of the lowest pole at $\mu^2 = -1/a$ does not vanish, so that the condition would imply the existence of a "tachyon" with negative mass-squared.

4. Factorization

Single Factorization

The real power of the multi-particle Veneziano formula depends on our being able to satisfy the factorization requirements stated in Section 1. We have to show that the residue at any pole is expressible as a finite sum of factored terms (Figure 9). Each side of Figure 9 represents

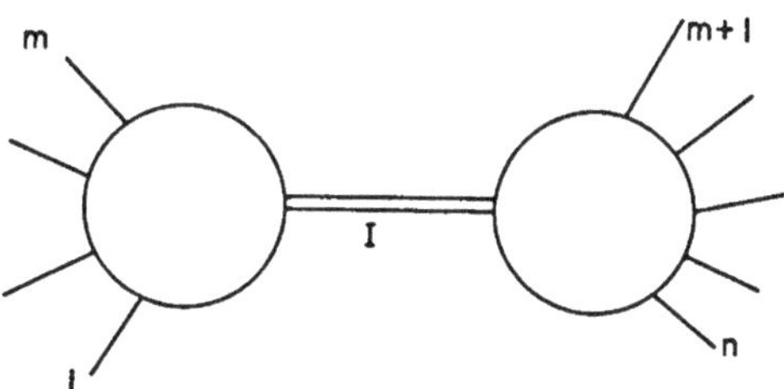

Figure 9. Factorization of the Multi-Particle Veneziano Formula.

one factor, which must be independent of the number of particles on the other side and of their momenta.

Once we have proved that this factorization requirement is met, we can construct amplitudes with spinning external particles by repeated factorization (Figure 10a). Such amplitudes can themselves be factorized, (Figure 10b), and the vertices will satisfy the symmetry requirements.

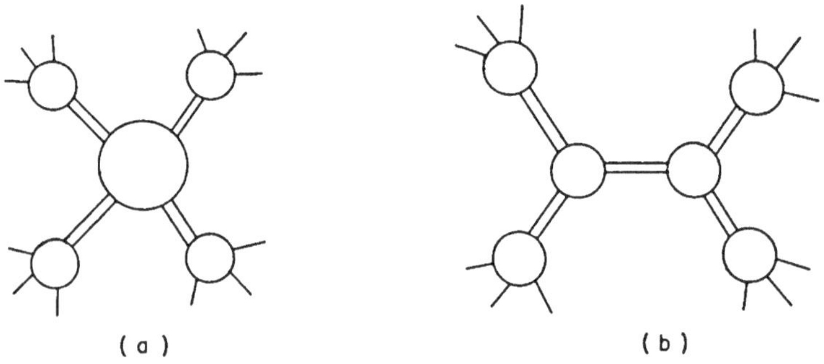

(a) (b)

Figure 10. Construction of general amplitudes by repeated factorization.

The demonstration that the factorization requirement is satisfied was given independently by Bardakci and Mandelstam and by Fubini and Veneziano.[9] We shall reproduce their results using the Koba-Nielsen formalism.

The formula (3.21) is more convenient than (3.14), since the exponents in the latter formula are expressed as products of two momenta. As in the last section, we set $z_1=0$, $z_{m+1}=1$, $z_n=\infty$. (Figure 9) The residue at the pole corresponding to the channel shown is zero when the cross-ratio $\{(z_m-z_1)(z_{m+1}-z_n)/(z_m-z_n)(z_{m+1}-z_1)\}$ vanishes. With our present choice of the three arbitrary z's, this cross-ratio is simply z_m. We shall therefore define

$$z_m = z.$$

The condition $z_m \leq z_{m+1}$ then becomes

$$z \leq 1.$$

In order to find the residue at the poles, we should like to obtain the z-dependence of the integrand in as explicit a form as possible. At the moment, one source of z-dependence is the series of inequalities

$$0 < z_2 < \ldots < z.$$

We therefore define new variables $y_i (i \leq m)$ by the formula

$$z_i = z\, y_i. \qquad 1 \leq i \leq m+1 \quad (y_m=1) \tag{4.1a}$$

Equation (3.21) then becomes

$$A=\int \frac{dy_2\ldots dy_{m-1}dz_{m+2}\ldots dz_{n-1}z^{-1}dz}{y_2(y_3-y_2)\ldots(y_{m-1}-y_{m-2})(1-y_{m-1})(1-z)(z_{m+2}-1)\ldots(z_{n-1}-z_{n-2})}\, z^{-s+\mu^2}$$

$$\times\prod_{j>i}{}'(y_j-y_i)^{-2p_ip_j}\prod{}'(y_{i+1}-y_i)\prod_{j>i}{}''(z_j-z_i)^{-2p_ip_j}\prod{}''(z_{i+1}-z_i)^{-\mu^2}$$

(4.2a)

$$\times\prod(z_j-zy_i)^{-2p_ip_j}\,(1-z)^{-\mu^2}\,(0=y_1\leq y_2\leq\ldots\leq y_m=1,\ 1=z_{m+1}\leq\ldots\leq z_n=\infty),$$

where

$$s = \left(\sum_{i=1}^{m} p_i\right)^2. \tag{4.2b}$$

In (4.2a) we have set our unit of mass so as to make the slope of the trajectories equal to 1, and we shall do in all subsequent equations. The product $\prod'$ is over all pairs from the left of Figure 9, the product $\prod''$ over all pairs to the right while, in the product $\prod$, the index i is to the left, j to the right. We should like to expand the last two factors in (4.2a) in powers of z, and this will be easier if the factor $z_j - zy_i$ is replaced by $1-zy_iy_j$. Let us therefore define

$$z_j = \frac{1}{y_j} \qquad m+2 \le j \le n \tag{4.1b}$$

following which we take out all factors of the y's which appear in denominators within the brackets. The result is

$$A = \int dy_2 \cdots dy_{m-1} dy_{m+2} \cdots dy_{n-1} I_1 I_2 \int_0^1 dz\; z^{-s+\mu^2-1} \prod (1-y_iy_jz)^{-2p_ip_j}$$

$$\times (1-z)^{-\mu^2-1}, \tag{4.3a}$$

where $I_1 = \dfrac{1}{y_2(y_3-y_2)\cdots(y_{m-1}-y_{m-2})(1-y_m)} \prod'_{j>i} (y_j-y_i)^{-2p_ip_j} (y_{i+1}-y_i)^{-\mu^2}$,

$$\tag{4.3b}$$

$$I_2 = \frac{1}{(1-y_{m+1})(y_{m+1}-y_{m+2})\cdots y_{n-1}} \prod''_{j>i} (y_i-y_j)^{-2p_ip_j} (y_i-y_{i+1})^{-\mu^2} \tag{4.3c}$$

$$0=y_1 \le y_2 \cdots \le y_{m-1} \le y_m = 1, \quad 1 = y_{m+1} \ge y_{m+2} \cdots \ge y_{n-1} \ge y_n = 0. \tag{4.3d}$$

The factors I_1 and I_2 depend only on the y's and p's from each half of the diagram. They are simply the n-point Veneziano formulas for the two halves of Figure 9, with the intermediate particle regarded as spinless, and with the arbitrary y's given the values:

$$I_1: \qquad y_1=0 \qquad y_m=1 \qquad y_I=\infty, \tag{4.4a}$$

$$I_2: \qquad y_n=0 \qquad y_{m+1}=1 \qquad y_{I'}=\infty \tag{4.4b}$$

The variables y_I and $y_{I'}$ in (4.4) denote the y's associated with the intermediate particle of Figure 9.

As expected, the amplitude A of (4.3a) has poles when s+b is equal to any positive integer, and the residue of the pole is simply the coefficient of z^n in the expression

$$\int dy_2 \cdots dy_{n-1} dy_{m+2} \cdots dy_{n-1}\; I_1 I_2 \prod_{\substack{i\leq m \\ j\geq m}} (1-y_i y_j z)^{-2p_i p_j} (1-z)^{-\mu^2-1}. \tag{4.5}$$

To evaluate (4.5) it is convenient to write it as the exponential of a logarithm. It is thus equal to

$$\int dy_2 \cdots dy_{m-1} dy_{m+2} \cdots dy_{n-1}\; I_1 I_2 \exp\left\{-\sum_{\substack{i\leq m \\ j\geq m}} 2p_i p_j\; \ell n(1-y_i y_j z) - (\mu^2+1)\ell n(1-z)\right\}$$

$$= \int dy_2 \cdots dy_{m-1} dy_{m+2} \cdots dy_{n-1}\; I_1 I_2 \exp\left\{\sum_{r=1}^{\infty} \Big(\sum_{\substack{i\leq m \\ j\geq m}} 2p_i p_j y_i^r y_j^r + \mu^2+1\Big) \frac{z^r}{r}\right\}.$$

For demonstrating factorization it is convenient to rewrite this expression in the form

$$\int dy_2 \cdots dy_{m-1} dy_{m+2} \cdots dy_{n-1}\, I_1 I_2 \exp\left\{\left(2 \sum_{r=1}^{\infty} P^{(r)} Q^{(r)} + \mu^2 + 1\right) \frac{z^r}{r}\right\}, \tag{4.6a}$$

where

$$P^{(r)} = \sum_{i \leq m} p_i y_i^{\,r}, \qquad Q^{(r)} = \sum_{j>m} p_j y_j^{\,r} \tag{4.6b}$$

The quantity $P^{(r)}$ depends only on variables from the left half of the diagram, $Q^{(r)}$ only on variables from the right. We remind the reader that we are still choosing $y_1=y_n=0$, $y_m=y_{m+1}=1$.

The coefficient of z^n in the exponential of (4.6a) will consist of a sum of terms of the form

$$c \prod_r \left\{ P^{(r)} Q^{(r)} + \mu^2 + 1 \right\}^{\kappa_r}, \tag{4.7a}$$

with

$$\sum r\kappa_r = n. \tag{4.7b}$$

There exists a term (4.7a) for each set of integers κ_r $(0 \leq r \leq n)$ which satisfy (4.7b). Since we are interested in making an angular-momentum analysis, we shall work in the centre-of-mass system of the channel under consideration, and shall write

$$P^{(r)} Q^{(r)} = P_0^{(r)} Q_0^{(r)} - \underset{\sim}{P}^{(r)} \cdot \underset{\sim}{Q}^{(r)}$$

By expanding (4.7a), we rewrite it as the sum of terms of the form

$$c \prod_{r,s} (-1)^{\lambda_r} (\underset{\sim}{P}^{(r)} \cdot \underset{\sim}{Q}^{(r)})^{\lambda_r} (P_0^{(s)} Q_0^{(s)})^{\mu_s} , \tag{4.8a}$$

with $\quad \sum r\lambda_r + \sum s\mu_s = n \qquad \text{if } \mu^2 + 1 = 0,$ (4.8b)

$$\sum r\lambda_r + \sum s\mu_s \leq n \qquad \text{if } \mu^2 + 1 \neq 0. \tag{4.8c}$$

The expression (4.8a) can then be written as the product of two Cartesian tensors

$$c \prod_{r,s} (\underset{\sim}{P}^{(r)})^{\lambda_r} (P_0^{(s)})^{\mu_s} \quad , \quad c \prod_{\lambda_r, \mu_s} (\underset{\sim}{Q}^{(r)})^{\lambda_r} (Q_0^{(s)})^{\mu_s} , \tag{4.9}$$

where the set of integers λ_r and μ_s still satisfy the restrictions (4.8b) or (4.8c). We have thus shown that the coefficient of z^n in (4.6a) can be expressed as a finite sum of terms, each of which is a product of one factor involving variables from the left of Figure 9, and another involving variables from the right. This is the result which we wished to prove, and we hope that the reader will by now be aware of its fundamental importance.

Each tensor in (4.9) can be resolved into components of definite angular momentum. There will be one component with $j = \sum_r \lambda r$, and several with lower values of j. Furthermore, there is one choice of integers λ_r and μ_r consistent with (4.8b) or (4.8c) where the quantity $\sum \lambda_r$ has its maximum value of n, namely $\lambda_1 = n$, $\lambda_r = 0$ $(r > 1)$, $\mu_s = 0$. The residue at the pole where $s+b = n$ will therefore contain one factored term with $j=n$ and several factored terms with $j < n$. In other words, the leading Regge trajectory is non-degenerate, the non-leading trajectories degenerate. The degeneracy increases rapidly as we increase the quantity n-j, since the number of solutions of (4.8b) or (4.8c) increases very rapidly with n. At large n, the degeneracy of the levels behaves roughly like $e^{\sqrt{n}}$.

We note that the degeneracy of the levels is somewhat reduced when the Veneziano supplementary condition $\mu^2+1 = 0$ is satisfied. The inequality (4.8c) then becomes replaced by the equality (4.8b), and the number of possible sets of integers λ_r and μ_s is reduced.

The large degeneracy of the levels at high energy corresponds physically to the large phase space which these levels represent. In fact, the increase of level density with energy is similar to that predicted by Hagedorn[10] from statistical considerations.

The factorization condition greatly reduces the choice of possible amplitudes which satisfy the requirements of Section 2. The more general amplitudes either do not factorize at all, or they lead to a spectrum whose density increases faster than the Hagedorn density at high energies. The simple n-point Veneziano formula is certainly not the only factorizable formula with a Hagedorn-like density at high energies, but it does appear to be the amplitude with the simplest spectrum. We do not necessarily seek the model with the absolutely simplest spectrum and, in particular, we should like to obtain one with a quark-like specturm. Nevertheless, if we demand that the spectrum be not too degenerate we obtain a model with a reasonable degree of uniqueness. In the following sub-section we shall observe that even the position of the leading trajectory can be fixed if we demand the absence of ghosts.

Ghosts and Ward Identities

Since the product $P^{(r)} Q^{(r)}$ in (4.7a) is to be taken in the Lorentz metric, the sign of some of the terms will be negative. In fact, the terms (4.8a) with $\Sigma\, \mu_s$ odd will be negative definite instead of positive definite; note that positive definiteness requires a factor $\prod (-1)^{\lambda_r}(-1)^{\mu_s}$, because all momenta in Figure 9 are directed inward. The Cartesian tensors in (4.9) with an odd number of factors $P_0^{(s)}$ will correspond to ghosts, or resonances with negative width, instead of particles. If we associate the vector $\underset{\sim}{P}^{(r)}$ with a vector particle, the fourth component $P^{(0)}$ will correspond to a ghost. The situation is analogous to that in a vector-meson theory where the fourth components have not been projected out.

Before we can conclude that our system possesses ghosts, we must ask whether all our states are linearly independent. More precisely, we must examine the factors corresponding to the two halves of Figure 8, which have the form

$$\int dy_2 \ldots dy_{m-1}\, I_1 \prod_{r,s} (\underset{\sim}{P}^{(r)})^{\lambda_r} (P_0{}^{(s)})^{\mu_s} \,. \tag{4.10}$$

Unless these factors are all linearly independent, we overcount the levels when we associate a different state with each set of integers (4.8b) and (4.8c). Furthermore, we may be able to use the linear dependences to express ghost states as a linear combination of particle states, and thereby to eliminate them.

We shall now show that linear dependences do in fact exist in our system. Such linear dependences were first found by Bardakci and Mandelstam,[9] Fubini and Veneziano,[9] and Brower and Weis.[11] The simplest linear dependence will enable us to eliminate the state corresponding to the factor $P_0^{(1)}$. It corresponds to the ordinary Ward identity, which enables us to eliminate the scalar photon in quantum electrodynamics. The other linear dependences may be regarded as more complicated Ward identities.

The easiest way of finding the linear dependences is to make a change of the integration variables in (4.10). The change which effects our purpose corresponds to an infinitesimal projective transformation which leaves the points 0 and 1 unchanged; we recall that the variables y_0 and y_m are equal to 0 and 1 respectively. Thus

$$y_i \to y_i + \epsilon\,(y_i - y_i^2). \tag{4.11}$$

The integrand in (4.10) does change under the transformation (4.11), since it is not quite projectively invariant. (The exponents do not satisfy (3.5b), as $\Sigma\, p_i$ is not zero). It turns out that the change in the integrand is a linear combination of terms which again have the form of the integrand of (4.10), with different values of λ_r and μ_s. The transformation (4.11) cannot change the integral (4.10), since it is merely a change of the variables of integration. We therefore conclude that a linear combination of terms of the form (4.10) vanishes.

To carry out this procedure, we write (4.3b) in the form

$$\{y_2(y_3-y_2)\ldots(1-y_m)\}^{-1}\, I' \;, \tag{4.12a}$$

where

$$I' = \prod_{j>i}{}' (y_j - y_i)^{-2p_i p_j} (y_{i+1} - y_i)^{-\mu^2} \tag{4.12b}$$

The factor $dy_2 \ldots dy_{m-1} \{y_2(y_3-y_2)\ldots(1-y_m)\}^{-1}$ does not change under the transformation (4.11). The factor I' transforms as follows:

$$I' \to I' + 2\epsilon P \{P^{(1)} - \tfrac{1}{2} P\} \tag{4.13a}$$

where

$$P = \sum_{i=1}^{m} p_i. \tag{4.13b}$$

Since we are working in the centre-of-mass system, P only has a zero component, so that

$$I' \to I' + 2\epsilon P_0 (P_0{}^{(1)} - \tfrac{1}{2} P_0) \tag{4.13c}$$

Finally, the quantities $\underset{\sim}{P}^{(n)}$ and $P_0{}^{(n)}$ transform according to the simple rule:

$$\underset{\sim}{P}^{(n)} \to \underset{\sim}{P}^{(n)} + \epsilon n (\underset{\sim}{P}^{(n)} - P^{(n+1)}) \tag{4.13d}$$

$$P_0{}^{(n)} \to P_0{}^{(n)} + \epsilon n (P_0{}^{(n)} - P_0{}^{(n+1)}). \tag{4.13e}$$

Hence

$$\int dy_2 \ldots dy_{m-1}\, I_1 \prod_{r,s} (\underset{\sim}{P}^{(r)})^{\lambda_r} (P_0{}^{(s)})^{\mu_s} \{ 2 P_0 (P^{(1)}{}_0 - \tfrac{1}{2} P_0) + \sum_r r\lambda_r (1 - \underset{\sim}{P}^{(r+1)} / \underset{\sim}{P}^{(r)}) + \sum_s s\mu_s (1 - P_0{}^{(s+1)} / P_0{}^{(s)} \} = 0. \tag{4.14}$$

The expression $\underset{\sim}{P}^{(r+1)}/\underset{\sim}{P}^{(r)}$ is interpreted in the sense that the factor $(\underset{\sim}{P}^{(r)})^{\lambda_r}$ becomes $(\underset{\sim}{P}^{(r)})^{\lambda_r - 1}\ \underset{\sim}{P}^{(r+1)}$. All terms of Equation (4.13) do have the form (4.10), with the λ's and μ's restricted by (3.8b) and (c), provided

$$\sum r\lambda_r + \sum s\mu_s + 1 = n \qquad \text{if } \mu^2+1 = 0^* \tag{4.15a}$$

$$\sum r\lambda_r + \sum s\mu_s + 1 \leq n \qquad \text{if } \mu^2+1 \neq 0. \tag{4.15b}$$

We thus conclude that a linear combination of terms of the form (4.10) vanish, as we wished to prove. At any pole, there is a separate identity for each set of integers λ_r and μ_s satisfying (4.15). The complete set of linear dependences was first given by Bardakci and Mandelstam.[9]

Let us examine the simplest of the linear dependences, which occurs when n=1. There are three expressions of the form (4.10), namely

$$\int dy_2 \ldots dy_{m-1}\ I_1\ \underset{\sim}{P}^{(1)},\ \int dy_2 \ldots dy_{m-1}\ I_1\ P_0^{(1)},\ \int dy_2 \ldots dy_{m-1}\ I_1. \tag{4.16}$$

The first has j=1, the last two j=0. The second term has an odd number of powers of $P_0^{(1)}$ and corresponds to a ghost. The only values of λ_r and μ_r in (4.14) which are compatible with (4.15) are λ_r=0, μ_r=0, so that the equation takes the simple form

$$2\ P_0 \int dy_2 \ldots dy_{m-1}\ I_1\ P_0^{(1)} - P_0^2 \int dy_2 \ldots dy_{m-1}\ I_1 = 0. \tag{4.17}$$

---

*If $\mu^2+1 = 0$, note that $P_0^2 = n-1$.

Equation (4.17) shows that the last two expressions (4.16) are multiples of one another. There is only one state with n=1, j=0, not two. Since one of the two states which occur before we make use of (4.17) is a ghost and the other a particle, it is not obvious whether the remaining state is a ghost or a particle. A more detailed examination shows that it is a particle, so that the ghost has been removed as required.

In general, the linear-dependence equation (4.14) enables us to eliminate any state with factors of $P_0^{(1)}$ in favor of a state where the total number of factors $P_0^{(r)}$ is reduced by one. Many ghosts can be eliminated, including all ghosts on the second trajectory, where there are no factors $P_0^{(r)}$, $r > 1$. More detailed examination shows that it is in fact the ghosts and not the particles which are removed from the second trajectory. However, we certainly do not have enough linear-dependence equations to remove all the ghosts. When n=2, for instance, we cannot remove the ghost which corresponds to the factor $P_0^{(2)}$.

In the special case $\mu^2+1=0$ the situation is transformed drastically. Virasoro then found a whole new set of linear dependences which result from the transformation.

$$y_i \rightarrow y_i + \epsilon\, y_i^m(1-y_i) \qquad \text{(m integral)} \qquad (4.18)$$

instead of (4.11).[12] The straightforward algebra is similar to that which we have given for the case m=1, and we shall not repeat it. We only mention that the factors $(y_{i+1}-y_i)^{\mu^2+1}$ do not transform suitably unless m=1, so that the new linear dependences do not exist unless $\mu^2+1=0$. The term $P_0P_0^{(1)}$ in the curly bracket of (4.14) is now replaced by $P_0 P_0^{(m)}$, and Equation (4.15a) becomes

$$\sum r\lambda_r + \sum s\,\mu_s + m = n.$$

We have sufficient linear-dependence equations to remove all the ghosts. As before, we have to verify that the ghosts and not the particles are eliminated in all cases. At the moment this has been done for $n \leq 3$ (S. Fubini and G. Veneziano, private communication).

It seems reasonable to hope that all the ghosts are eliminated when $\mu^2 + 1 = 0$. We pointed out in the previous section that this equation is precisely the Veneziano supplementary condition, which thus appears to be playing the fundamental role of allowing a ghost-free theory. The theory is still not physically acceptable, as it has one particle with $\mu^2 = -1$. In any case it does not lead to a realistic spectrum of particles, and so far no-one has constructed a model with a quark-like spectrum which is free of ghosts. We nevertheless regard it as most significant that a factorized Veneziano model without ghosts can be constructed.

Multiple Factorizations

We may now factorize a second time as in Figure 11.[13,14] To do so we redefine the y_i's $(0 \leq i \leq m)$ so as to

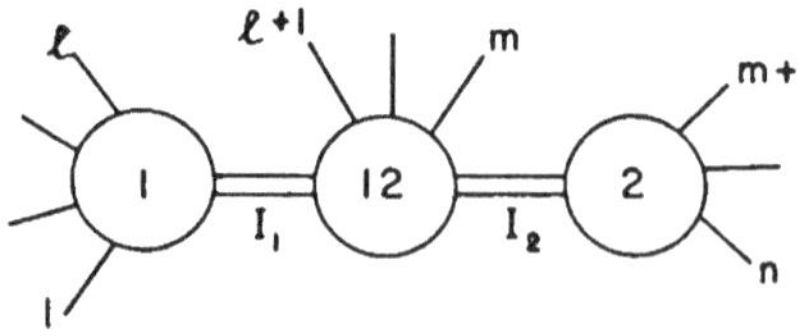

Figure 11. Twice-factorized multi-particle Veneziano amplitude.

make $y_0=0$, $y_{\ell+1}=1$, $y_I=\infty$; the definition is simply $y_i' = y_i/y_{\ell+1}$. We treat the new variables y_i' $(0 \le i \le m)$ in exactly the same way as we treated the variables z in the first factorization. The factor I_2 in (4.6a) does not depend on the y_i''s $(0 \le i \le m)$, and will be unaffected. The treatment of the factor I_1 is similar to the treatment of the entire integrand in the first factorization, but will be slightly more complicated since the mass of the particle I is not μ. The exponential factor also depends on the y's and must be included in our analysis.

We shall not go through the details here, but shall simply quote the result:

$$A=\int dy_2\cdots dy_{\ell-1}dy_{\ell+2}\cdots dy_m dy_{m+2}\cdots dy_n I_1 I_{12} I_2 \int_0^1 dz_1 z_1^{-s_1+\mu^2-1}\int_0^1 dz_2 z_2^{-s_2+\mu^2-1}$$

$$\times\, y^{-s_2+\mu_2} \exp\left(2\sum_{r=1}^{\infty}\frac{1}{r}\left\{\left(P^{(r)}K^{(r)}+\mu^2+1\right)z^r_1+\left(\bar{K}^{(r)}Q^{(r)}+\mu^2+1\right)z^r_2+P^{(r)}Q^{(r)}(yz_1z_2)^r\right\}\right)$$

(4.19a)

where

$$0 = y_1 \le y_2 \le \cdots \le y_\ell=1,\quad 1=y_{\ell+1}\ge \cdots \ge y_m=y,\quad 1=y_{m+1}\ge \cdots \ge y_n=0,$$

(4.19b)

$$I_{12}=\frac{1}{(1-y_{\ell+1})(y_{\ell+1}-y_{\ell+2})\cdots(y_{m-1}-y)y}\prod_{i>j}''' (y_j-y_i)^{-2p_ip_j}$$

$$\times\prod_i''' y_i^{-2p_ip_I}(y_i-y_{i+1})^{-\mu^2}\, y_m^{-\mu^2}, \qquad (4.19c)$$

$$K^{(r)} = \sum_{i=\ell+1}^{m} p_i y_i^{\,r} \qquad \bar{K}^{(r)} = \sum_{i=\ell+1}^{m} p_i (y/y_i)^r, \tag{4.19d}$$

and s_1 and s_2 are the channel variables associated with the left-hand and right-hand resonances of Figure 10. The quantities I_1 and P, I_2 and Q, refer to the left-and right-hand sides of Figure 11, and they are defined in the same way as before. The product $\prod'''$ in I_{12} is over the variables from the middle of Figure 11, so that I_{12} is just the Veneziano integrand for this portion of the diagram with $y_{I_1} = \infty$, $y_{I_2} = 0$. We could have replaced our variables $y_i(\ell+1 \leq i \leq m)$ by y/y_i. They would then increase from y to 1 as i increased from ℓ+1 to m, and the definitions of $K^{(r)}$ and $\bar{K}^{(r)}$ would be interchanged.

From Equation (4.19) we can find the multi-particle Veneziano amplitude for a process where two adjacent particles are excited. By modifying the procedure we can define an amplitude for a process where two non-adjacent particles are excited.[15] We begin by redefining the y's by a projective transformation in such a way that three y's adjacent to the new factorized channel are equal to 0,1 and ∞. The method is then the same as before; the result has a slightly more complicated form than for adjacent particles. We can also perform triple and further factorizations. It then becomes somewhat cumbersome to express the results in our present notation. In the following section we shall develop a new formalism which simplifies the treatment of our spectrum of resonances.

5. The Operator Formalism

Principles of the Formalism

In the present section we shall develop formal methods which greatly simplify calculations involving the spectrum of resonances which we have found. Amplitudes with three or more excited legs would appear quite complicated if they were expressed in the notation used up till now, whereas they can be written down much more concisely in the new formalism. Another application is to summation over intermediate states, which again is greatly simplified. Furthermore, the factorizability of our amplitude is much more evident in the formalism, and it may well be useful in enabling us to modify our amplitudes without losing factorization.

We have seen that there is one level associated with each component of each Cartesian tensor of the form

$$(P^{(r)})^{\lambda_r} \tag{5.1}$$

where $P^{(r)}$ is a four-vector. If we begin by considering the simplest case where $\mu^2+1 = 0$, the λ's are restricted by the condition

$$\sum r\lambda_r = n \tag{5.2}$$

at the nth resonance. We introduce a four-dimensional harmonic-oscillator operator a_r for each r, and we associate the resonance (5.2) with the state

$$\prod_r \frac{(a^\dagger_r)^{\lambda_r}}{(\lambda_r)^{\frac{1}{2}}} \mid 0 \rangle. \tag{5.3}$$

The operator a_r has four independent components $a_{r,\mu}$, but we shall suppress the indices μ in our notation. Since we are using a Lorentz metric and we require the oscillators to be real for μ=1,2 and 3, the operator $a^\dagger_r$ is really minus the adjoint of a_r. Note that the "vacuum" in our present notation corresponds to the scalar particle on the leading trajectory. The exponents λ_r in (5.3) satisfy the relation (5.2), so that each state corresponds to a definite pole position n.

The analogy of the spectrum of resonances to a harmonic-oscillator system was first pointed out by Bardakci and Mandelstam.[9] The operator formalism was proposed independently by Fubini, Gordon and Veneziano,[14] by Nambu and by Susskind.[16]

We now observe that the exponential in (4.6a) can be written as follows (when $\mu^2+1 = 0$):

$$\exp\left\{ (2 \Sigma\, P^{(r)}Q^{(r)}) \frac{z^r}{r} \right\} = \langle 0| \exp\left\{ \Sigma \frac{\sqrt{2}\, Q^{(r)} a_r}{\sqrt{r}} \right\} z^R \exp\left\{ \Sigma \frac{\sqrt{2}\, P^{(r)} a^\dagger_r}{\sqrt{r}} \right\} | 0 \rangle, \tag{5.4a}$$

where

$$R = \Sigma\, r a^\dagger_r\, a_r \tag{5.4b}$$

Equation (5.4) is easily proved from the identities:

$$\exp\,(fa)\; z^{a^\dagger a} = z^{a^\dagger a} \;\exp\,(faz), \tag{5.5a}$$

$$\exp\,(fa)\,\exp\,(ga^\dagger) = \exp\,(ga^\dagger)\,\exp\,(fa)\,\exp\,(fg). \tag{5.5b}$$

It follows at once from (5.2) and (5.3) that the eigenvalues of the operator R are simply the pole positions of the state in question. Insertives (5.4) in (4.6), multiplying by $z^{-\alpha(s)-1}$ and integrating, we obtain the following expression for the amplitude:

$$A = \langle 0 \mid G(Q,a)\; D(R,s)\; G(P,a^\dagger) \mid 0 \rangle\,, \tag{5.6}$$

where

$$G(P,a^{\dagger}) = \int dy_2 \cdots dy_{m-1} I_1 \exp\left\{\Sigma \frac{\sqrt{2}\, P^{(r)} a^{\dagger}_r}{\sqrt{r}}\right\} \tag{5.7a}$$

$$G(Q,a) = \int dy_{m+2} \cdots dy_{n-1}\, I_2 \exp\left\{\Sigma \frac{\sqrt{2}\, Q^{(r)} a_r}{\sqrt{r}}\right\} \tag{5.7b}$$

$$D(R,s) = \int_0^1 dz\, z^{R-\alpha(s)-1} = \left\{R-\alpha(s)\right\}^{-1} . \tag{5.7c}$$

One can now insert complete sets of intermediate states $|\, C \rangle\langle C \,|$ and $|\, C'\rangle\langle C' \,|$ in (5.6). The expression $\langle C' |\, G(P,a^{\dagger})\, |\, 0 \rangle$ may be interpreted as the vertex function between the harmonic-oscillator state $\langle C' \,|$ and the incoming m-particle state; the expression $\langle 0 \,|\, G(Q,a) |\, C \rangle$ is similarly interpreted. The middle factor $\langle C | \left\{ R-\alpha(s) \right\}^{-1} | C' \rangle$ is simply equal to $\delta_{CC'}\, \left\{n-\alpha(s)\right\}^{-1}$, and may be interpreted as the "propagator" between the states $\langle C \,|$ and $|\, C' \rangle$. Equation (5.6) is thus the mathematical expression corresponding to Figure 9. It makes evident the fact that poles are present when $\alpha(s)=n$, that the residues factorize, and that the spectrum of resonances corresponds to the states of our system of oscillators.

It is also worth noting that the states $G(P,a^{\dagger})|\, 0 \rangle$ are superpositions of the "coherent states" of our oscillator Hilbert space; such states are much used in quantum opics. From the present point of view the state $G(P,a^{\dagger})\, |\, 0 \rangle$ represents a state of m scalars in the sense that it is a superposition of oscillator states, each weighted with the vertex function corresponding to its coupling with the scalars.

In drawing a diagram for the vertex $G(P,a^\dagger)$ (Figure 12), it is necessary to place a dot above the

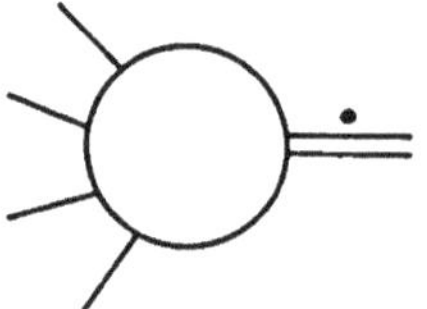

Figure 12. The Vertex $G(P,a^\dagger)$

excited leg in order to indicate that the particles in the portion which has been removed from Figure 9 are numbered from top to bottom. More precisely, the dot indicates that the upper external particle in the removed portion had been assigned the value $y_{m+1}=1$, the lower particle the value $y_n=0$. If we had numbered the particles in the reverse order, the states of our oscillator system would correspond to a different set of basic states of the degenerate resonances. To take a simple example, the particle with j=1 on the leading trajectory correspond to a Cartesian tensor

$$\underset{\sim}{p}^{(1)} = \Sigma\, \underset{\sim}{p}_i y_i \quad (\ell+1 \le i \le n).$$

If the particles on the right of Figure 8 had been numbered from bottom to top, the values 0 and 1 of the y's would be interchanged, with the value ∞ remaining unchanged. The two sets of y's are related by the simple equation

$$y_i' = 1-y_i,$$

so that

$$\underset{\sim}{P}^{(1)'} = - \underset{\sim}{P}^{(1)} .$$

since $\sum \underset{\sim}{p}_i = 0$ in the centre-of-mass system. With the new set of basic states, the state vector corresponding to the vector particle changes sign. Since reversing the order of a series of scalar particles is equivalent to applying the change-conjugation operator, this result shows that the lowest vector particle has C=-1. In general, placing the dot in Figure 12 below the excited line corresponds to charge-conjugating the basic states.

When μ^2+1 is not equal to zero, the spectrum of resonances is larger than that of our system of oscillators. However, instead of modifying the latter system, we shall modify the propagator. The factor $(1-z)^{-\mu^2-1}$ is not expanded as the exponential of a logarithm as in (4.6a), but is left unchanged. Equations (5.6), (5.7a) and (5.7a) are taken over unaltered, but (5.7c) is replaced by the equation

$$D(R,s) = \int_0^1 dz(1-z)^{-\mu^2-1} z^{R-\alpha(s)-1} = B\left\{ R-\alpha(s), -\mu^2 \right\} . \tag{5.7d}$$

The propagator now has poles when $\alpha(s)$ is an integer which is at least equal to the eigenvalue of R. Each state of our oscillator system corresponds to a series of resonances with $n \geq R$.

One can easily extend this method to the doubly-factorized amplitude (Figure 11). By using the identities (5.5), one can show that expression (4.19d) is equal to

$$\langle 0|\ G(Q,a)\ D(R,s_2)\ \Gamma(K,a,a^\dagger)\ D(R,s_1)\ G(P,a^\dagger)|\ 0\ \rangle \tag{5.8}$$

where

$$\Gamma(K,a,a^\dagger) = \int dy_2\dots dy_n \quad I_{12}\ \exp\left\{\Sigma\frac{\sqrt{2}\ \overline{K}^{(r)}a^\dagger_r}{\sqrt{r}}\right\} y^R \exp\left\{\Sigma\frac{\sqrt{2}\ K^{(r)}a_r}{\sqrt{r}}\right\}. \tag{5.9}$$

The quantities I_{12}, K and $\overline{K}$ are defined in (4.19c) and (d), and the y's satisfy the inequalities

$$1=y_1 \geq y_2\dots\geq y_n= y.$$

The amplitude corresponding to the operator Γ is shown in Figure 13(a). In the particular case

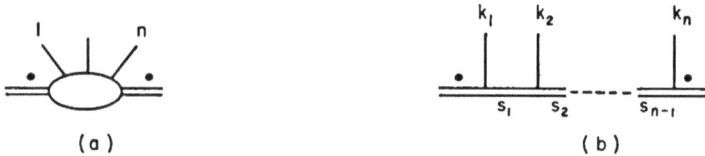

Figure 13. The vertex operator Γ.

where n=1, so that there are two excited legs and one scalar leg, we obtain a three-particle vertex operator given by the formula

$$V(k,a,a^\dagger) = \exp\left\{k\sum_r\frac{\sqrt{2}\ a^\dagger_r}{\sqrt{r}}\right\}\ \exp\left\{k\ \Sigma\frac{\sqrt{2}\ a_r}{\sqrt{r}}\right\}, \tag{5.10}$$

k being the momentum of the scalar particle.

Once we have an expression for the vertex operator V, we can write down a completely factorized formula for Γ (Figure 13(b).) Thus

$$\Gamma(K,a,a^\dagger) = V(k_n,a,a^\dagger)\, D(R,s_{n-1})\, V(k_{n-1},a,a^\dagger)\, D(R,s_2)\ldots V(k_1,a,a^\dagger). \tag{5.11}$$

If the particle on the left of Figure 14(b) is a scalar and that on the right is an arbitrary state S, we obtain the vertex of Figure 12, so that

$$\langle S \mid G(P,a^\dagger) \mid 0 \rangle = \langle S \mid \Gamma(k,a,a^\dagger) \mid 0 \rangle .$$

On inserting Equation (5.11) for Γ, we obtain a general formula for a coherent state:

$$G(P,a^\dagger) \mid 0 \rangle = V(k_n,a,a^\dagger)\, D(R,s_{n-1})\, V(k_{n-1},a,a^\dagger)\ldots D(r,s_1) \times \exp\left\{ \sum_r \frac{\sqrt{2}\, a^\dagger{}_r}{\sqrt{r}} \right\} 0 \rangle. \tag{5.12}$$

We may also take the vacuum-expectation value of (5.11), in which case we obtain a completely factorized form for the (n+2)-point Veneziano amplitude. There are no integrations left in this formula or in Equations (5.11) and (5.12). Such expressions are often most useful for handling Veneziano amplitudes. On the other hand, we must warn the reader that the simplication is purely formed, since the operators act on a complicated Hilbert Space.

Factorization according to Equations (5.11) or (5.12), which correspond to Figure 13(b), is sometimes known as factorization in the multi-peripheral configuration.

Projective Transformations

It is often useful to change the variables of integration in the coherent states

$$G(P,a^\dagger) \mid 0 \rangle$$

and, in particular, to make projective transformations. We saw in the previous section that a sub-class of such transformations leads to linear-dependence equations, and we obviously wish to study linear dependences within the operator formalism. Another application is to obtain a "twist" operator, i.e., an operator which reverses order of the unexcited legs in Figure 12. Such an operator will prove to be most useful, and it is again a particular case of a projective transformation.

We define the following three operators, corresponding to the three possible transformations:

$$\tfrac{1}{2}\epsilon L_3:\ y' = y-\epsilon y \qquad \epsilon L_+:\ y'=y-\epsilon y^2 \qquad \epsilon L_-:\ y'=y+\epsilon. \tag{5.13}$$

The commutators between the operators have the same form as those between the generators of the three-dimensional rotation group. The operators were first introduced by Gliozzi,[17] though we have adopted the notation of Fubini and Veneziano.[18]

The vertex operators are now redefined as follows:

$$G(P,a^\dagger) = \int dy_2 \ldots dy_{m-1} \{y_2(y_3-y_2)\ldots(1-y_m)\}^{-1} H(P,a^\dagger) \tag{5.14a}$$

where $H(P,a^\dagger) = I_1' \, (y_m-y_1)^{-P^2} \exp\left\{\sum \frac{\sqrt{2}\,P^{(r)} a_r^\dagger}{\sqrt{r}}\right\}$ (5.14b)

and the function I_1' is defined in (4.12b). The extra factor $(y_m-y_1)^{-P^2}$ has no effect when $y_m=1$, $y_1=0$, the values which occur in the definition of G. It does simplify the effect of the transformations (5.13) on H. In fact, application of the operators (5.13) to the state $H(P,a^\dagger)|0\rangle$ is equivalent to applications of the following operators:

$$\tfrac{1}{2} L_3(P) = P^2 - \sum_r r\, a_r^\dagger a_r = P^2 - R, \tag{5.15a}$$

$$L_+(P) = \sqrt{2}\, P\, a_1 - \sum_r \sqrt{r(r+1)}\ \ a^\dagger_r\, a_{r+1}, \tag{5.15b}$$

$$L_-(P) = \sqrt{2}\, P\, a^\dagger_1 + \sum_r \sqrt{r(r+1)}\ \ a^\dagger_{r+1}\, a_r. \tag{5.15c}$$

It follows from (5.15) that the operators satisfy the correct commutation relations.

We can also study the finite projective transformations generated by the L's. Let us consider, for example, the transformation $\exp\,(\frac{1}{2}\, L_3\, \zeta.)$ If

$$y' = \exp\,(\tfrac{1}{2}\, L_3\, \zeta).\ y$$

it follows from (5.13) that y satisfies the differential equation

$$\frac{dy'}{d\zeta} = -\, y'.$$

On inserting the boundary condition that $y' = y$ where $\zeta = 0$, we find:

$$\exp(\tfrac{1}{2}\, L_3\, \zeta\,) \longrightarrow y' = y\ell^{-\zeta}\ . \tag{5.16a}$$

Similarly

$$\exp(L_-\zeta) \longrightarrow y' = y+\zeta\ . \tag{5.16b}$$

The transformation

$$A(P) = -\tfrac{1}{2}\, L_3(P) + L_+(P) \tag{5.17a}$$

is particularly important, since we notice from (4.11) that this is the infinitesimal transformation corresponding to linear dependences. We might therefore investigate finite transformations generated by (5.17a), but we shall be more interested in the finite transformation generated by the adjoint $A^{\dagger}$, defined by*

$$A^{\dagger}(-P) = -\tfrac{1}{2}L_3(P) - L_-(P). \tag{5.17b}$$

We find that

$$\exp(A^{\dagger}\zeta) \rightarrow y'=ye^{\zeta} + 1-e^{\zeta}. \tag{5.18}$$

From (5.16), we observe that this transformation can be expressed as a product of finite transformations involving L_3 and L_-:

$$\exp(A^{\dagger}\zeta) = \exp\{L_-(1-e^{\zeta})\}\ \exp\{-\tfrac{1}{2}L_3\,\zeta\}\ .$$

This equation can be expressed in a slightly simpler form by writing $e^{\zeta}= 1-z$:

$$(1-z)^{A^{\dagger}(-P)} = \exp(zL_-)(1-z)^{-\frac{1}{2}L_3}\ . \tag{5.19}$$

Equation (5.19), which will be most useful in the sequel depends only on the group structure of the L's, and is valid whether the L's are defined by (5.13) or (5.15). It was first found independently by Kaku and Thorn[19] and by Gliozzi,[17] to whose papers the reader is referred if he desires a proof which does not make explicit use of projective transformations.

---

*Note that the operator $A^{\dagger}(-P)$ is really the adjoint of A(P), since the adjoint of a is $-a^{\dagger}$.

Linear Dependences

We are now equipped to study linear dependences within our present formalism. In the last section we saw that the state

$$A\ G(P,a^\dagger)|\ 0\ \rangle$$

vanishes. We proved this theorem by using the definitions (5.13) for the L's, and noting that we were simply changing the integration variables in G. Before we investigate the result further, let us rederive it purely within the operator formalism. We shall apply the operator A, with the L's defined by (5.15), to the expression (5.12) for the coherent states. By commuting the operator all the way to the right, we shall show that it annihilates the state.

We thus have to commute the operator A through the operator $V(k_m,a,a^\dagger)\ D(R,s_{n-1})$. We can commute A through V by means of the formula:

$$A(P)\ V(k,a,a^\dagger) = V(k,a,a^\dagger)\ \left\{A(P-k) + \mu^2\right\}\ , \tag{5.21a}$$

where $\mu^2=k^2$, the square of the mass of the external particle. We note that P-k is just the total momentum of the particles to the right of the vertex under consideration, so that the operator A on the right-hand side of (5.20) depends on the appropriate momentum. Further, $s_{n-1}=(P-k)^2$. We can now commute the operator A(P-k) through D; after a little algebra we obtain the result:

$$A(P-k)\ D\left\{R,\ (P-k)^2\right\} = D\left\{R+1,\ (P-k)^2\right\} A(P-k)-\mu^2 D\left\{R,(P-k)^2\right\}. \tag{5.21b}$$

On combining (5.18) and (5.19), we find:

$$A(P)\ V(k,a,a^\dagger)\ D\left\{R,(P-k)^2\right\} = V(k,a,a^\dagger)\ D\left\{R+1,\ (P-k)^2\right\} A(P-k). \tag{5.22}$$

By repeated use of (5.22) we can commute the operator A through all the factors of (5.12), until it ultimately reaches the vacuum, which it annihilates. The proof we have just given is due to Thorn;[20] the linear-dependence operator A was discovered independently by him, by Chiu, Matsuda and Rebbi,[21] and by Gliozzi.[17]

Let us now interpret the linear-dependence equation within the operator formalism. All matrix elements in which we are interested are taken between states of the form $G(P,a^\dagger)\mid 0\rangle$, which we shall call real states $\mid R\rangle$. The fundamental linear-dependence equation is

$$A\,G(P,a^\dagger)\mid 0\rangle = 0 \tag{5.23a}$$

or

$$A\mid R\rangle = 0. \tag{5.23b}$$

The real states therefore do not span the complete Hilbert space of oscillator states, but they are restricted by the condition (5.23). It is sometimes convenient to rewrite (5.23) as a condition on the states $D(R,s)\,G(P,a^\dagger)\mid 0\rangle$ or $D(R,s)\mid R\rangle$. If we multiply (5.23) on the left by $D(R+1, s)$ and apply (5.21b), we find

$$(A+\mu^2)\,D(R,s)\,G(P,a^\dagger)\mid 0\rangle = 0, \tag{5.24a}$$

or

$$(A+\mu^2)\,D(R,s)\mid R\rangle = 0. \tag{5.24b}$$

We thus notice that any state $\mid S\rangle$ of the form $\{A^\dagger(-P)+\mu^2\}\mid C\rangle$, where C is an arbitrary state in the oscillator Hilbert space, satisfies the relation

$$\langle S \mid D(R,s) \mid R \rangle = 0. \qquad (5.25)$$

The states $\mid S \rangle$ do not contribute to the summation over intermediate states in (5.6) and similar equations, and they will be called spurious states. Since our physical Hilbert space of resonances is obtained by factorizing expression such as (5.6), we can conclude that the Hilbert space of oscillators is too large to represent the physical space of resonances.

The simplest spurious state is obtained by applying the operator $A^{\dagger}(-P) + \mu^2$ to the "vacuum" state. It is thus equal to

$$\{ A^{\dagger}(-P) + \mu^2 \} \mid 0 \rangle = (-P^2 - \sqrt{2}\, Pa^{\dagger}_1 + \mu^2) \mid 0 \rangle$$

$$= - \sqrt{2}\, Pa^{\dagger}_1 \mid 0 \rangle,$$

since $P^2 = \mu^2$ for the state $\mid 0 \rangle$. This is the ghost state whose elimination we discussed in the previous section. It has a slightly different appearance in the present context owing to our somewhat unorthodox treatment of the Hilbert space with its multi-pole propagators.

It is often necessary to modify the propagator (5.7d) so that it automatically satisfies the equation

$$\langle S \mid \tilde{D}(R,s) = 0, \qquad (5.2b)$$

where S is a spurious state. Such a propagator would not give a redundant spectrum of states in a factorization, even if the matrix element to be factorized is not taken between real states. To quote an instance where a modified propagator is useful we may mention the calculation of the trace of $\tilde{D}\,\Gamma$, which will be required in the final section. The formula (5.9) for Γ is obtained by factorizing the expression (5.8), and we therefore know that matrix elements of the form $\langle R \mid D\,\Gamma\,D \mid R\rangle$ are correctly given. On the other hand, matrix elements such $\langle S \mid D\,\Gamma\,D \mid S\rangle$ are not uniquely determined from the factorization of (5.8) and, if we use a propagator with a redundant spectrum, we may sum over unphysical states when taking the trace.

A modified propagator $\tilde{D}$ with the correct properties was proposed by Thorn[20] and altered somewhat by Brower and Weis.[22] They made use of a projection operator

$$P_S(P) = \left\{ A^{\dagger}(-P) + \mu^2 \right\} \frac{1}{A(P)\left\{ A^{\dagger}(-P) + \mu^2 \right\}} \quad A(P) \,. \tag{5.27a}$$

Commuting P_S with D changes it into its adjoint, as may be shown from (5.21):

$$D(R,P^2)\, P_S(P) = P_S{}^{\dagger}(-P)\, D(R,P^2).$$

The modified propagator is then defined by the equation:

$$\tilde{D}(R,P^2) = D(R,P^2)\,\{ 1-P_S(P) \} = \{ 1-P_S{}^{\dagger}(-P) \}\, D(R,P^2). \tag{5.28}$$

It follows at once from its definition that $\tilde{D} \mid R \rangle = D \mid R \rangle$, $\langle R \mid \tilde{D} = \langle R \mid D$. The propagator $\tilde{D}$ may thus be used instead of D in all equations up to (5.12). On the other hand, $\langle S \mid \tilde{D} = \tilde{D} \mid S \rangle = 0$, where $\mid S\rangle$ is any spurious state, so that $\tilde{D}$ has no redundant spectrum.

Brower and Weis have rewritten (5.25a) in a form which involves no reciprocals:

$$P_S(P) = \sum_{\ell=0}^{\infty} (-1)^{\ell} \binom{-m^2-1}{\ell}^{-1} \left\{ {-A^{\dagger}(-P)-m^2 \atop \ell} \right\} \left\{ {-A(P) \atop \ell} \right\} \qquad (5.27b)$$

Their expression follows from the general formula

$$(AB)^{-1} = \sum_{\ell=0}^{\infty} (-1)^{\ell} (\ell+1)^{-2} \binom{-c-1}{\ell}^{-1} \binom{B-1}{\ell} \binom{A-1}{\ell},$$

provided [A,B] = - A-B-c, where c is a c-number.

It is not difficult to show that only a finite number of terms of (5.27b) contribute to each pole of $\tilde{D}$, so that the residues of the poles of $\tilde{D}$ are well-defined. Away from the poles it is not obvious that the expressions (5.27) and (5.28) are meaningful, and Gliozzi has given arguments to show that they are not.[23] When a modified propagator is used in a factorization, it is therefore only implied that the poles of the amplitude are correctly given by the factorized expression. The amplitude can be constructed from its poles by making use of the known analyticity properties

Since the operator A, when acting on any real state, gives zero, it follows that any finite transformation $e^{\zeta A}$ leave a real state unchanged. We shall later make use of (5.19) in this connection, and we therefore write the result in the form

$$\langle R \mid (1-z)^{-A^{\dagger}(-P)} = \langle R \mid, \qquad (5.29a)$$

for any z. Alternatively, if we have a modified propagator $\tilde{D}$, we may write

$$\tilde{D}\, V(k,a,a^\dagger)\,(1-z)^{-A^\dagger(-P)} = 0, \tag{5.29b}$$

as a consequence of (5.28). Equation (5.29a), which was first obtained by Chiu, Matsuda and Rebbi,[21] may be regarded as a kind of gauge invariance which follows from the Ward identities.

The Twist Operator

Another useful operator which can be constructed from, the L's is the "twist" operator of Caneschi, Schwimmer and Veneziano.[24] The purpose of this operator is to reverse the order of the unexcited particles in Figure 12. From Equations (5.14a) and (4.12b) we notice that reversal of the order of the particles has two effects; it reverses the signs of the factors $y_j - y_i$, and it reverses the inequalities of the integration variables:

$$0 = y_1 \leq y_2 \dots \leq y_m = 1.$$

Apart from a phase factor $(-1)^{-P^2}$, the required twist operator Ω is therefore the operator which changes the variables y_i to $1-y_i$. We can express this operator as a function of the L's from (5.13):

$$\Omega = (-1)^{-P^2} e^{L_-} e^{\frac{1}{2}i\pi L_3} = (-1)^{-P^2} e^{L_-} (-1)^{-\frac{1}{2}L_3} = e^{L_-}(-1)^R = (-1)^R e^{-L_-}. \tag{5.30}$$

Equation (5.30) is due independently to Kaku and Thorn[19] and to Gliozzi.[17] We notice from (5.30) that $\Omega^2 = 1$, an identity which would be expected, since two applications of the twist operator restores the original diagram.

Application of the twist operator enables one to perform factorizations in configurations other than Figure 13(b). The operator twists the entire portion of the diagram to the left of it. Thus the diagram of Figure 14 is represented by the function

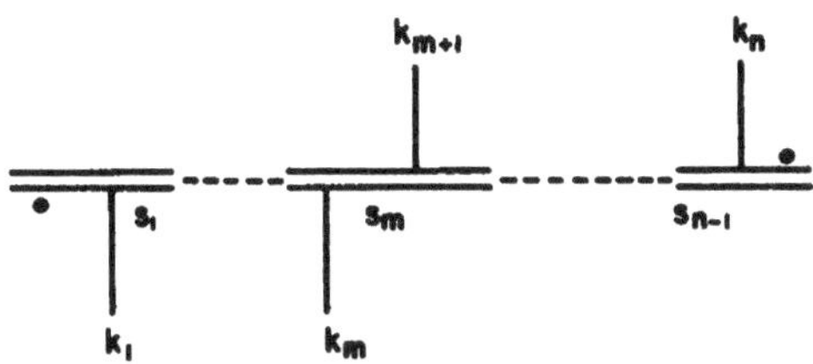

Figure 14. Multiperipheral diagram with one twist.

$$V(k_n,a,a^\dagger)\ D(R,s_{n-1})\ldots V(k_{m+1},a,a^\dagger)\ D(R,s_m)\ \Omega\ V(k_m,a,a^\dagger)\ldots V(k_1,a_1a^\dagger). \tag{5.31}$$

Notice that the dot on the left is now at the bottom, since, if the diagram is joined to a vertex G on the left, the operator Ω will reverse the ordering of the particles of G. By applying the twist operator several times, we can obtain a diagram with upper and lower scalar legs in any combination. The amplitude which the diagram represents depends only on the order of the external particles; different methods of drawing a diagram with the same ordering of the external particles. correspond to factorizing in different channels.

We can easily commute twist operators through linear-dependence operators from the formulas[20]

$$\Omega A = -A\,\Omega , \tag{5.32}$$

$$\Omega A^{\dagger} = -L_3 \Omega , \tag{5.33}$$

which are most easily proved by considering the operators as performing transformations on the y's.

By making use of the identity (5.19), it is possible to reduce the number of Ω's in a twisted diagram. Thus

$$D\Omega = \int dz(1-z)^{-\mu^2-1}\, z^{R-\alpha(s)-1}\,\Omega ,$$

and $z^R\Omega = z^R e^{L_-}(-1)^R$ from (5.30)

$= e^{zL_-}(-z)^R$ using an identity similar to (5.5a)

$$= (1-z)^{A^{\dagger}(-P)} \left(\frac{-z}{1-z}\right)^R (1-z)^{P^2}, \text{ from (5.19).} \tag{5.34}$$

The factor $(1-z)^{A^{\dagger}(-P)}$ can be commuted to the left through vertex operators by using the adjoint of Equation (5.20). To commute it through the factors z'^R in the propagators we make use of the following relation, which can be derived from (5.19):

$$z'^R(1-z)^{A^{\dagger}} = (1-zz')^{A^{\dagger}} \left\{\frac{z'(1-z)}{1-zz'}\right\}^R \left(\frac{1-zz'}{1-z}\right)^{P^2}. \tag{5.35}$$

If the factor $(1-z)^{A^{\dagger}(-P)}$ meets a second twist operator one can eliminate it using (5.33). One can then perform similar manipulations on the new twist operator. The factor involving $A^{\dagger}$ can ultimately be moved all the way to the left, when it becomes unity if the particles at the end of the chain in Figure (14) are real (Equation (5.29a)).

When matrix elements are evaluated with the aid of (5.34), the usual power series in z, such as the series on the right of Equation (5.4a), becomes replaced by a power series in $\frac{z}{1-z}$, which diverges if $z > \frac{1}{2}$. In certain calculations the summation can be performed analytically at a later stage but, in general, we must interpret formulas involving (5.34) in the sense that they give correctly only the residues at the poles in s.

Another formula which can be proved in a similar way to (5.34) is

$$\langle R \mid \Omega^{\dagger} D\Omega \mid R' \rangle = \langle R \mid D \mid R' \rangle, \tag{5.36}$$

if $\langle R \mid$ and $\mid R \rangle$ are real states. Equation (5.36) must obviously be true, for it states that reversal of the order of the particles on both sides of Figure (9) leaves the amplitude unchanged. It is interesting to note that the direct verification requires the assumption that the states $\langle R \mid$ and $\mid R' \rangle$ are real, as it is necessary to use Equation (5.29a)

The Sciuto Vertex

Knowledge of the propagators D and $\tilde{D}$, the vertex V, and the twist operator Ω, enables us to factorize in the configurations of Figure 13(b) and Figure (14). In order to factorize in a more general configuration such as Figure (10b), we require the vertex operator for three excited particles, which was originally found by Sciuto.[25] The method is to perform a third factorization on the vertex Γ, given in Equation (5.9) and Figure 13(a). It is first necessary to change the variables of integration.

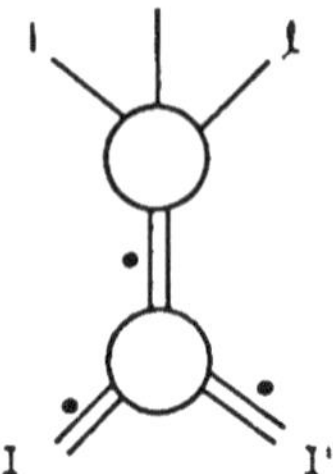

Figure 15. Factorization of the vertex Γ.

As written in (5.1), the fixed points are $y_1=1$, $y_I=\infty$, $y_{I'}=0$, and the variable y_ℓ has been set equal to y. To redefine our variables as in our original factorization, we have to set $y_\ell=0$, $y_I=1$, $y_{I'}=\infty$. (Note that we are reading the particles in counter-clockwise order from ℓ to I'). The new variables are thus

$$y_i = 1-\frac{y}{y_i}.$$

Let us also set

$$z = y_i' = 1-y; \tag{5.37}$$

this variable then corresponds to the z in our original calculation. Finally, by analogy to (4.2a), we again make a change of variables

$$y_i'' = \frac{y_i'}{z} = \frac{1}{z}\left(1-\frac{1-z}{y_i}\right)$$

or

$$y_i = \frac{1-z}{1-zy_i''}, \quad 1 \leq i \leq \ell-1, \tag{5.38a}$$

$$y = 1-z, \quad (\text{from } (5.37)) \tag{5.38b}$$

After substituting the change of variables (5.38) in (5.9), we can factorize as we did in the previous section. We thus obtain the three-particle vertex, as shown in Figure 15. Note that the dot on the intermediate line is on the right, since it is always on the side of increasing y's.

We shall not quote Sciuto's result here, but instead we shall give the formula for the symmetric vertex of Caneschi and Schwimmer.[26] The vertex shown in Figure 15 is unsymmetric in the position of the dots. By applying the twist operator $\Omega^{\dagger}$ to the leg 1

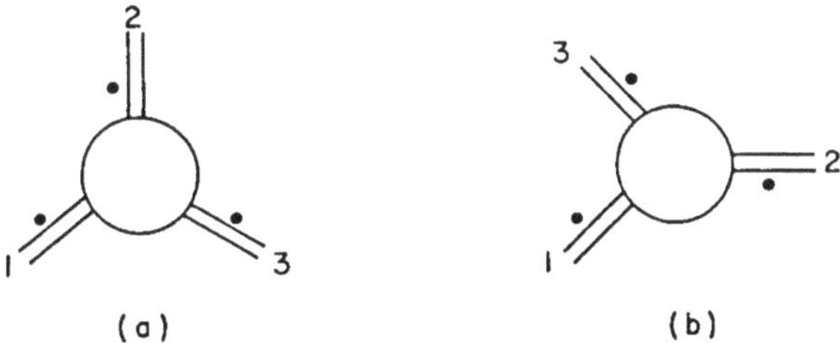

Figure 16. Unsymmetric and symmetric Sciuto vertices.

of the vertex (Figure 16), (so that it acts to the left in the corresponding mathematical expression), we obtain a completely symmetric vertex. To quote the formula it is convenient to associate creation operators $a^{\dagger}_r$, $b^{\dagger}_r$, $c^{\dagger}_r$, with the legs 1, 2 and 3 respectively. We then write the vertex as a matrix element between the state $\langle\lambda, \mu, \nu|$ and the vacuum, where $\langle\lambda|$, $\langle\mu|$ and $\langle\nu|$ are the states in our oscillator space corresponding to particles 1, 2 and 3. The formula is then:

$$\mathcal{W}(p_1,p_2,p_3,\lambda,\mu,\nu)\equiv\langle\lambda,\mu,\nu|W(p_1,p_2,p_3;a^{\dagger},b^{\dagger},c^{\dagger})|0\rangle$$

$$=\langle\lambda,\mu,\nu|\exp\{p_3a^{\dagger}+p_1b^{\dagger}+p_2c^{\dagger}+(a^{\dagger},c^{\dagger})_-+(c^{\dagger},b^{\dagger})_-+(b^{\dagger},a^{\dagger})_-\}|\rangle, \tag{5.39a}$$

where

$$(\alpha,\beta)_- = \sum_{r,s=1}^{\infty} (-1)^s \sqrt{\frac{s}{r}} \binom{r}{s} a_r b_s . \tag{5.39b}$$

The expression (5.39) is fairly compact and, by using it, one can factorize a diagram in any configuration such as Figure (10b). However, it is important to recognize one limitation of the formula; it only gives the correct matrix element $\langle \lambda,\mu,\nu | D_1 D_2 D_3 W | 0 \rangle$ if the states λ,μ and μ are real. The matrix element does not vanish if one of these states is spurious. We must therefore use modified propagators $\tilde{D}$ throughout. With presently available techniques the manipulation of the projection operators P_S is considerably more complicated than it is for factorization in the configurations of Figures (13b) or (14). As a matter of fact, the unsymmetric vertex originally proposed by Sciuto is simpler in this respect,[27] since it does vanish if one of the legs 1 or 2 is spurious and the other two are real.

Gross and Schwarz[28] have recently introduced a formula for a modified twisted propagator which may be helpful in this respect. The formula is as follows:

$$\mathcal{D}(p) = \int_0^1 dz\, (1-z)^{A^\dagger(-P)-1} (-1)^R \left(\frac{z}{1-z}\right)^{R-\alpha(s)-1} (1-z)^{A(P)-1} . \tag{5.40}$$

They obtained their result by using Equations (5.30) for the twist operator and Equations (5.28) and (5.27b) for the modified propagator; they then used an integral representation for the hypergeometric function. Once the formula is obtained, it can be proved by noting that it gives the same matrix elements between real states as the unmodified twisted propagator, and that the expressions $(A^{\dagger}+\mu^2)\mathcal{U}$ or $\mathcal{U}(A+\mu^2)$ vanish, since they are equal to integrals of a total derivative.

From the formula for the Sciuto vertex, it has been shown directly that we can pass from one configuration of Figure 17 to the other without changing the result.[27,28,29] By performing such transformations within more complicated diagrams, we can pass from any given configuration of resonances to any other. We thus verify the fundamental duality requirement that the sum of the contributions of all sets of resonances to a scattering amplitude is independent of the channels in which one chooses to factorize.

Figure 17. Duality transformations with Sciuto vertices.

By repeated factorization of the n-point scalar amplitude, it is possible to generalize the Sciuto formula to a process with any number of excited particles.[15,30] As in Equation (5.9), the formula involves the usual Koba-Nielsen integration; the integrand is the product of a factor I_1 and an exponential factor which contains creation operators corresponding to the excited particles. Gross and Schwarz[28] have obtained similar results by joining symmetric Sciuto vertices with their modified twisted propagator.

6. Veneziano-type Quark Models

Meson-Meson and Meson-Baryon Amplitudes

The models presented in the previous sections do fulfil most of the consitency requirements, except for the presence of ghosts, (if $\mu^2 \neq -1$) or of a negative-mass particle (if $\mu^2=1$). On the other hand, the spectrum of resonances which they predict does not agree with that observed in nature. A realistic model will have to possess a spectrum similar to that predicted by the quark model. One of the most appealing features of the Veneziano theory is that it does enable us to understand how a quark model may predict the correct spectrum, even though quarks need not exist as actual particles. Every Veneziano-type amplitude with quarks of spin $\frac{1}{2}$ which has so far been constructed possesses ghosts, so that a consistent model with all the required features should be regarded as a potentiality rather than an actuality. Nevertheless, since the models possess so many attractive properties, and since they enable us to reconcile features of hitherto completely different approaches, we may hope that they can be modified so that their defects will be overcome.

The simplest dual-resonance model without spin or isotopic spin appears to be the ordinary Veneziano model which we have treated in the foregoing sections. All other models so far constructed possess a more complicated spectrum of resonances. In the simplest dual-resonance quark model, therefore, the scattering amplitude will be the direct product of an ordinary Veneziano amplitude and a matrix in spin and SU(3) space. Models of this type have been constructed by Mandelstam and by Bardakci and Halpern.[31] They possess too much symmetry; all $Q\,\overline{Q}$ states are degenerate, for instance. They may well serve as starting-points for more complicated models, however, in which the Veneziano amplitude itself is modified.

It follows from our construction that the scattering amplitude is an eigenfunction of the spin and SU(3) crossing matrices. In this respect our present model resembles several models previously proposed, and a method for finding eigenfunctions of the crossing matrix was given by Capps.[32] His construction may be represented by the duality diagrams proposed independently by Harari, by Rosner, by Matsuoka, Ninomiya and Sawada, and, in a rather different form, by Neville.[33]

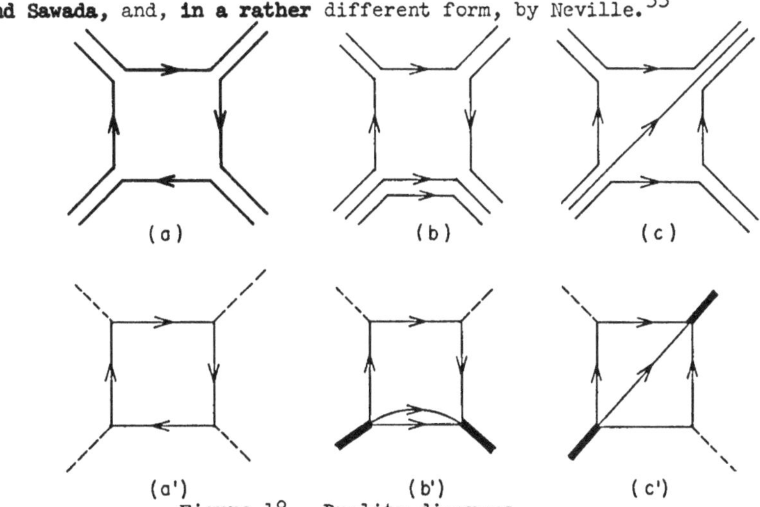

Figure 18. Duality diagrams.

Figure 18(a) is a duality diagram for meson-meson scattering. The quark lines represent delta functions in spin and SU(3) space; the external and internal mesons are pictured as composed of two quarks. There will be a second duality diagram in which the quarks circulate in the opposite direction, and further duality diagrams in which pairs of adjacent mesons are interchanged.

The amplitude corresponding to our duality diagrams is clearly an eigenfunction of the crossing matrix. Moreover, the multiplets of internal resonances are the same as the multiplets of external particles, and each vertex is symmetric in its three particles. It is probable that th eigenfunction of the crossing matrix constructed from duality diagrams is the only eigenfunction with all the required properties. We can therefore maintain that the quark model follows from our simplified hypotheses. We do not have to make additional assumptions such as the absence of exotic resonances. Of course, we cannot predict that the quarks form a unitary triplet with spin $\frac{1}{2}$. It would be possible to construct models with quarks of any spin and with an internal symmetry group of SU(n).

The spin and SU(3) duality diagrams must now be combined with the Veneziano amplitude which describes the orbital motion of the quarks. A diagram such as Figure 18(a) is associated with the eight-point Veneziano diagram with the external particles in the same cyclic order. One can then obtain a four-meson diagram by factorization, as in Figure 10(a). The external mesons of the amplitude so constructed will possess arbitrary orbital angular momentum, besides their spin angular momentum and their SU(3) quantum numbers. Once the amplitude has been constructed, the external quarks may be discarded. The concept of quarks with space, spin and SU(3) degrees of freedom is thus most useful in obtaining our amplitude, even though there need be no physical particles corresponding to quarks.

Amplitudes for meson-baryon scattering may be constructed in a similar way. Two duality diagrams have been drawn in Figures 18(b) and (c). If s and u are the two meson-baryon channels and t the meson-meson channel, Figure 18(b) will correspond to an s t term, Figure 18(c) to the s u term.

The duality diagram of Figures 18(a) (b) and (c) can alternatively be drawn as in Figures 18(d), (e) and (f); this was the method employed by Neville. The external particles are meson or baryons, and the vertices contain the appropriate SU(3)-and γ-matrices. Amplitudes for the scattering of meson or baryons with any orbital angular momentum can again be obtained from the n-point amplitude by factorization.

The introduction of SU(3) into our model by the above method introduces no difficulties, but the spin degree of freedom introduces extra ghosts. Such ghosts are a general feature of factorized Veneziano-type models with fermions, and they occur in any amplitude with baryons, whether or not it is constructed from quarks. The problem of ghosts in fermion amplitudes is the main problem connected with the Veneziano model which faces us at the moment. If we construct our scattering amplitude from quarks, the problem exists in meson amplitudes as well.

The basic reason behind the difficulty is the indefinite nature of the Lorentz metric. We recall that the ghosts in the ordinary factorized Veneziano model without quarks originated from this source too. In order that our amplitude be invariant under Lorentz transformations and reflections, it is necessary that we introduce two quarks, corresponding to the $(\frac{1}{2},0)$ and $(0,\frac{1}{2})$ representations of the Lorentz group. The quarks will thus have four spin degrees of freedom in all. Meson composed of such quarks will always be associated with parity doublets. The normal-parity mesons, corresponding to those predicted by the ordinary quark model, will be particles or resonances with real coupling constants. The abnormal-parity mesons will be ghosts.

Let us investigate in more detail the spin multiplets which the model predicts. By combining two $(\frac{1}{2},0)+(0,\frac{1}{2})$ representations of the Lorentz group, we obtain the following representations:

$(1,0)+(0,1)$	M=1,	J=1,	P= + or -		
$(0,0)$ (twice)	M=0	J=0	P= + or -		
$(\frac{1}{2},\frac{1}{2})$ (twice)	M=0	J=1	P= + or -	J=0	P= + or - .

The quoted values for the angular momentum and parity apply to the lowest particles on the Regge trajectory, where there is no orbital motion. The quantity M is the Toller quantum number, which is only a good quantum number when s=0. We notice that the expected vector and pseudo-scalar mesons are accompanied by a second vector-pseudo-scalar pair, and by two axial-vector-scalar pairs. The latter two pairs are ghosts. The two normal-parity particles have the same value of C (+ 1 for the pseudo-scalars, -1 for the vectors); the abnormal-parity particles have opposite values of C.

It will probably be possible to construct a model with symmetry breaking in which the extra trajectories lie below the main trajectory. Whether the ghosts can actually be removed within the framework of the narrow-resonance approximation is a question which is as yet unanswered. Taking $\mu^2=-1$ removed the ghosts in the original model, but it does not remove the new ghosts. The difficulty may be indicative of the fact that a consistent narrow-resonance model is impossible. In that case a solution of the problem would involve unitarization of the Veneziano amplitude. The extra trajectories might thereby be eliminated, or they might cease to give rise to inconsistencies. The presence of the M=1- pair of trajectories is in fact indicated by high-energy photoproduction experiments, but the results of such experiments may possibly be explained by cuts in the j-plane.

Both in the original Veneziano model and in the quark model, the ghosts are connected with the parity doublets predicted at s=0 by Mac-Dowell doubling or, more generally, by Toller doubling. Whenever the Toller quantum number M is non-zero and, in particular, for all fermion channels, where it is half-integral, the trajectories must appear in parity doublets at s=0. The sign of the residues at this point implies that one trajectory is ghost-like, though there may be a change of sign before we come to values of s at which resonances occur. Carlitz and Kisslinger[34] have proposed a model in which Mac-Dowell or Toller doubling is avoided by the presence of cuts in the j-plane. By using their ideas, Bardakci and Halpern[35] have removed the leading-trajectory ghosts from the Veneziano quark model, but at the expense of losing factorization.

Our model possesses one undetermined mixing parameter for the vector and one for the pseudo-scalar trajectory, since, as long as the trajectories are degenerate, the model cannot predict what linear combination of degenerate states corresponds to the physical particle. The parameters might be fixed so as to minimize the residues of the poles due to unphysical particles relative to those due to physical particles in meson-meson scattering.

We then find that the total contribution from the ghosts is about one-half that from the physical particles, while the contribution from the second vector and pseudo-scalar particles is very small. We also find that the coupling constants involving physical mesons satisfy the conditions of $SU(6)_W$ symmetry.

The situation with the meson-baryon system is similar; there is one mixing parameter to fix for each of the two trajectories and, if we adjust them to minimize the coupling to unphysical particles, the meson-baryon coupling constants also satisfy the $SU(6)_W$ relations. Furthermore, subject to a remark which we shall make later, the ratio of the MMM and $MB\bar{B}$ coupling constants is that predicted by vector-meson universality. Since $SU(6)_W$ and vector-meson universality are sufficient to predict all MMM and $MB\bar{B}$ coupling constants in terms of a single parameter, we conclude that the ratios between such constants are correctly predicted by our model. The concepts of $SU(6)_W$ symmetry and vector-meson universality were of course familiar before the Veneziano model was proposed, but we can now begin to understand how such concepts, together with the quark model, may be related to the assumptions of S-matrix theory.

The Regge parameters predicted by our model can be compared with experiment and, where this comparison has been made, the agreement is reasonably satisfactory.[36] More detailed work in this respect is probably premature until the ghost problem is better understood.

Exotic Resonances

While there are no exotic intermediate states in meson-meson and meson-baryon scattering, such intermeidate states do exists in baryon anti-baryon scattering. The duality diagram for that process is shown in Figure 19(a). It will be noticed that the resonances in the t-channel

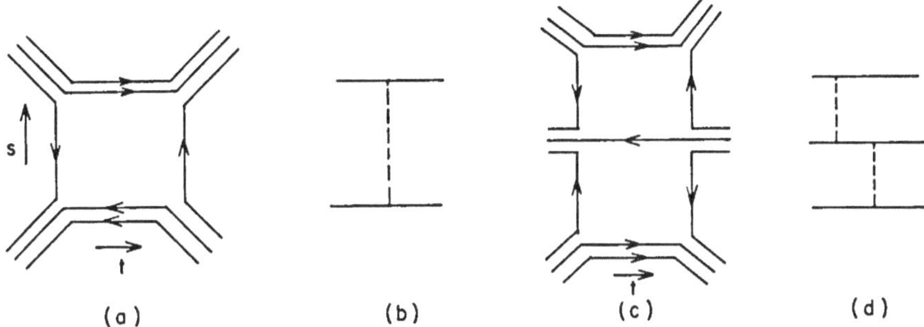

Figure 19. Duality diagrams with exotic resonances.

are exotic mesons consisting of two quarks and two anti-quarks. Rosner[37] was the first to draw attention to such exotic mesons, and their presence is a necessary feature of our model. By viewing Figure 19(a) from the s-channel, we can see that the diagram must be present. The intermediate state is then an ordinary meson (Figure 19(b)), and the $B\bar{B}M$ coupling is required for consistency with the meson-baryon amplitude.

One can employ similar reasoning to prove the existence of exotic baryons consisting of four quarks and one anti-quark. A diagram for the process $B\bar{B}B \to B\bar{B}B$ is shown in Figure 19(c). Such a diagram must be present in order to represent the sequence of exchanges shown in Figure 19(d). The resonances in the t-channel of Figure 19(c) and the exotic baryons under discussion. It is evident that exotic mesons and baryons with an arbitrarily large number of quarks and anti-quarks are present in the model.

A model with exotic resonances would not conflict with experiment provided that the masses of such resonances were sufficiently high. The resonances could decay with large Q-value and no centrifugal barrier, and they would not appear experimentally as narrow resonances. The intercept of the trajectory must therefore depend on the total quark number (the number of quarks plus the number of anti-quarks). If this is the case it is necessary to re-examine the factorization properties, since the discussion of Section 4 assumed that all trajectories had the same intercept. One can show that the model is factorizable provided that the intercept of the trajectory is a polynomial function of the total quark number.[38] The degeneracy of the resonances increases with the order of the polynomial and, in the limit of a polynomial of infinite order, the factorization properties are lost.

In a model where the spectrum of resonances is not too degenerate, the intercept must be a polynomial function of low degree in the total quark number. A linear function is not acceptable, since the square of the mass of a highly exotic particle would be proportional to the total quark number, and we would have an infinite sequence of stable particles with increasing strangeness and isotopic spin. The simplest model is therefore one in which the intercept is a quadratic function of the total quark number. If we provisionally accept this model, we can use the meson and baryon masses to fix two of the three constants of the polynomial. By setting the quadratic term equal to zero, we obtain a mass of 1.6-1.8 BeV for the lightest exotic meson. This is a lower limit; the quadratic term, which must be present, will increase this mass.

Non-planar Multi-particle Veneziano Amplitude

The meson-baryon diagrams shown in Figure 18 are planar diagrams. We now ask the question whether we should include non-planar diagrams, such as Figure 20(a), as well. It is certainly possible to construct a

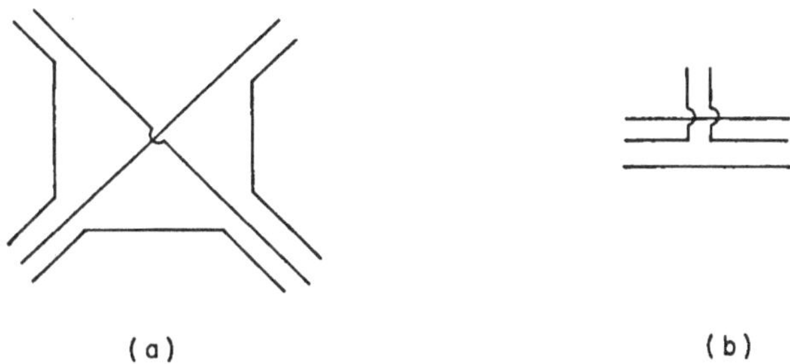

Figure 20. Non-planar duality diagrams.

model where such non-planar diagrams are absent. This model has one or two drawbacks. First, the baryon states are not limited to states symmetric in the three quarks. In a model with only symmetric states, any quark in the intermediate state should be able to go into any quark in the initial or final states. The duality diagram of Figure 18(b) would therefore necessarily be accompanied by Figure 20(a). The presence of non-symmetric quark states in the planar-diagram model was noticed by Mandula, Rebbi, Slansky, Weyers and Zweig,[39] who found an $\ell=0$ $\underset{\sim}{70}$ together with the expected $\ell=0$ $\underset{\sim}{56}$ and $\ell=1$ $\underset{\sim}{70}$. The coupling of this state to the meson-baryon system was about one-quarter as strong as that of the other states.

Another feature of the planar-diagram model is that it does not possess vector-meson universality. The vector meson can interact with either of the outer quarks of the baryon, but it cannot interact with the middle quark as in Figure 20(b), since that diagram is excluded by hypothesis. The strength of the coupling of vector mesons to baryons is thus two-thirds of the strength of their coupling to other mesons.

At the present time we cannot exclude the planar-diagram model on the basis of these two defects, but it is possible to construct an alternative model where the defects are not present. Before we explain how this may be done, we must mention a third difference between the present model and the non-relativistic harmonic-oscillator quark model, namely the degeneracy of the baryon spectrum on the leading trajectory. In the Veneziano model which we have used hitherto, the leading trajectory possesses no "orbital" degeneracy. It has a finite degeneracy associated with the spin and SU(3) degrees of freedom. In this respect the model is simpler than the harmonic-oscillator quark model, where the degeneracy of resonances on the leading trajectory increases without limit as the angular momentum is increased. Thus, in a model with scalar quarks, the states with j=0 and j=1 are not degenerate. The state with j=2 is doubly degenerate, since it can be composed of one quark in a D-state and two in S-states, or of two quarks in P-states and one in an S state. It is evident how the degeneracy of states of higher j increases.[40]

The alternative model which we shall now propose resembles the harmonic-oscillator quark model as regards the features just discussed. The baryon states are symmetric in the three quarks, and vector-meson universality holds. On the other hand, the baryon spectrum on the leading trajectory has a degeneracy which resembles that of a system of three quarks bound by harmonic forces. The model involves a generalization of the multi-particle Veneziano formula. We recall that a duality diagram must be associated with a Veneziano diagram with the same ordering of the external particles, and we have pointed out that the duality diagram of Figure 20(a) must be present in a symmetric model. This diagram may be redrawn as in Figure 21(a); more complicated diagrams will have a similar structure. It will be slightly more convenient to use the Neville form of the diagrams as in Figure 21(b). We wish to

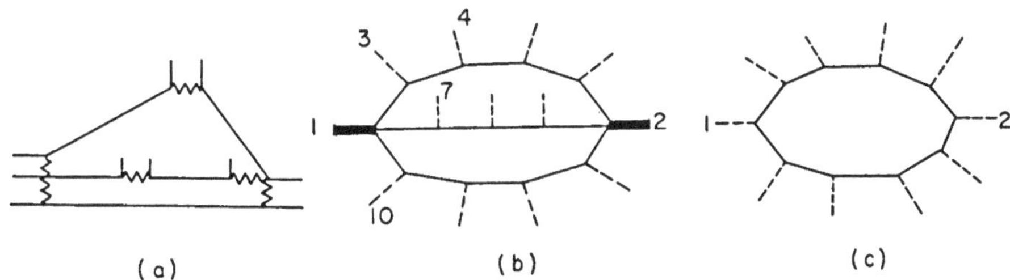

Figure 21. Non-planar and planar Veneziano diagrams.

obtain a new multi-particle Veneziano formula with resonances in all channels of Figure 21(a) or (b) which consist of adjacent particles. Thus, in Figure 21(b), we require resonances in meson channels such as 34, or in baryon channels such as 3 1 10 and 3 1 7 10, but not in channels such as 47 or 1 4 7 10. The required generalization was given by Mandelstam;[41] the formulation of the model which we shall present here is due to Thorn.[42]

We shall obtain our formula by analogy with that for the ordinary planar amplitude, where the y's associated with the points 1 and 2 are given the values 0 and ∞ (Figure 21 c). The y's for the particles along the top line then have positive values in increasing numerical order from 0 to ∞; those for the particles along the bottom line have negative values, again in increasing numerical order from 0 to ∞. In the non-planar diagram (Figure 21(b)), we assign positive y's, in increasing numerical order, to the particles on the top line. The y's associated with particles on the middle and bottom lines will be assigned values of the form $re^{2i\pi/3}$ and $re^{4i\pi/3}$ respectively, where the r's are again in increasing numerical order. Apart from the ranges of integration of the y's, the formula will be almost identical to the Koba-Nielsen formula for planar diagrams:

$$A=\int \frac{dy_1\ldots[dy_a]\ldots[dy_b]\ldots[dy_c]\ldots dy_n(y_b-y_a)(y_c-y_b)(y_a-y_c)(y_2-y_1)}{\prod_A(y_i-y_{i1})}$$

$$\times\prod_{i>j}(y_i-y_j)^{-2p_ip_j}\prod_A(y_i-y_{i1})^{-\mu^2}(y_2-y_1)^{3\mu^2-2M^2} \quad . \qquad (6.1)$$

The product $\prod_A$ is over all adjacent pairs y_i, y_{i1} in Figure 21(b); it is of course analogous to the product of factors $(y_{i+1}-y_i)$ in the formula for planar diagrams. The new formula has an extra factor $(y_2-y_1)^{3\mu^2-2M^2+1}$, which is due to the fact that the vertices 1 and 2 are four-particle vertices and which is necessary for projective invariance.

In our exposition we have fixed the values of y_1 and y_2 in (6.1) at 0 and infinity, but the formula can be generalized to any specification of the three arbitrary y's.[39] The integrand of (6.1) remains unchanged, and the rules for fixing the range of the y's can be stated in a projectively invariant manner.

The factorization properties of the generalized amplitude were initially investigated by Mandelstam,[41] who showed that the meson trajectories had precisely the same degeneracy as in planar amplitudes, but that the leading baryon trajectory had a degeneracy similar to that of the harmonic-oscillator quark model. The discussion of the factorization properties was completed by Frye, Lee and Susskind[43] (for $\mu^2=-1$) and by Thorn (for arbitrary μ^2)[42]. These authors generalized the operator formalism presented in the previous section to the new amplitude.

By associating the non-planar Veneziano diagrams with non-planar duality diagrams, we can construct a quark model which avoids the defects of the planar-diagram model. For elastic meson-baryon scattering, the two diagrams are shown in Figures 18(e) and (f). The s t term, corresponding to Figure 18(e), is exactly the same as previously since, if all the meson lines of Figure 21(b) are attached to one of the three quark lines, the formula (6.1) is identical to the ordinary Koba-Nielsen formula. The s u term, corresponding to Figure 18(f), is not the same as the ordinary Veneziano amplitude. If we ignore spin and isotopic-spin, it is as follows:

$$\int_0^1 dx\, x^{-\alpha(s)-1}\,(1-x)^{-\alpha(u)-1}\,\{1-x(1-x)\}^{\frac{1}{2}(s+u-M^2-\mu^2)}\quad . \tag{6.2}$$

7. Higher-order Feynman-like Diagrams

Formula for the Single-Loop Planar Diagram

In any consistent theory the Veneziano model can only be regarded as a first approximation, which one would wish to generalize, at least approximately, to systems with resonances of finite width. In other words, the

Veneziano amplitude should be regarded as the Born term of our theory. A very appealing suggestion for including further effects has been made by Kikkawa, Sakita and Virasoro,[44] who construct a series of Feynman-like diagrams which correspond to the higher-order corrections. It is unlikely that a straight perturbation approach can be applied to strong-interaction physics, even if the concept of the perturbation series is altered by the removal of elementary particles. In fact, the integration over the large phase space of intermediate states does appear to give rise to a divergence. Nevertheless, by appropriately modifying the perturbation treatment the divergence can be avoided, and the series of Feynman-like diagrams may well motivate a semi-perturbation approach to the problem.

Kikkawa, Sakita and Virasoro based their formulas for the higher-order diagrams at least partly on conjecture. One can actually deduce the formulas for the Feynman-like diagrams, however, since higher-order terms can always be calculated from lower-order terms by unitarity, as was pointed out by several groups.[45]

To obtain a diagrammatic representation of the higher-order terms, we represent the Born term by the duality diagrams of Figure 18(a) or 18(a′). The solid lines may be regarded as quarks, or they may be regarded as the boundary of a rubber sheet which represents the intermediate resonances (Virasoro, Susskind). In any case Figures 18(a) and (a′) are definitely Born diagrams, not loops. The system contains no actual quarks, and any $Q\bar{Q}$ intermediate state represents a meson. By stretching out the $Q\bar{Q}$ pair or the rubber sheet, we obtain the diagrams of Figure 2.

If we now combine two diagrams of the form of Figure 18(a) by unitarity, we obtain Figure 22(a) or (b). We can stretch out the $Q\bar{Q}$

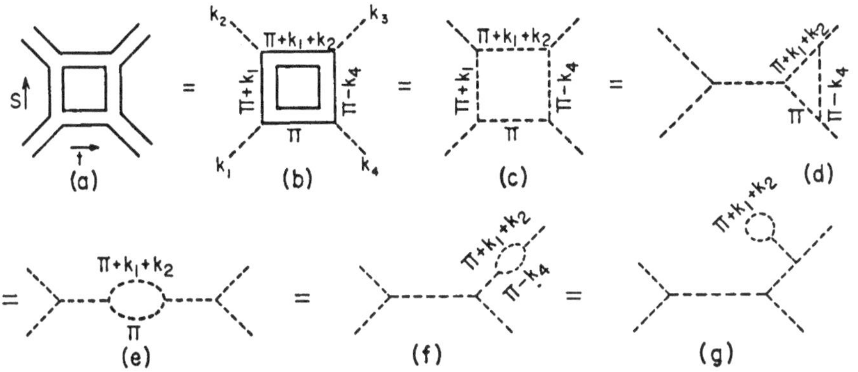

Figure 22. Alternative forms of the single-loop planar diagram.

pair or the rubber sheets to obtain any of the diagrams shown in Figures 22(c) to 22(g), as well as several others. These diagrams have two-particle intermediate states (two sheets of rubber, or two $Q\bar{Q}$ pairs). Just as with the Born term, all conventional single-loop Feynman diagrams (Figures 22(c) to 22(g)) are combined into one diagram (Figure 22(b)) in the Veneziano model. The duality transformation of Figure 17 allows us to pass from one conventional diagram to another.

As in the ordinary Feynman theory, the amplitude will be in the form of an integral over the loop momentum Π (Figure 22). The integrand will have poles in all channels which are represented by the dotted lines of Figure 22. There are six such channels, two of which are the external channels s and t, and the other four the internal-line channels with momenta Π, $\Pi+k_1$, $\Pi+k_1+k_2$ and $\Pi-k_3$. The Feynman integrand will itself be an integral over Koba-Nielsen variables.

Before we outline the derivation of the formula from unitarity we shall describe the form of the result. The original conjecture of Kikkawa, Sakita and Virasoro, made before the unitarity calculations were performed, came close to the correct formula, though there were some differences.

In the ordinary Feynman theory with spinless particles, the integrand is obtained simply by opening the loop at any point and adding a propagator $(p^2-\mu^2)^{-1}$. If we carried out the same procedure for Veneziano-type Feynman diagrams, the result would depend on the point at which the loop was opened. If we opened the loop of Figure 22(c) at the line with momentum $\Pi+k_1$, the resulting amplitude would lose the t-channel double pole of Figure 22(e) whereas, if we opened the loop at the line with momentum Π, the amplitude would retain the double pole. One can retain all the poles by repeating the external lines (Figure 23), as may be shown by applying the duality transformation of Figure 17. In order that

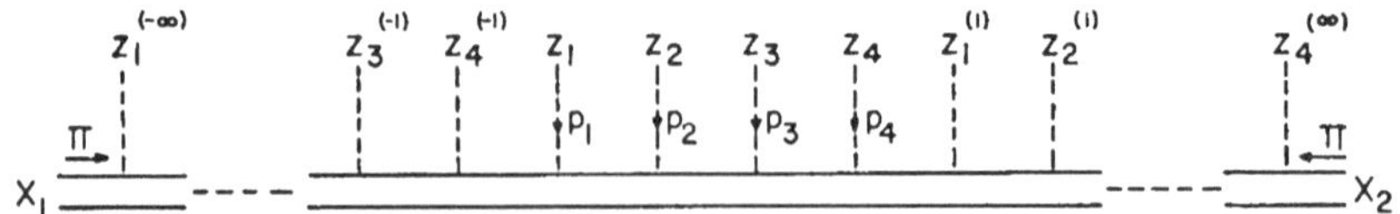

Figure 23. Opened loop of a Feynman-Like Veneziano diagram.

the diagram be symmetric in the lines 1, 2, 3, 4, it is necessary to repeat the lines an infinite number of times.

The Koba-Nielsen variables z in Figure 23 cannot all be independent. The t-channel pole in Figure 22(d) arises from the region of integration of the Koba-Nielsen variables where the cross-ratio $(z_1-z_2)(z_4^{(-1)}-z_3)/\{(z_1-z_3)(z_4^{(-1)}-z_2)\}$ vanishes. The second t-channel pole in Figure 22(e) arises from the region of integration where the cross-ratio $(z_3-z_4)(z_2-z_1^{(1)})/\{(z_3-z_1^{(1)})(z_2-z_4)\}$ vanishes. We do not desire an additional pole where the cross-ratio $(z_1^{(1)}-z_2^{(1)})(z_4-z_3^{(1)})/\{(z_1^{(1)}-z_3^{(1)})(z_4-z_2^{(1)})\}$ vanishes, otherwise the amplitude would have multiple t-channel poles of increasing order. It is therefore necessary to ensure that the cross-ratio $(z_1^{(1)}-z_2^{(1)})(z_4-z_3^{(1)})/\{(z_1^{(1)}-z_3^{(1)})(z_4-z_2^{(1)})\}$ vanishes only where the cross-ratio $(z_1-z_2)(z_4^{(-1)}-z_3)/\{(z_1-z_3)(z_4^{(-1)}-z_2)\}$ vanishes. This will be the case if the variables $z^{(r)}$ are related to $z^{(r-1)}$ by a given projective transformation P.

Our Veneziano integrand will thus be a function of seven variables: the four variables z_1, z_2, z_3 and z_4, and the three variables associated with the projective transformation. As we explained in Section 3, we can specify the projective transformation by the two points x_1 and x_2 which it leaves invariant, and by the invariant quantity w, so that the seven variables can be chosen as z_1, z_2, z_3, z_4, x_1, x_2 and w. The variables $z_i^{(r)}$ can then be obtained by applying the projective transformation P to z_i r times.

The Veneziano integrand will itself be invariant under an arbitrary projective transformation Q. Under such a transformation, the projective transformation P transforms according to (3.11). As we saw in Section 3, this implies that x_1 and x_2 transform in the ordinary way under Q, while w remains invariant. As in the case of the n-point Born diagram, we can choose the projective transformation Q so as to set three of the variables z_1, z_2, z_3, z_4, x_1 and x_2 equal to preassigned values. The amplitude for the four-point single-loop diagram will thus be an integral over four Koba-Nielsen variables (including the variable w); the n-point amplitude will similarly be an integral over n variables.

The Koba-Nielsen variables for the lines at the end of the chain in Figure (23), which are associated with the loop momenta Π and $-\Pi$, will be obtained by applying the projective transformation P an infinite number of times. Such variables will be left unchanged by the projection P, since $PP^{(\infty)}(y_i) = P^{(\infty)}(y_i)$. They will therefore simply be the points which we have called x_1 and x_2. Note that $x_1^{(r)} = x_1$, $x_2^{(r)} = x_2$.

Notice that the concept of projective transformations occurs twice when discussing the Koba-Nielsen integrand for a loop. A <u>particular</u> transformation P occurs in the definition of the variables of integration. In this respect the integrand for a loop differs from that for a tree. The loop and tree integrands resemble one another in being invariant under a <u>general</u> projective transformation Q.

We can now write down the formula for the Feynman integrand corresponding to Figure 22, but with n internal lines:

$$A=\int \frac{dz_1\ldots[d\tilde{z}_a]\ldots[d\tilde{z}_b]\ldots[d\tilde{z}_c]\ldots dz_n dx_1 dx_2 dw(\tilde{z}_b-\tilde{z}_a)(\tilde{z}_c-\tilde{z}_b)(\tilde{z}_a-\tilde{z}_c)}{(z_2-z_n^{(-1)})(z_3-z_1)\ldots(z_n-z_{n-2})(z^{(1)}-z_{n-1})(x_2-x_1)^2(z_1-x_1)(x_2-z_n)(z_1^{(1)}-x_1)^{-1}(x_2-z_n^{(-1)})^{-1}}$$

$$\times(1-w)\prod_{r=1}^{\infty}(1-w^r)^{-4}\prod_{\substack{r=0\\ j>i}}^{\infty}\prod_{\substack{r=1\\ i,j}}^{\infty}\left\{\frac{(z_j^{(r)}-z_i)(z_{j+1}^{(r)}-z_{i-1})}{(z_j^{(r)}-z_{i-1})(z_{j+1}^{(r)}-z_i)}\right\}^{-\alpha(s_{ij})-1}$$

$$\times\prod_i\left\{\frac{(z_i-x_1)(z_{i+1}-x_2)}{(z_i-x_2)(z_{i+1}-x_1)}\right\}^{-\alpha(s_i)-1} \qquad (7.1)$$

If $j=n$, $z_{j+1}^{(r)} = z_j^{(r+1)}$ and, if $i=1$, $z_{i-1}= z_n^{(-1)}$. The variables which we have denoted by $\tilde{z}$, and which are omitted from the integration, may be any of the z's or x's. The z's and x's are cyclically ordered in the sequence

$$x_1,\ldots,z_n^{(-1)},\ z_1,\ z_2,\ldots z_n,z_1^{(1)},\ldots,x_2,x_1. \qquad (7.2)$$

Note that this cyclic ordering is consistent with the identification $x_1=z^{(-\infty)}$, $x_2=z^{(\infty)}$. The range of integration of w is determined by the condition (7.2). In particular, w is always between 0 and 1 and if any of the cross-ratios involving the x's vanish, w is zero.

Let us now discuss the factors of (7.1). As usual, there is one cross-ratio for each channel of Figure (23), and the cross-ratio is raised to the power $-\alpha(s)-1$, where s is the channel in question. We never repeat the same cross-ratio more than once. For instance, the cross-ratio $(z_j^{(r)}-z_i^{(s)})(z_{j+1}^{(r)}-z_{i-1}^{(s)})/\{(z_j^{(r)}-z_{i-1}^{(s)})(z_{j+1}^{(r)}-z_i^{(s)})\}$ is equal to

$(z_j^{(r-s)}-z_i)(z_{j+1}^{(r-s)}-z_{i-1})/\{(z_j^{(r-s)}-z_{i-1})(z_{j+1}^{(r-s)}-z_i)\}$, and it is not included separately in (7.1). Similarly, the cross-ratio $(z_i^{(r)}-x_1)(z_{i+1}^{(r)}-x_2)/\{(z_i^{(r)}-x_2)(z_{i+1}^{(r)}-x_1)\}$ is equal to $(z_i-x_1)(z_{i+1}-x_2)/\{(z_i-x_2)(z_{i+1}-x_1)\}$ and is not repeated. The second product of cross-ratios in (7.1) is over the internal-line channels of Figures (22) and (23); there are n cross-ratios in all. The first product is over the external-line channels. There are an infinite number of such cross-ratios, since the external channels of Figure (23) may have any number of external lines. It can be shown from the condition of cyclic ordering that the cross-ratio associated with a channel containing n or more lines never vanishes. Accordingly there are no poles associated with such channels. Kikkawa, Sakita and Virasoro omitted most of the cross-ratios which never vanish from their original conjectured formula, but their presence follows from the unitarity calculation.

The volume element which forms the first factor of (7.1) possesses factors $(z_{i+2}-z_i)$ in the denominator, as is usual with Koba-Nielsen formulas expressed in terms of cross-ratios. It is projectively invariant, since each explicit variable of integration occurs in two net factors of the denominator, whereas every other variable occurs in zero net factors. Also, when a particular cross-ratio vanishes and one treats the variables as on pp.30-31, the volume element remains finite. The simplest possible volume element with these properties is that which occurs in (7.1).

The factor $\prod(1-w^r)^{-4}$ cannot be deduced from plausibility arguments, but it results from the unitarity calculation. It was not included in the original conjecture of Kikkawa, Sakita and Virasoro. This factor gives rise to a divergence in the integral at $w=1$; we shall discuss the significance of the divergence later in this section.

Finally, the factor $1-w$ in (7.1) is the correction factor due to the existence of linear dependences. It is interesting that this factor is precisely analogous to the Feynman term in the perturbation theory of the Yang-Mills and gravitational fields,[46] and that it arises from precisely the same source. Feynman found that the Ward identity did not prevent the circulation of scalar Yang-Mills quanta, on vector gravitons, in closed loops. The uncorrected perturbation formulas would thus violate unitarity. He showed that unitarity could be restored by including diagrams with closed loops of fictitious scalar particles (in the Yang-Mills theory) or fictitious vector particles (in gravitation).

In the present problem, the difficulty arises from the fact that the vertex Γ (Figure 13) does not vanish if the particles on the ends are spurious. Hence, if such a vertex is closed into a loop, spurious particles will circulate. The prescription for correcting for the spurious particles, originally found by Bardakci, Halpern and Shapiro,[45] is to multiply the Veneziano integrand by w and subtract the result from the original integrand. It is not difficult to show that multiplication of (7.1) by w is equivalent to decreasing all the exponents $\alpha(s_i)$, which correspond to the internal-line channels of Figure (22), by 1. In other words, from the uncorrected result we subtract the contribution due to a loop with fictitious internal trajectories one unit below the real trajectories. The fictitious trajectories resemble Feynman's fictitious scalar or vector particles.

While the factors of (7.1) which depend on w are new to the loop amplitude, the other factors are the analogs of corresponding factors in the formula for tree diagrams.

The Koba-Nielsen formula for a closed loop can be written in a form similar to Equation (3.21) for the Born term. The expression is as follows:

$$A=\int\frac{dz_1\ldots[d\tilde{z}_a]\ldots[d\tilde{z}_b]\ldots[d\tilde{z}_c]\ldots dz_n dx_1 dx_2 dw(\tilde{z}_b-\tilde{z}_a)(\tilde{z}_c-\tilde{z}_b)(\tilde{z}_a-\tilde{z}_c)}{(z_2-z_1)(z_3-z_2)\ldots(z_n-z_{n-1})(z_1^{(1)}-z_n)(x_2-x_1)^2(z_1-x_1)(z_1^{(1)}-x_1)^{-1}}(1-w)$$

$$\times\prod_{r=1}^{\infty}(1-w^r)^{-4}\,w^{-\alpha(\Pi^2)-1}\lim_{N\to\infty}\prod_{\substack{r=0\\ j>i}}\prod_{\substack{r=1\\ i,j}}^{N}(z_j^{(r)}-z_i)^{-2p_ip_j}\prod_i\left\{(z_i-x_1)^{2\Pi p_i}(z_i-x_2)^{2\Pi p_i}\right\}$$

$$\times\prod_{i=1}^{n-1}(z_{i+1}-z_i)^{-\mu^2}\left\{(z^{(1)}-z_n)(z-x_1)(z^{(1)}-x_1)^{-1}\right\}^{-\mu^2}. \tag{7.3}$$

Again, the formula for the loop bears a close resemblance to the formula for the tree. The factors whose exponents contain products of momenta are just those which would have been expected. The factors with exponents μ^2 are necessary for projective invariance. Two net factors must contain a term z_i ($1\leq i\leq n$), whereas the net number of factors with a term $z_1^{(1)}$ or x_1 must be zero.

The equivalence of formulas (7.1) and (7.3) can be proved by reasoning similar to that used for tree diagrams, though the algebra is somewhat more complicated. We shall not give the details here.

It is often convenient to choose our fixed $\tilde{z}$'s to be $x_1=0$, $x_2=\infty$, $z_1=1$. As we pointed out in Section 3, the projective transformation P is then multiplicative: $z_i^{(n)} = w^{-n} z_i$. In conformity with our usual notation we shall use y's for integration variables defined in this way. Equation (7.3) then becomes:

$$A=\int \frac{dy_2 \ldots dy_n dw}{(y_2-1)(y_3-y_2)\ldots(y_n-y_{n-1})(w^{-1}y_1-y_n)w}(1-w)\prod_{r=1}^{\infty}(1-w^r)^{-4}\, w^{-\alpha(\Pi^2)-1}$$

$$\times \lim_{N\to\infty} \prod_{\substack{r=0\\ j>i}} \prod_{\substack{r=1\\ i,j}}^{N} (w^{-r}y_j-y_i)^{-2p_ip_j} \prod_i y_i^{-2\Pi p_i} \prod_{i=1}^{n-1}(y_{i+1}-y_i)^{-\mu^2}(w^{-1}z_1-z_n)^{-\mu^2} w^{-\mu^2}. \tag{7.4}$$

The y's and w satisfy the inequalities:

$$y_1 = 1 \leq y_2 \leq \ldots \leq y_n \leq w^{-1} y_1 \,. \tag{7.5}$$

Equations (7.3) and (7.4) are of course completely equivalent. If we are not particularly interested in projective invariance, (7.4) is often simpler to work with. On the other hand the factorization properties of A in the external channels are more evident in (7.3) and, moreover, this form of the equation gives us more insight into the generalization to multi-loop diagrams, since the simplification leading to (7.4) only exists for single-loop diagrams.

The integration over the Feynman momentum in our amplitudes can be performed analytically before the integration over the Koba-Nielsen variables, which may therefore be regarded as playing the role of Feynman parameters. Needless to say, the remaining integral over the Koba-Nielsen variables is more complicated then the conventional integrals over Feynman parameters. For a loop with n external particles there are n Koba-Nielsen variables as opposed to the (n-1) Feynman parameters of the conventional loop.

Kikkawa, Sakita and Virasoro also gave a prescription for calculating an n-loop planar amplitude. They were able to sum the leading asymptotic contributions from all such amplitudes, and to show that their result had a Regge asymptotic form with a non-linear trajectory.

Derivation from Unitarity

We now outline the derivation of the formulas of the previous subsection. The complete results were first derived by Bardakci, Halpern and Shapiro.[45] The formula without the linear-dependence factor was derived simultaneously with, and independently of, these authors by Kikkawa, Sakita, Veneziano and Virasoro[45] and by Amati, Bouchiat and Gervais.[45] The last-named authors used the operator formalism of Section 5; their method is probably the simplest and most powerful, and it can be used for more complicated problems. The calculation of Amati, Bouchiat and Gervais was completed by Thorn.[20]

A direct unitarity calculation of an n-point single-loop diagram would consist in joining an (m+2)-point diagram with an (n-m+2) point diagram, and thereby calculating the imaginary part of the n-point diagram. The procedure involves a summation over two-particle intermediate states. A simpler, indirect procedure, involving a summation over a one-particle intermediate state, is to use the tree theorem: if a tree diagram (such as Figure 13(b)) is closed onto itself to form a loop, and if the resulting function is used as a Feynman integrand, the integral will possess the correct unitarity properties. The appropriate propagator must be inserted at the point where the loop is joined, and a summation must be carried out over all one-particle states.

Applied to our present problem, the tree theorem states that the Feynman integrand A for the single planar loop can be calculated by taking the trace of the operator $\tilde{D}\Gamma$, where D is the modified propagator (5.28), and Γ is the vertex (5.9) or (5.11). The trace is of course to be taken over the Hilbert space of the harmonic oscillators.

It is important that the modified propagator $\tilde{D}$ be used in the calculation, since it projects out the spurious states. The matrix element of Γ between spurious states is not zero and, if we were to calculate the trace without projecting out the spurious states, we would sum over such states as well as over real states. To indicate the method of calculation we shall begin by using the ordinary propagator D in our formulas. We shall then show how one can correct for the presence of spurious states.

From Equations (5.11), (5.10) and (5.7d), we find that we have to evaluate the following trace:

$$\int_0^1 dx_n \cdots \int_0^1 dx_1 (1-x_n)^{-\mu^2-1} \cdots (1-x_1)^{-\mu^2-1} x_n^{-\alpha(s_n)-1} \cdots x_1^{-\alpha(s_1)-1}$$

$$\times \operatorname{Tr}\left(x_n^R \exp\left\{ p_n \quad \Sigma \frac{\sqrt{2}\, a^\dagger_r}{r} \right\} \quad \exp\left\{ p_n \, \Sigma \frac{\sqrt{2}\, a_r}{r} \right\} x_{n-1}^R \cdots x_1^R \right.$$

$$\left. \times \exp\left\{ p_1 \, \Sigma \frac{\sqrt{2}\, a^\dagger_r}{r} \right\} \quad \exp\left\{ p_1 \quad \Sigma \frac{\sqrt{2}\, a_r}{r} \right\} \right) . \qquad (7.6)$$

To avoid confusion with the Koba-Nielsen variables we have replaced the z's by x's.

The trace (7.6) can be evaluated fairly simply by making use of coherent states, defined (for a single oscillator) as follows:

$$|\zeta\rangle = \exp \zeta a^{\dagger} \mid 0 \rangle. \tag{7.7a}$$

It is almost trivial to verify the properties:

$$a \mid \zeta \rangle = \zeta \mid \zeta \rangle, \tag{7.7b}$$

$$\exp (\zeta' a^{\dagger}) \mid \zeta \rangle = \mid \zeta + \zeta' \rangle, \tag{7.7c}$$

$$x^{a^{\dagger}a} \mid \zeta \rangle = \mid x \zeta \rangle, \tag{7.7d}$$

$$\langle \zeta' \mid \zeta \rangle = \exp (\zeta'^{*} \zeta) \tag{7.7e}$$

The coherent states from an overcomplete, continuous, non-orthogonal, normalizable basis, and the sum over states is given by the formula

$$\sum \mid n \rangle\langle n \mid = \frac{1}{\pi}\int d(\mathrm{Re}\ \zeta)\, d(\mathrm{Im}\ \zeta) \exp (-|\zeta|^{2}) \mid \zeta \rangle\langle \zeta \mid. \tag{7.7f}$$

Note that the integration in (7.7f) is a two-dimensional integration over the complex plane, not a one-dimensional integration over a contour.

In our present problem the coherent states will be of the form $| \zeta_1 \ldots \zeta_r \ldots \rangle$, where the variable ζ_r corresponds to the oscillator a_r. We can thus write

$$\mathrm{Tr}\ M = \left\{ \prod_{r=1}^{\infty} \frac{1}{\pi}\int d(\mathrm{Re}\ \zeta_r) d(\mathrm{Im} \zeta_r) \exp(-|\zeta_r|^{2}) \right\} \langle \zeta_1 \ldots \zeta_r \ldots | M | \zeta_1 \ldots \zeta_r \ldots \rangle. \tag{7.8}$$

After substituting the operator within the large brackets of (7.6) for M we can find the matrix element using (7.7b)-(7.7e). The result is

$$\mathrm{Tr}\ M = \left\{ \prod_{r=1}^{\infty} \frac{1}{\pi}\int d(\mathrm{Re}\ \zeta_r) d(\mathrm{Im} \zeta_r) \exp [-|\zeta_r|^{2}(1-w^{r})] \right\}$$

$$\times \exp\left\{ \sum_{i,r} \zeta_r \frac{\sqrt{2}\ p_i y_i^{-r}}{\sqrt{r}} \sum_{i,r} \zeta_r^{*} \frac{\sqrt{2}\ p_i w\ y_i^{r}}{\sqrt{r}} + \sum_{j>i} \frac{2 p_i p_j y_j^{-r} y_i^{r}}{r} \right\}, \tag{7.9a}$$

where

$$y_i = \prod_{k=1}^{i-1} (x_i)^{-1} \qquad w = \prod_{k=1}^{n} x_i . \tag{7.9b}$$

To simplify the ζ-integration, we complete the square within the exponent. Thus

$$\mathrm{Tr}\ M=\left\{\prod_{r=1}^{\infty} \frac{1}{\pi}\int d(\mathrm{Re}\ \zeta_r)d(\mathrm{Im}\ \zeta_r)\ \exp\ [-|\zeta_r|^2\ (1-w^r)]\right\}$$

$$\times \exp\left\{ \sum_{i,j,r} \frac{2p_i p_j (y_j{}^{-1} y_i w)^r}{r(1-w^r)} + \sum_{j>i} \frac{2p_i p_j y_j{}^{-r} y_i{}^r}{r} \right\} . \tag{7.10}$$

The ζ-integration of the factor $\exp[-|\zeta_r|(1-w^r)]$ gives us the factor $\prod (1-w^r)^{-4}$ in (7.4). The factor $(1-w^r)$ in the second exponential of (7.10) can be expanded in powers of w; the exponent is then equal to a sum of logarithms. On substituting (7.10) in (7.6) and making the change of variables (7.9b), we obtain the result (7.4).

Now let us indicate how this result can be corrected for linear dependences. By (5.11) and (5.28), the trace to be evaluated is

$$\mathrm{Tr}\left\{ D(R,s_n)(1-P_S)V\ D(R,\ s_{n-1})(1-P_S)\ldots D(R,s_1)(1-P_S)V \right\}. \tag{7.11}$$

Using the formula (5.27a) for P_S, and the relation (5.22), which enables us to commute the operator A through the operator $V\ D(R,s_i)$, we can show at once that

$$(1-P_S)V\ D(R,s_{n-1})(1-P_S)=V\ D(R,s_{n-1})(1-P_S).$$

All factors $(1-P_S)$ except one can thus be removed from (7.11), and we have to subtract the expression

$$\mathrm{Tr}\left\{D(R,s_n)\ P_S V\ D(R,s_{n-1})\ V\ldots D(R,s_1)\ V\right\}$$

from the trace calculated with neglect of linear dependences. The factor A in the expression for P_S is next commuted through each factor V D(R,s) in turn. By (5.22), it changes the factor to V D(R+1,s). Ultimately, when the factor A has been commuted right round the cycle, the trace becomes:

$$\mathrm{Tr}\left\{D(R+1,\ s_n)\ A(A^{\dagger}+m^2)\ [A(A^{\dagger}+m^2)]^{-1}V\ D(R+1,s_{n-1})\ldots D(R+1,s_1)V\right.$$

$$= \mathrm{Tr}\left\{D(R+1,\ s_n)\ V\ D(R+1,\ s_{n-1})\ldots D(R+1,s_1)\ V.\right. \qquad (7.12)$$

The expression (7.12) is exactly the same as the trace calculated without regard to linear dependences, except that each term $\alpha(s_i)$ in the exponents of the D's is replaced by $\alpha(s_i)-1$. From (7.4) and (7.9b), it is at once evident that such a change in the α's is equivalent to multiplying the Veneziano integrand by a factor w. We thereby obtain the linear-dependence correction factor quoted above, and the derivation of (7.4) has been completed.

Non-Planar Diagrams

We emphasized earlier in this section that all conventional fourth-order perturbation diagrams are included in a single Veneziano diagram. On the other hand there exist other Veneziano diagrams which have no analog in ordinary perturbation theory. The diagrams which we have examined up to this point were obtained by combining two st Veneziano Born terms by unitarity in the t-channel. If we combine two su diagrams, or an st and a su diagram, by unitarity in the t-channel, we obtain a new, non-planar single-loop diagram. An example is shown in Figure (24a)-(24c) where we have combined two su diagrams by unitarity in the

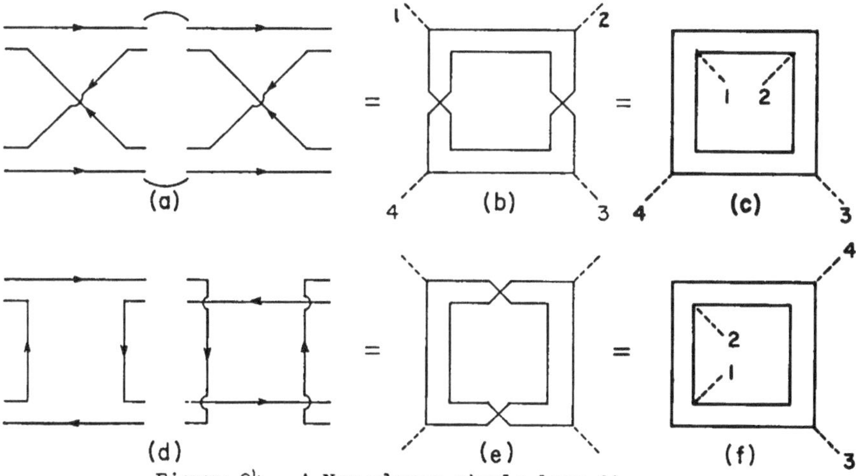

Figure 24. A Non-planar single-loop diagram.

t-channel. Another way of constructing the diagram is shown in Figure (24d)-(24f). Here we combine two st diagrams by unitarity in the t-channel, as we did with the planar diagram, but the quarks in the two diagrams circulate in opposite directions.

When we quote the prescription for calculating non-planar amplitudes we shall see that Figures 24(c) and 24(f), which can be deformed into one another without crossing meson lines attached to the same quark line, are different forms of the same diagram. The imaginary part of the amplitude contains contributions from Fig.24(a) (t-channel) and Fig.24(d) (s-channel), as well as from several other processes which lead to the same diagram. The only non-planar single-loop four-point diagram which is different from Figure 24 is shown in Figure 25.*

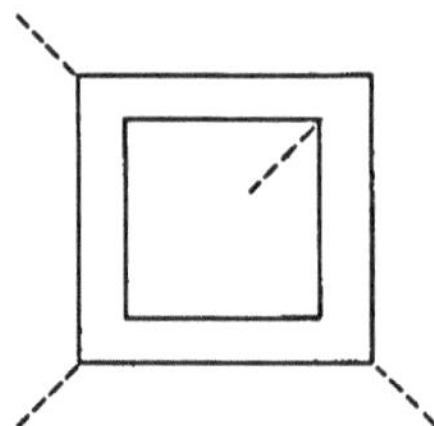

Figure 25. Another non-planar single-loop diagram.

The rules for calculating amplitudes corresponding to non-planar diagrams were first conjectured by Kikkawa, Klein, Sakita and Virasoro.[47] Again their conjecture resembled the correct result in its general form, though it did not agree with it exactly. A unitarity calculation was performed by Galli, Gallardo and Susskind for the case $\mu^2 = -1$. The result of this calculation approximated further to the correct result, but it was still not completely correct, since Galli, Gallardo and Susskind did not take linear dependences into account and, in fact, their method of factorization introduced a new set of redundant states. A complete unitarity calculation was performed by Kaku and Thorn.[19,49] They used a method analogous to that of Amati, Bouchiat and Gervais, except that the vertex Γ

*In a theory without a conserved net quark number, we can also have a non-orientable diagram with only one twist.[29]

was replaced by a twisted vertex. With the use of techniques explained in Section 5, they were able to reduce the twist operators to a single factor of the form $e^{u A^\dagger}$. They then employed similar techniques to commute the operator $e^{u A^\dagger}$ right round the trace and replace it by $e^{u' A^\dagger}$ with $u' < u$. At the same time, the variables of integration in the propagators were changed. By repeating the procedure an infinite number of times, Kaku and Thorn got rid of the operators $A^\dagger$ completely. They could then calculate the trace in the same way as Amati, Bouchiat and Gervais.

The result found by Kaku and Thorn resembles Equation (7.4); by projective invariance it can be rewritten in a form similar to (7.3). Instead of the factor 1-w in (7.4) or (7.3), we now have a factor $(1-w)^2$*. The other and more crucial respect in which the formula for a non-planar loop differs from that for a planar loop lies in the range of integration of the y's or z's. The variables associated with the mesons attached to the inner and outer quark lines of the loop are separately arranged in cyclic order, but no restriction is made on the variables associated with the two sets of mesons relative to one another.

One further restriction must be made on the range of integration. It is not very difficult to see that the integrand of (7.4) remains invariant under the change of variables

$$y_{i,I}' = w^{-1} y_{iI}, \tag{7.13a}$$

$$\Pi' = \Pi + \sum_I p_i. \tag{7.13b}$$

---

*The calculation of Galli, Gallardo and Susskind, with its extra redundant states, possessed no factor 1-w. A calculation performed without eliminating the usual spurious states leads to a single factor 1-w.

where the variables $y_{i,I}$ are those associated with the mesons attached to the inner loop of Figures 24(c) or (f) or Figure 25, and $\sum_I p_i$ represents the sum of the momenta of the mesons attached to the inner loop. More generally, the integrand of (7.2) remains invariant under the change of variables:

$$z_{i,I}' = z_{i,I}^{(1)} , \tag{7.14a}$$

$$\Pi' = \Pi + \sum_I p_i . \tag{7.14b}$$

If the loop of Figure 24 is opened and the external lines repeated indefinitely (Figure 26), the transformation (7.13) or (7.14) corresponds to moving the lower meson lines through one cycle. We certainly would

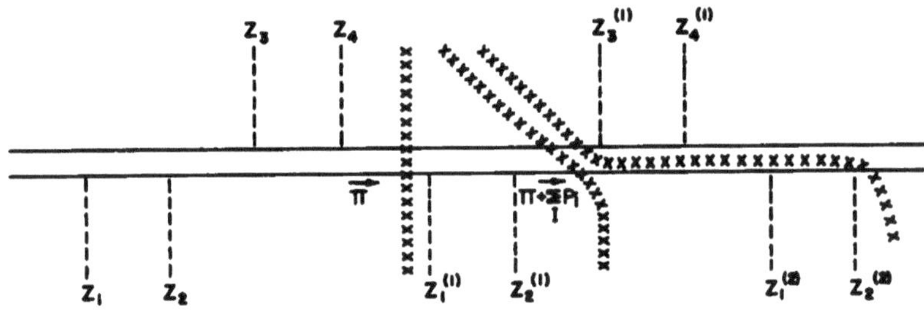

Figure 26. Infinite opened loop corresponding to Figure 24.

not expect to integrate over the entire range of variables, since we would then be integrating a periodic function over an infinite number of periods. The correct prescription is either to integrate over one period of the Feynman momentum Π :[19]

$$\Pi_1 < \Pi < \Pi_1 + \sum p_i , \tag{7.15}$$

or to integrate over one period of the Koba-Nielsen variables:

$$z_{i,1} < z_i < z_{i,1}^{(1)} \quad \text{for any one } z_i. \tag{7.16}$$

The amplitude for a non-planar diagram can also be written in a form analogous to (7.1); again there is one cross-ratio for each channel of Figure 26, and equal cross-ratios are not repeated. Now, however, there are an infinite number of channels involving the loop momentum Π, the channel variables s being of the form $(s+m \sum_I p_i)^2$. Examples of such channels have been denoted by crossed lines in Figure 26.

From the unitarity calculation we can confirm that the amplitude represented in Fig.24(c) or (f) has imaginary parts corresponding to Fig.24(a) (t-channel and Fig.24(d) (s-channel). If the loops of Figure 24(c) or Figure 24(f) are opened at the bottom, the resulting Born diagrams are the same in both cases, since the cyclic ordering of the particles is the same, (Figure 27).

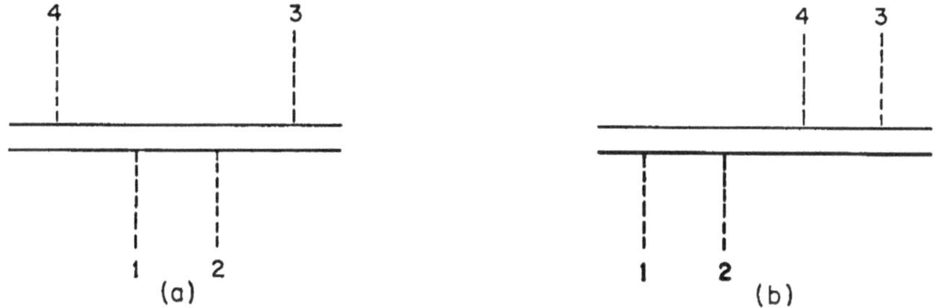

Figure 27. Opened loops corresponding to Figures 24(c) and (f).

The amplitude will possess poles in the channels shown in Figure 27(a), as well as those shown in Figure 27(b). When the integral over the Feynman momentum is performed, the poles corresponding to Figure 27(a) will give the unitarity contribution of Figure 24(a), while those corresponding to Figure 27(b) will give the unitarity contribution of Figure 24(f).

One interesting point about non-planar diagrams is that they give rise to cuts in the j-plane. Figure 24(b) is obtained by combining two su terms by unitarity in the t-channel, and the amplitude corresponding to it would be expected to possess angular-momentum cuts in the t-channel. This has been confirmed by Kikkawa.[50] Figure 24(b) is considerably simpler than any conventional Feynman diagram with j-plane cuts, and it might be hoped that the calculation of the effects of such cuts will now become more feasible.

The non-planar loops of Figures (24) and (25) have been constructed from planar tree diagrams. In addition one can construct loops from the non-planar baryon amplitudes given in the previous section. There exist baryon loops with external mesons, as well as loops with external baryons. A prescription for calculating such diagrams has been given by Thorn.[42]

The Divergence Problem

The factor $\prod_{r=1}^{\infty} (1-w^r)^{-4}$ in (7.1), (7.3) and (7.4) is highly divergent near w=1, the upper limit of integration. At the time of writing physicists are not agreed on the reason for the divergence. It may simply indicate that we have used a bad method to calculate the amplitude. However, it is very plausible that the divergence is due to the rapid growth of the number of resonances with energy. The imaginary part of an amplitude due to two-particle intermediate states may well increase exponentially as the energy is increased. Since the large number of resonances at high energy corresponds to the large volume of phase space, one can associate the divergence with the exponentially increasing volume of phase space which occurs in the integration over intermediate states. Once the imaginary part of a scattering amplitude increases exponentially, the real part is not determined by the analytic properties. It would of course be possible to test this conjecture for the origin of the divergence by calculating the imaginary part of the loop amplitude, but so far the calculation has been to complicated to perform.

If the reason for the divergence is the exponentially increasing level density, one would have to modify the perturbation approach. Instead of calculating the exact imaginary part in n^{th} order and letting n approach infinity, one would calculate an amplitude whose imaginary part was correct up to the mth threshold. One would then sum the perturbation series (in principle!) so as to obtain a unitary amplitude. Finally, one would let m to infinity. The amplitude as given by (7.1), (7.2) or (7.4) appears to be almost tailor made for this procedure since, by expanding the function $\Pi(1-w^r)^{-4}$ in powers of w and keeping powers up to w^m, one does obtain an amplitude with a correct imaginary part up to the mth threshold, and at the same time one avoids the divergence. The limiting procedure outlined above is no doubt more complicated than a straight perturbation expansion, but it would in any case be somewhat surprising if the physics of strong interactions yielded to an unmodified perturbation treatment. In particular, it is unlikely that effects due to the large volume of phase space at high energy, such as the diffraction peak, could be calculated by perturbation theory.

A very interesting suggestion for treating the divergence problem has been made by Neveu and Sherk and amplified by Burnett, Gross, Neveu, Scherk and Schwarz.[51] They show that the divergence can be removed by subtracting a counter term, which can be interpreted as a renormalization. However, it is always possible to obtain a finite scattering amplitude with an exponentially increasing imaginary part. The crucial question is that of uniqueness of the amplitude or equivalently, of the asymptotic behavior at high energy. Until such questions have been answered, we believe that any procedure for making the amplitude finite by subtraction should be treated with a certain amount of caution.

References

1. For a further discussion of infinitely rising Regge trajectories, see S. Mandelstam, 1966 Tokyo Summer Lectures in Theoretical Physics, Part 11 (Syokabo, Tokyo and Benjamin, New York, 1967).
2. For calculations performed in this spirit before the discovery of the Veneziano formula, see: M. Ademollo, H. R. Rubinstein, G. Veneziano and M. A. Virasoro, Phys. Rev. Letters 14, 1402 (1967); Phys. Letters 27B, 99 (1968); Phys. Rev. 176, 1904 (1968); S. Mandelstam, Phys. Rev. 166, 1539 (1968); C. Schmid, Phys. Rev. Letters 20, 628 (1968); C. Schmid and J. Yellin, Phys. Letters 27B, 19 (1968); Phys. Rev. 182, 1449 (1969).
3. R. Dolen, D. Horn and C. Schmid, Phys. Rev. Letters 19, 402 (1967); Phys. Rev. 166, 1768 (1968).
4. G. Veneziano, Nuovo Cimento 57A, 190 (1968).
5. M. A. Virasoro, Phys. Rev. 177, 2309 (1969); S. Mandelstam, Phys. Rev. 183, 1374 (1969).
6. K. Bardakci and H. Ruegg, Phys. Letters 28B, 342 (1968); M. A. Virasoro, Phys. Rev. Letters 22, 37 (1969).
7. K. Bardakci and H. Ruegg, Phys. Rev. 181, 1884 (1969); H. M. Chan and S. T. Tsou, Phys. Letters 28B, 485 (1969); C. G. Goebel and B. Sakita, Phys. Rev. Letters 22, 257 (1969); Z. Koba and H. B. Nielsen, Nuclear Physics B10, 633 (1969).
8. Z. Koba and H. B. Nielsen, Nuclear Physics B12, 517 (1969).
9. K. Bardakci and S. Mandelstam, Phys. Rev. 184, 1640 (1969); S. Fubini and G. Veneziano, Nuovo Cimento 64A, 811 (1969).

10. R. Hagedorn, Nuovo Cimento 56A, 1027 (1968).

11. R. C. Brower and J. H. Weis, L.R.L. report no. UCRL 19220, (1969).

12. M. A. Virasoro, to be published.

13. K. Bardakci, M. B. Halpern and J. Shapiro, Phys. Rev. 185, 1910 (1969).

14. S. Fubini, D. Gordon and G. Veneziano, Phys. Letters 29B, 679 (1969).

15. L. P. Yu, to be published.

16. Y. Nambu, to be published; L. Susskind, Phys. Rev. D 1, 1182 (1970).

17. F. Gliozzi, Lettere al Nuovo Cimento 2, 846 (1969).

18. S. Fubini and G. Veneziano, to be published.

19. M. Kaku and C. B. Thorn, to be published.

20. C. B. Thorn, Phys. Rev., to be published.

21. C. B. Chiu, S. Matsuda and C. Rebbi, Phys. Rev. Letters 23, 1526 (1969).

22. R. Brower and J. H. Weis, Lettere al Nuovo Cimento 3, 285 (1970).

23. F. Gliozzi, private communication to C. B. Thorn.

24. L. Caneschi, A. Schwimmer and G. Veneziano, Phys. Letters 30B, 356 (1969).

25. S. Sciuto, Lettere al Nuovo Cimento 2, 411 (1969).

26. L. Caneschi and A. Schwimmer, Lettere al Nuovo Cimento 3, 213 (1970).

27. M. Kaku and L. P. Yu, to be published.

28. D. J. Gross and J. H. Schwarz, to be published.

29. D. J. Gross, A. Neveu, J. Scherk and J. H. Schwarz, to be published.

30. G. Carbone and S. Sciuto, Lettere al Nuovo Cimento 3, 246 (1970); J. M. Kosterlitz and D. A. Wray, to be published.

31. S. Mandelstam, Phys. Rev. 184, 1625 (1969); K. Bardakci and M. B. Halpern, Phys. Rev. 183, 1456 (1969).

32. R. H. Capps, Phys. Rev. 168, 1731 (1968).

33. H. Harari, Phys. Rev. Letters 22, 562 (1969);

J. Rosner, Phys. Rev. Letters 22, 689 (1969); T. Matsuoka, K. Ninomiya and S. Sawada, Prog. Theor. Phys. 42, 56 (1969); D. Neville, Phys. Rev. Letters 22, 494 (1969).

34. R. Carlitz and M. Kisslinger, Phys. Rev. Letters 24, 186 (1970).

35. K. Bardakci and M. B. Halpern, Phys. Rev. Letters 24, 428 (1970).

36. S. Mandelstam, Phys. Rev., to be published.

37. J. Rosner, Phys. Rev. Letters 21, 950 (1968).

38. S. Mandelstam, Phys. Rev., to be published.

39. J. Mandula, C. Rebbi, R. C. Slansky, J. Weyers and G. Zweig, Phys. Rev. Letters 22, 1147 (1969).

40. The degeneracy of the harmonic-oscillator quark states has been emphasized by P. G. O. Freund and R. Waltz, to be published.

41. S. Mandelstam, Phys. Rev., to be published.

42. C. B. Thorn, private communication.

43. G. Frye, C. W. Lee and L. Susskind, to be published.

44. K. Kikkawa, B. Sakita and M. A. Virasoro, Phys. Rev. 184, 1701 (1969).

45. K. Bardakci, M. B. Halpern and J. A. Shapiro, Phys. Rev. 185, 1910 (1969); K. Kikkawa, B. Sakita, G. Veneziano and M. A. Virasoro, note added in proof to reference 44; D. Amati, C. Bouchiat and J. L. Gervais, Lettere al Nuovo Cimento 2, 399 (1969).

46. R. P. Feynman, Acta Physica Polon. 24, 697 (1963).

47. K. Kikkawa, S. A. Klein, B. Sakita and M. A. Virasoro, to be published.

48. J. C. Gallardo, E. Galli and L. Susskind, Phys. Rev. D 1, 1189 (1970).

49. C. B. Thorn, to be published.

50. K. Kikkawa, Phys. Rev. 187, 2249 (1969).

51. A. Neveu and J. Scherk, to be published; T. Burnett, D. J. Gross, A. Neveu, J. Scherk and J. H. Schwarz, to be published.

Dynamic and Algebraic Symmetries

Steven Weinberg
Massachusetts Institute of Technology
Cambridge, Massachusetts

CONTENTS

1. INTRODUCTION

In George Orwell's novel, 1984, the hero is employed by the Ministry of Truth in the task of re-writing history. Any lecturer or author who attempts to summarize progress in an area of theoretical physics is faced with a similar task. The actual historical development of a physical theory is always confused by false starts, theoretical misapprehensions, experimental errors, and the play of personalities. To make sense out of all this, one has to go back and re-write history according to one's best understanding of the underlying logic of the subject. The result is not good history, but it sometimes makes sense, in a way that history rarely does.

The history of current algebra is particularly complicated, because at least three connected developments have been going on at the same time: (1) We have been learning about the currents of the weak interactions. We may recall here the realization by Feynman and Gell-Mann, and Marshak and Sudarshan, that the β-decay interaction has the V, A form; the suggestion by Feynman and Gell-Mann that the vector current is conserved; the derivation of the pion lifetime by Goldberger and Treiman; the explanation of the Goldberger-Treiman relation in terms of a "partially conserved axial-vector current" by Gell-Mann and Levy, Bernstein, Fubini, Gell-Mann, and Thirring, and others; the suggestion by Gell-Mann that the V and A densities form an $SU(2) \times SU(2)$ or $SU(3) \times SU(3)$ algebra; the development by Fubini and Furlan and Gell-Mann and Dashen of a technique for saturating the current commutation relations at infinite momentum; and the calculation of g_A/g_V by Adler and Weisberger. (2) As more and more was learned about the weak interaction currents, it became increasingly clear that the strong interactions must possess an underlying approximate symmetry, chiral SU(2) × SU(2) or SU(3) × SU(3), whose currents are to be identified with the V and A currents of the weak interactions. Here one can mention the remark of Nambu, that a vanishing pion mass would allow an exactly conserved axial-vector current and hence a chiral-invariant Lagrangian; the conjecture of Goldstone, proved by Goldstone, Salam, and myself, that an invariance of the Lagrangian which is not an invariance of the vacuum requires the presence of particles of zero mass; the development of soft pion theorems by Nambu and his co-workers; and the re-interpretation of the Adler-Weisberger sum rule as a formula for pion scattering lengths by Tomozawa and myself. (3) Finally, while all this was going on, the techniques developed to deal with these problems were being ploughed back into hadron electrodynamics, and used to derive sum rules, such as those of Drell and Hearn and Cabibbo and Radicati, which could actually have been derived ten years earlier without waiting for the development of current algebra.

In order to give these developments a logical coherence that historically they lacked, I will concentrate on one aspect of the subject, the exploration of symmetries of the Lagrangian which are not symmetries of Hilbert space. Symmetries of physical Hilbert space, are manifested through algebraic conditions on the S-matrix, and are therefore called algebraic symmetries. Symmetries of the Lagrangian which are not symmetries of Hilbert space are manifested through low energy theorems for massless bosons, and are called dynamic symmetries. I will first discuss the simplest and best-known example of an interplay between algebraic and dynamic symmetries, that of electrodynamics, where exact results can be obtained and our discussion need not be confused by the need for approximations. Next, I will discuss general local dynamic symmetries, and finally, will apply the results obtained to the chiral invariant theory of pions. By concentrating on the algebraic and dynamic symmetries of the strong and electromagnetic interactions, I will be neglecting almost all aspects of the weak interactions. I will also not discuss the transformation properties of the terms in the Lagrangian which break chiral symmetry, and hence will not be able to go into the problem of pion-pion scattering.

Implicit in these lectures is the assumption that the subject can be divided along these lines, and in particular, that the symmetry properties of the strong interactions can be understood without reference to the weak interactions. This approach cannot yet be said to be absolutely validated by experimental evidence, and has recently been subject to a vigorous attack by the "weak PCAC" group at Rockefeller. However, the interpretation of current algebra in terms of a dynamic symmetry of the strong interactions has an internal consistency, and, if I may say so, beauty, which merits at least an attempt at a coherent, ahistorical, presentation.

2. HADRON ELECTRODYNAMICS

A. Gauge Invariance of the First Kind

Any Lagrangian which respects the conservation of charge will automatically be invariant under gauge transformations of the first kind, which transform any field ψ according to the rule $\psi(x) \to e^{ie\Lambda}\psi(x)$. (2.A.1)

Here e is the charge destroyed by ψ, and Λ is an arbitrary real number. In particular, the electromagnetic field A^μ carries no charge, so it is invariant under such transformations:

$$A^\mu(x) \longrightarrow A^\mu(x) \quad . \tag{2.A.2}$$

From the invariance of the Lagrangian under the transformation (2.A.1), (2.A.2), we can readily derive information about the electromagnetic current. For Λ infinitesimal, we have $0 = \delta L = \sum_\psi \left[\frac{\partial L}{\partial \psi} ie\Lambda\psi + \frac{\partial L}{\partial(\partial_\mu\psi)} ie\Lambda\partial_\mu\psi\right]$.

By using the Euler-Lagrange equations $\frac{\partial L}{\partial \psi} = \partial_\mu\left(\frac{\partial L}{\partial(\partial_\mu\psi)}\right)$ we find $0 = i\Lambda\sum_\psi \partial_\mu\left[\frac{\partial L}{\partial(\partial_\mu\psi)} e\psi\right]$.

Thus gauge invariance of the first kind is entirely equivalent to the conservation of current: $0 = \partial_\mu J^\mu$ (2.A.3)

where $J^\mu \equiv -i\sum_\psi \frac{\partial L}{\partial(\partial_\mu\psi)} e\psi$. (2.A.4)

From (2.A.3), it follows that $\frac{dQ}{dt} = 0$ (2.A.5)

where Q is the charge operator, $Q \equiv \int J^0 d^3x$ (2.A.6)

(The reason for this particular normalization of J^μ and Q will be made clear in Section D.) The time-independence of Q implies that Q commutes with the Hamiltonian, and hence also with the S-matrix, $[Q, S] = 0$. (2.A.7)

My purpose in boring you with this extremely familiar material is to present one example of an <u>algebraic symmetry</u>, an invariance principle of the Lagrangian which is manifested in algebraic conditions on the S-matrix.

B. Gauge Invariance of the Second Kind

Now suppose that the Lagrangian is also invariant under gauge transformations of the second kind, in which Λ in Eq. (2.A.1) is an arbitrary function of space-time position:

$$\psi(x) \to e^{ie\Lambda(x)}\,\psi(x) \qquad (2.B.1)$$

and A_μ undergoes the transformation

$$A_\mu(x) \longrightarrow A_\mu(x) + \partial_\mu \Lambda(x) \qquad (2.B.2)$$

The change in the Lagrangian for Λ infinitesimal is now

$$\begin{aligned} \delta L = {} & \sum_\psi \left[\frac{\partial L}{\partial \psi}\, ie\Lambda\psi + \frac{\partial L}{\partial(\partial_\mu \psi)}\, ie\,\partial_\mu(\Lambda\psi) \right] + \\ & + \frac{\partial L}{\partial A_\mu}\,\partial_\mu \Lambda + \frac{\partial L}{\partial(\partial_\nu A_\mu)}\,\partial_\nu\partial_\mu\Lambda \\ = {} & -J^\mu \partial_\mu \Lambda + \frac{\partial L}{\partial A_\mu}\,\partial_\mu\Lambda + \frac{\partial L}{\partial(\partial_\nu A_\mu)}\,\partial_\nu\partial_\mu\Lambda \end{aligned}$$

so the Lagrangian is invariant under (2.B.1) and (2.B.2) if and only if

$$\frac{\partial L}{\partial A_\mu} = J^\mu \qquad (2.B.3)$$

$$\frac{\partial L}{\partial(\partial_\nu A_\mu)} = -\frac{\partial L}{\partial(\partial_\mu A_\nu)} \qquad (2.B.4)$$

Thus, what we learn by adding gauge invariance of the second kind to gauge invariance of the first kind is a set of prescriptions for coupling the electromagnetic potential to charged fields. Taken together, Eqs. (2.B.3) and (2.B.4) require that the Lagrangian be formed out of various fields ψ, their "gauge-covariant derivatives"

$$D_\mu \psi \equiv \partial_\mu \psi - ie A_\mu \psi \qquad (2.B.5)$$

and the gauge-invariant derivative of the electromagnetic field

$$F_{\mu\nu} \equiv \partial_\mu A_\nu - \partial_\nu A_\mu \,. \qquad (2.B.6)$$

Under gauge transformations of the second kind, $D_\mu \psi$ behaves like ψ,

$$D_\mu \psi(x) \longrightarrow e^{ie\Lambda(x)}\, D_\mu \psi(x) \qquad (2.B.7)$$

and $F_{\mu\nu}$ is invariant $F_{\mu\nu}(x) \longrightarrow F_{\mu\nu}(x)$. (2.B.8)

Thus any Lagrangian constructed solely from ψ , $D_\mu \psi$, and $F_{\mu\nu}$ is invariant under gauge transformations of the second kind if it is invariant under gauge transformations of the first kind.

We saw in the previous section, that the invariance of the Lagrangian with respect to gauge transformations of the first kind led to the existence of a conserved current J^μ and thence to an algebraic condition (2.A.7) on the S-matrix. Is the same true for gauge invariance of the second kind? For any gauge function $\Lambda(x)$, we certainly can construct a conserved current

$$J^\mu_\Lambda \equiv -i \sum_\psi \frac{\partial L}{\partial(\partial_\mu \psi)} e \Lambda \psi - \frac{\partial L}{\partial(\partial_\mu A_\nu)} \partial_\nu \Lambda$$

or

$$J^\mu_\Lambda = J^\mu \Lambda - \frac{\partial L}{\partial(\partial_\mu A_\nu)} \partial_\nu \Lambda \qquad (2.B.9)$$

Note that this is conserved, because (2.B.3), (2.B.4) , and the Euler-Lagrange equation for A_μ allow (2.B.9) to be written

$$J^\mu_\Lambda = \partial_\nu \left\{ \frac{\partial L}{\partial(\partial_\nu A_\mu)} \Lambda \right\} \qquad (2.B.10)$$

and conservation of J^μ_Λ then follows from the antisymmetry in μ and ν of the quantity in brackets. It thus appears that we can construct an infinite class of charges ,

$$Q_\Lambda \equiv \int J^0_\Lambda \, d^3x \, , \qquad (2.B.11)$$

which are conserved if $\Lambda(\infty)$ is bounded. However, this variety is an illusion, for (2.B.10) and (2.B.8) give

$$J^0_\Lambda = \nabla \cdot (\mathcal{E} \Lambda) \qquad (2.B.12)$$

where

$$\mathcal{E}^i \equiv \frac{\partial L}{\partial(\partial_i A_0)} \qquad (2.B.13)$$

If Λ vanishes at large distances, then Q_Λ itself vanishes, while if Λ approaches some finite constant λ at infinity, then Q_Λ is just λQ, not a new "charge".

The moral of this little exercise is that gauge invariance of the second kind, although a legitimate symmetry of the Lagrangian, does not lead to any new algebraic properties of the S-matrix, but only tells us how the electromagnetic field may enter in the Lagrangian. Such invariance principles are called dynamic symmetries.

How are we to use a dynamic symmetry like gauge invariance of the second kind? One approach is to use the symmetry to construct a Lagrangian, and then use the Lagrangian in a perturbative calculation. For instance, in quantum-electrodynamics one writes

$$L_{QED} = -\bar{\psi}[\gamma^\mu D_\mu + m]\psi - \tfrac{1}{4}F^{\mu\nu}F_{\mu\nu} \qquad (2.B.14)$$

and then expands in the charge e carried by the electron field ψ. This Lagrangian is not unique, for gauge invariance of the second kind would not forbid us from adding an additional term

$$L_{PAULI} = -\mu\bar{\psi}\sigma^{\mu\nu}\psi F_{\mu\nu}, \qquad (2.B.15)$$

a possibility to which I shall return later on. However there is no doubt that the perturbative calculations based on (2.B.14), without any Pauli term, do a really spectacular job of accounting for the observed properties of photons and electrons.

Unfortunately, this direct approach fails us when we try to deal with hadrons, whose interactions are too strong to allow the use of perturbation theory. In order to use the invariance of the Lagrangian under gauge invariance of the second kind to say something useful about the electromagnetic interactions of hadrons, we need to first use the invariance of the Lagrangian to derive properties of the electromagnetic current J^μ, and then use these properties to derive soft-photon theorems which are valid to all orders in the strong interactions. This procedure is the essence of what has come to be called current algebra.

C. The LSZ Theorem - A Review

The current algebra program described above is usually implemented by manipulation of time-ordered products of current and field operators. In order to make contact with physics, we need a theorem, originally proved by Lehmann, Symanzik, and Zimmerman,[1] which relates the residues of the poles in the Fourier transforms of matrix elements of time-ordered products to physical matrix elements. The LSZ proof was restricted to certain simple fields, such as Dirac or Klein-Gordon fields, whose free-field equations are known. The derivation given here is completely general, and makes no mention of the spin of the fields involved.

We consider a matrix element

$$M(p) = \int d^4x\, e^{-ip\cdot x} \langle 0 | T\{ A(x)\, B(y)\, C(z) \cdots \} | 0 \rangle \qquad (2.C.1)$$

Here $A, B, C, \cdots$ are arbitrary local operators (not necessarily canonical fields) , $|0\rangle$ is the vacuum state, p^μ is any momentum four-vector, and T denotes a time-ordered product, in which the operators A , B , C , $\cdots$ are arranged from left to right in order of decreasing time argument. (A minus sign is inserted for each permutation of fermion fields.)

Let us suppose that $A(x)$ has the correct quantum numbers to destroy a set of single particle states $|\mathbf{p},a\rangle$ of mass m_A. That is,

$$\langle 0|A(x)|\mathbf{p},a\rangle \neq 0 .$$

Here a stands for all the discrete indices, including helicity, which distinguish the various degenerate single-particle states of mass m_A destroyed by A. Let us consider the behaviour of M as p^μ approaches the A mass shell,

$$p^2 \longrightarrow -m_A^2 \quad , \quad p^0 > 0 \quad . \tag{2.C.2}$$

Among the various terms which contribute to M, there is one specially interesting time-ordering in which $A(x)$ stands furthest to the left. Isolating this term, we have

$$M(p) = \int d^4x\, e^{-ip\cdot x}\langle 0|A(x)T\{B(y),C(z),\cdots\}|0\rangle\,\theta(x^0 - \max\{y^0,z^0,\cdots\}) + \\ + \text{O.T.} \tag{2.C.3}$$

Here θ is the usual step function $[\theta = 1$ for positive argument, and $\theta = 0$ for negative argument$]$ and O.T. is an elegant piece of notation standing for "other terms". We can insert a complete set of intermediate states between A and the remaining time-ordered product, so that

$$M(p) = \sum_n \int d^4x\, e^{-ip\cdot x}\langle 0|A(x)|n\rangle\langle n|T\{B(y),C(z),\cdots\}|0\rangle\,\theta(x^0 - \max\{y^0,z^0,\cdots\}) + \\ + \text{O.T.} \tag{2.C.4}$$

The x-integral can now be carried out, because

$$\langle 0|A(x)|n\rangle = \langle 0|A(0)|n\rangle\, e^{ip_n\cdot x}$$

and

$$\int d^4x\, e^{i(p_n-p)\cdot x}\, \theta(x^0-t) = \frac{-i(2\pi)^3}{p_n^0-p^0-i\epsilon}\, \delta^3(\mathbf{p}_n-\mathbf{p})\, e^{-i(p_n^0-p^0)t}$$

(As usual, ϵ is a positive infinitesimal.) The contribution of a single-particle state of mass m_A in (2.C.4) thus has a pole at $p^0 = \sqrt{\mathbf{p}^2+m_A^2}$ while no other state $|n\rangle$ and none of the "other terms" arising from other time-orderings, can have such a pole. Thus the pole dominates in the limit (2.C. 2) , and we have

$$M(p) \to \frac{-i(2\pi)^3}{\sqrt{\mathbf{p}^2+m_A^2}-p^0-i\epsilon} \sum_a \langle 0|A(0)|\mathbf{p},a\rangle\langle \mathbf{p},a|T\{B(y),C(z),\cdots\}|0\rangle$$

This can also be written in a more familiar form:

$$M(p) \to \frac{-2i(2\pi)^3\sqrt{\mathbf{p}^2+m_A^2}}{p^2+m_A^2-i\epsilon} \sum_a \langle 0|A(0)|\mathbf{p},a\rangle\langle \mathbf{p},a|T\{B(y),C(z),\cdots\}|0\rangle \tag{2.C.5}$$

(The factor $2(2\pi)^3\sqrt{\mathbf{p}^2+m_A^2}$ appears here because we use a non-covariant normalization

$$\langle \mathbf{p}',a'|\mathbf{p},a\rangle = \delta^3(\mathbf{p}'-\mathbf{p})\,\delta_{a'a}$$

Both matrix elements in (2.C.5) are therefore proportional to $\sqrt{2(2\pi)^3\sqrt{\mathbf{p}^2+m_A^2}}$, and the result is Lorentz invariant.) In exactly the same way, we can show that, in the limit (2.C. 2) ,

$$M(-p) \longrightarrow \frac{-2i(2\pi)^3\sqrt{\underline{p}^2+m_A^2}}{p^2+m_A^2-i\epsilon} \times$$

$$\times \sum_a \langle 0|T\{B(y),C(z),\cdots\}|\underline{p},\bar{a}\rangle\langle \underline{p},\bar{a}|A(0)|0\rangle$$

$$(2.C.6)$$

the sum now running over the single particle states of mass m_A created by $A(x)$, which of course are the antiparticles of the particles $|\underline{p},a\rangle$. By repeating the steps that led to (2.C.5) and (2.C.6), we can build up any physical matrix element out of the residues of the poles of Fourier transforms of time-ordered products. The only tricky point to watch out for is the appearance of disconnected parts, which arise when a field operator destroys a particle created by some other operator. Also note that the multiparticle states built up on the left or right are respectively out states or in states, because the poles arise from integrals which diverge at these poles in the limits $x^0 \to +\infty$ or $x^0 \to -\infty$ respectively.

D. Charge Non-Renormalization

As our first exercise in current algebra, we will prove the famous theorem of Ward and Takahashi,[2,3] that the strong interactions do not renormalize particle charges.

The renormalized charge of a particle is a parameter which describes its response to an external slowly-varying electromagnetic field. According to Eq. (2.B.3), the response of a particle to an external field which varies so slowly that no appreciable momentum is transferred to or from the particle is described by a matrix element

$$\langle \underset{\sim}{p}', \lambda' | J^\mu(0) | \underset{\sim}{p}, \lambda \rangle$$

where λ , λ' are helicity indices, all other discrete indices being suppressed. (This is only strictly true if we treat electromagnetic effects only to first order; in higher orders there is an electromagnetic renormalization of the electromagnetic fields. However, this radiative correction, though infinite, is small, and in any case enters as a common factor $Z_3^{1/2}$ in all charges.) For $\underset{\sim}{p} \to 0$, parity and rotational invariance give

$$\langle \underset{\sim}{0}, \lambda' | J^i(0) | \underset{\sim}{0}, \lambda \rangle = 0 \tag{2.D.1}$$

$$\langle \underset{\sim}{0}, \lambda' | J^0(0) | \underset{\sim}{0}, \lambda \rangle = e_1 \, \delta_{\lambda\lambda'} \, (2\pi)^{-3} \tag{2.D.2}$$

with e_1 a constant, defined as the renormalized charge. (In a box of volume V , the factor $(2\pi)^3$ would be replaced with V , so that the charge density would be e_1/V .) By boosting from rest to momentum $\underset{\sim}{p}$, we have then

$$\langle \underset{\sim}{p}, \lambda' | J^\mu(0) | \underset{\sim}{p}, \lambda \rangle = e_1 \, p^\mu \, \delta_{\lambda'\lambda} \, / (2\pi)^3 E \tag{2.D.3}$$

(A factor $E = \sqrt{\underset{\sim}{p}^2 + m^2}$ appears in the denominator instead of m because of the non-covariant normalization of our states.) We shall see that e_1 is the "charge" which appears in the formulas for emission, absorption, and scattering of soft photons.

The unrenormalized charge e is a parameter associated with a particular <u>field,</u> rather than a particle; it is defined by the gauge transformation rule

$$\psi(x) \longrightarrow e^{ie\Lambda(x)}\psi(x) \qquad (2.D.4)$$

The fundamental theorem we wish to prove here is that a particle $|p,\lambda\rangle$, which is associated with a field $\psi(x)$ in the sense that

$$\langle 0|\psi(x)|p,\lambda\rangle \neq 0 \qquad (2.D.5)$$

must have a renormalized charge e_1 equal (aside from higher order electromagnetic effects) to the unrenormalized charge e of the field.

The crucial link between the properties of the field and the particle is provided by the commutation relations of J^0 with ψ . (This can be called paleo-current-algebra.) According to the definition (2.A.4) , the charge density is

$$J^0 = -i\sum_n \pi_n e_n \psi_n \qquad (2.D.6)$$

The sum runs over all fields ψ_n appearing in the Lagrangian, e_n is the charge for the field ψ_n , and π_n is the canonical conjugate to ψ_n:

$$\pi_n \equiv \frac{\partial L}{\partial(\partial_0\psi_n)} \qquad (2.D.7)$$

The canonical commutation relations give

$$[\pi_n(\underset{\sim}{x},t), \psi_m(\underset{\sim}{y},t)]_\mp = -i\delta_{nm}\delta^3(\underset{\sim}{x}-\underset{\sim}{y})$$

$$[\psi_n(\underset{\sim}{x},t), \psi_m(\underset{\sim}{y},t)]_\mp = [\pi_n(\underset{\sim}{x},t), \pi_m(\underset{\sim}{y},t)]_\mp = 0 \qquad (2.D.8)$$

the $\mp$ signs denoting commutation relations (for bosons) or anti-commutation relations (for fermions) . It follows that, for both boson and fermion fields, J^0 and ψ have the commutator

$$[J^0(\underline{x},t), \Psi_n(\underline{y},t)] = -e_n \Psi_n(\underline{y},t)\,\delta^3(\underline{x}-\underline{y}) \ . \qquad (2.D.9)$$

If the field Ψ is any one of the canonical fields Ψ_n, or is built up as a product of such fields (for instance we might form a deuteron field from the product of a proton and a neutron field) then

$$[J^0(\underline{x},t), \Psi(y,t)] = -e\Psi(\underline{y},t)\,\delta^3(\underline{x}-\underline{y}) \ , \qquad (2.D.10)$$

where e is the sum of the unrenormalized charges of the canonical fields entering in Ψ.

One way to prove our theorem is to note that since the charge operator Q commutes with the Hamiltonian and is rotationally invariant, it may be diagonalized within the single-particle subspace, i.e.,

$$Q|\underline{p},\lambda\rangle = e'|\underline{p},\lambda\rangle \ .$$

It follows then that

$$(2\pi)^3\,\delta^3(\underline{p}'-\underline{p})\,\langle \underline{p}',\lambda'|\,J^0(0)|\,\underline{p},\lambda\rangle =$$

$$= \int d^3x\,\langle \underline{p}',\lambda'|\,J^0(\underline{x},0)|\,\underline{p},\lambda\rangle =$$

$$= \langle \underline{p}',\lambda'|\,Q\,|\,\underline{p},\lambda\rangle = e'\,\delta^3(\underline{p}'-\underline{p})\,\delta_{\lambda'\lambda}$$

and therefore

$$\langle \underline{p},\lambda'|\,J^0(0)|\,\underline{p},\lambda\rangle = (2\pi)^{-3}\,e'\,\delta_{\lambda'\lambda}$$

Comparing with (2.D.3), we see that e' is just the renormalized charge e_1 :

$$Q|\underset{\sim}{p},\lambda\rangle = e_1|\underset{\sim}{p},\lambda\rangle \qquad (2.D.11)$$

By the same reasoning, and since the vacuum expectation value of J^μ vanishes by Lorentz invariance, the charge Q must annihilate the vacuum state $|0\rangle$:

$$Q|0\rangle = 0 \qquad (2.D.12)$$

But integration of (2.D.10) over $\underset{\sim}{x}$ yields

$$[Q,\psi] = -e\psi \qquad (2.D.13)$$

Taking the matrix element of (2.D.13) between the vacuum and a single-particle state, and using (2.D.11) and (2.D.12), we have then

$$-\langle 0|\psi|\underset{\sim}{p},\lambda\rangle e_1 = -e\langle 0|\psi|\underset{\sim}{p},\lambda\rangle$$

and so, if $\langle 0|\psi|\underset{\sim}{p},\lambda\rangle$ does not vanish, we must have

$$e_1 = e \qquad (2.D.14)$$

as was to be proven.

It will be instructive to consider another proof, using the LSZ theorem of the last section, which is closer in spirit to the original work of Ward and Takahashi and to the subsequent developments in current algebra. We consider the time-ordered product $T\{\psi(x),\psi^\dagger(y),J^\mu(z)\} \equiv$

$$\begin{aligned}&\theta(x^0-y^0)\,\theta(y^0-z^0)\,\psi(x)\psi^\dagger(y)J^\mu(z) + \theta(x^0-z^0)\,\theta(z^0-y^0)\,\psi(x)J^\mu(z)\psi^\dagger(y) +\\ &+\theta(z^0-x^0)\,\theta(x^0-y^0)\,J^\mu(z)\psi(x)\psi^\dagger(y) \pm \theta(y^0-x^0)\,\theta(x^0-z^0)\,\psi^\dagger(y)\psi(x)J^\mu(z) \pm\\ &\pm\theta(y^0-z^0)\,\theta(z^0-x^0)\,\psi^\dagger(y)J^\mu(z)\psi(x) \pm \theta(z^0-y^0)\,\theta(y^0-x^0)\,J^\mu(z)\psi^\dagger(y)\psi(x)\end{aligned}$$

the sign $\pm$ being + for bosons and − for fermions. Since the divergence of $J^\mu(z)$ vanishes, the divergence of this time-ordered product receives contributions only from the z^0 -dependence of the θ -functions:

$$\frac{\partial}{\partial z^\mu} T\{\psi(x), \psi^+(y), J^\mu(z)\} = \theta(x^0-y^0)\left[\delta(y^0-z^0)\psi(x)[J^0(z),\psi^+(y)]\right]+$$

$$+\ \theta(x^0-y^0)\,\delta(x^0-z^0)\,[J^0(z),\psi(x)]\,\psi^+(y)$$

$$\pm\ \theta(y^0-x^0)\left[\delta(x^0-z^0)\psi^+(y)[J^0(z),\psi(x)] + \delta(y^0-z^0)[J^0(z),\psi^+(y)]\,\psi(x)\right] =$$

$$= \delta(x^0-z^0)T\{[J^0(z),\psi(x)],\psi^+(y)\} + \delta(y^0-z^0)T\{\psi(x),[J^0(z),\psi^+(y)]\} \qquad (2.D.15)$$

Using (2.D.10) and its adjoint, this gives

$$\frac{\partial}{\partial z^\mu} T\{\psi(x),\psi^+(y), J^\mu(z)\} =$$

$$= -e\left[\delta^4(z-x)-\delta^4(z-y)\right]T\{\psi(x),\psi^+(y)\} \qquad (2.D.16)$$

By taking the vacuum-expectation-value, multiplying by $\exp(-ip'\cdot x + ip\cdot y)$, integrating over x and y and setting $z=0$, we obtain the generalized Ward-Takahashi identity

$$i(p-p')_\mu \int d^4x\, d^4y\; e^{-ip'\cdot x}\, e^{ip\cdot y} \left\langle T\{\psi(x),\psi^+(y), J^\mu(0)\}\right\rangle_0 =$$

$$= -e\int d^4y\; e^{ip\cdot y}\left\langle T\{\psi(0),\psi^+(y)\}\right\rangle_0 +$$

$$+\ e\int d^4x\; e^{-ip'\cdot x}\left\langle T\{\psi(x),\psi^+(0)\}\right\rangle_0 \qquad (2.D.17)$$

Translation invariance gives

$$\left\langle T\{\psi(0),\psi^+(y)\}\right\rangle_0 = \left\langle T\{\psi(-y),\psi^+(0)\}\right\rangle_0$$

so by setting $y=-x$, we see that the second term in

(2.D.17) is the same as the last, except that it has opposite sign and contains p in place of p'. Letting p' approach p thus gives the original Ward identity.

$$-i\int d^4x\, d^4y\, e^{-ip\cdot(x-y)} \langle T\{\psi(x),\psi^+(y),J^\mu(0)\}\rangle_0 =$$

$$= e\frac{\partial}{\partial p^\mu}\int d^4x\, e^{-ip\cdot x}\langle T\{\psi(x),\psi^+(0)\}\rangle_0 \qquad (2.D.18)$$

According to the argument of Sec. 2C, when p^μ approaches the mass-shell $p^2 = -m^2$, the matrix elements are dominated by the poles: $\int d^4x\, d^4y\, e^{-ip\cdot(x-y)}\langle T\{\psi(x),\psi^+(y),J^\mu(0)\}\rangle_0 \to$

$$\to \left[\frac{-i(2\pi)^3 2\sqrt{\mathbf{p}^2+m^2}}{p^2+m^2-i\epsilon}\right]^2 \sum_{\lambda\lambda'} \langle 0|\psi(0)|\mathbf{p},\lambda\rangle\langle\mathbf{p},\lambda|J^\mu(0)|\mathbf{p},\lambda'\rangle\langle\mathbf{p},\lambda'|\psi^+(0)|0\rangle\,;$$

$$\int d^4x\, e^{-ip\cdot x}\langle T\{\psi(x),\psi^+(0)\}\rangle_0 \to$$

$$\to \frac{-i(2\pi)^3 2\sqrt{\mathbf{p}^2+m^2}}{p^2+m^2-i\epsilon}\sum_\lambda \langle 0|\psi(0)|\mathbf{p},\lambda\rangle\langle\mathbf{p},\lambda|\psi^+(0)|0\rangle$$

The leading term on the right-hand-side of Eq. (2.D.18) comes from the derivative of the singular part

$$\frac{\partial}{\partial p^\mu}(p^2+m^2-i\epsilon)^{-1} \longrightarrow -\frac{2p^\mu}{(p^2+m^2-i\epsilon)^2}$$

Comparing coefficients of this double pole, we have then

$$(2\pi)^3 2\sqrt{\mathbf{p}^2+m^2}\sum_{\lambda\lambda'}\langle 0|\psi(0)|\mathbf{p},\lambda\rangle\langle\mathbf{p},\lambda|J^\mu(0)|\mathbf{p},\lambda'\rangle\langle\mathbf{p},\lambda'|\psi^+(0)|0\rangle =$$

$$= ep^\mu\sum_\lambda \langle 0|\psi(0)|\mathbf{p},\lambda\rangle\langle\mathbf{p},\lambda|\psi^+(0)|0\rangle\,.$$

Comparing now with (2.D.3) , we see again that

$$e_1 = e$$

Perhaps I should stress again that all of the above is valid for any spin.

E. Zero Mass Photons

As a second "soft photon" theorem, even more important than the non-renormalization of charge, one can show that the photon mass vanishes. The Euler-Lagrange equations for A^0 read

$$\underset{\sim}{\nabla} \cdot \underset{\sim}{\mathcal{E}} = J^0 \qquad (2.E.1)$$

where

$$\mathcal{E}^i \equiv \frac{\partial L}{\partial(\partial_i A_0)} \qquad (2.E.2)$$

(If L involves $F_{\mu\nu}$ only in the kinematic term $-\frac{1}{4} F_{\mu\nu} F^{\mu\nu}$, the $\underset{\sim}{\mathcal{E}}$ is just the usual electric field.) Taking the matrix element of (2.E.1) between single-particle states gives

$$i(\underset{\sim}{p}-\underset{\sim}{p}') \cdot \langle \underset{\sim}{p}', \lambda' | \underset{\sim}{\mathcal{E}}(0) | \underset{\sim}{p}, \lambda \rangle = \langle \underset{\sim}{p}', \lambda' | J^0(0) | \underset{\sim}{p}, \lambda \rangle$$

and letting $\underset{\sim}{p}'$ approach $\underset{\sim}{p}$, we have

$$i(\underset{\sim}{p}-\underset{\sim}{p}') \cdot \langle \underset{\sim}{p}', \lambda' | \mathcal{E}(0) | \underset{\sim}{p}, \lambda \rangle \longrightarrow \frac{e}{(2\pi)^3} \delta_{\lambda'\lambda} \qquad (2.E.3)$$

Thus, as long as $e \neq 0$, there must be a singularity in this matrix element at $\underset{\sim}{p}' = \underset{\sim}{p}$. Normally, the necessary singularity comes from a photon propagator, with

$$\langle \underset{\sim}{p}', \lambda' | \underset{\sim}{\mathcal{E}}(0) | \underset{\sim}{p}, \lambda \rangle \longrightarrow \frac{-ie(\underset{\sim}{p}-\underset{\sim}{p}')\delta_{\lambda'\lambda}}{(2\pi)^3 (\underset{\sim}{p}-\underset{\sim}{p}')^2} \quad \text{for } \underset{\sim}{p} \to \underset{\sim}{p}' .$$

However, if the photon mass were not zero, there would be no way that this matrix element could acquire a singularity of this type.

This proof is not quite rigorous, because the renormalization of charge by higher-order electromagnetic effects could allow $e = 0$ even if the physical charge of the particle does not vanish. However, even if the above argument does not really prove that the photon mass is necessarily zero, it certainly makes a good prima facie case for a zero photon mass.

F. Lorentz Invariance - A Review

In what follows, we shall be making heavy use of the Lorentz transformation properties of single-particle states.[4,5] These ought to be familiar, but a brief review may be useful in refreshing your memory, and it will serve to establish the notation we shall use.

The homogeneous Lorentz group consists of real 4×4 matrices $\Lambda^\mu{}_\nu$, with

$$\Lambda^\mu{}_\nu \Lambda_\rho{}^\nu = \delta^\mu_\rho \quad , \qquad (2.F.1)$$

indices being raised and lowered, as always, with the Minkowski metric $\eta_{\mu\nu}$. These transformations are represented on Hilbert space by unitary operators $U[\Lambda]$, with

$$U[\Lambda_1]\, U[\Lambda_2] = U[\Lambda_1 \Lambda_2] \qquad (2.F.2)$$

In particular, a four-vector like the electromagnetic current J^μ is transformed into

$$U[\Lambda]\, J^\mu(0)\, U^{-1}[\Lambda] = \Lambda_\nu{}^\mu J^\nu(0) \qquad (2.F.3)$$

A single-particle state of momentum $\underset{\sim}{p}$ and spin z-component σ is defined by "boosting" from rest

$$|\underset{\sim}{p},\sigma\rangle \equiv \sqrt{\frac{m}{E}}\; U[L(\underset{\sim}{p})]\, |\underset{\sim}{0},\sigma\rangle \qquad (2.F.4)$$

where $E \equiv \sqrt{\underset{\sim}{p}^2+m^2}$, and $L(\underset{\sim}{p})$ is the Lorentz boost

$$
\begin{aligned}
L^i{}_j(\underset{\sim}{p}) &= \delta^i_j + \frac{p^i p^j}{m(E+m)} , \\
L^i{}_0(\underset{\sim}{p}) &= L^0{}_i(\underset{\sim}{p}) = \frac{p^i}{m} , \\
L^0{}_0(\underset{\sim}{p}) &= \frac{E}{m} .
\end{aligned}
\tag{2.F.5}
$$

(The factor $\sqrt{\frac{m}{E}}$ appears so that $|\underset{\sim}{p},\sigma\rangle$ will have the conventional non-covariant norm.) If we operate on these states with a general Lorentz transformation $U[\Lambda]$, we find that

$$
\begin{aligned}
U[\Lambda]|\underset{\sim}{p},\sigma\rangle &= \sqrt{\frac{m}{E}}\, U[\Lambda L(\underset{\sim}{p})]|\underset{\sim}{0},\sigma\rangle = \\
&= \sqrt{\frac{m}{E}}\, U[L(\Lambda\underset{\sim}{p})]\, U[L^{-1}(\Lambda\underset{\sim}{p})\Lambda L(\underset{\sim}{p})]\, |\underset{\sim}{0},\sigma\rangle
\end{aligned}
\tag{2.F.6}
$$

But $L^{-1}(\Lambda\underset{\sim}{p})\Lambda L(\underset{\sim}{p})$ gives a particle of momentum zero first a momentum $\underset{\sim}{p}$, then $\Lambda\underset{\sim}{p}$, and then zero, so it is a rotation, called the Wigner rotation. For any rotation R , we have

$$
U[R]|\underset{\sim}{0},\sigma\rangle = \sum_{\sigma'} D^{(j)}_{\sigma'\sigma}[R]\, |\underset{\sim}{0},\sigma'\rangle , \tag{2.F.7}
$$

where $D^{(j)}[R]$ is the usual unitary representation of the rotation group for spin j . Using (2.F.7) and (2.F.4) in (2.F.6) , we have then the transformation rule

$$
U[\Lambda]|\underset{\sim}{p},\sigma\rangle = \sqrt{\frac{E(\Lambda\underset{\sim}{p})}{E(\underset{\sim}{p})}}\ \sum_{\sigma'} D^{(j)}_{\sigma'\sigma}[L^{-1}(\Lambda\underset{\sim}{p})\Lambda L(\underset{\sim}{p})]\, |\Lambda\underset{\sim}{p},\sigma'\rangle \tag{2.F.8}
$$

We will occasionally encounter the generators for infinitesimal rotations and boosts. A rotation by an infinitesimal angle $|\underset{\sim}{\epsilon}|$ about the axis $\hat{\underset{\sim}{\epsilon}}$ may be written

$$
R_{\underset{\sim}{\epsilon}} = 1 + i\, \underset{\sim}{\epsilon}\cdot \underset{\sim}{J} \tag{2.F.9}
$$

where

$$\begin{aligned} (J_k)^i{}_j &= -i\epsilon_{ijk} \\ (J_k)^i{}_0 &= (J_k)^0{}_i = (J_k)^0{}_0 = 0 \end{aligned} \qquad (2.F.10)$$

A boost from rest to an infinitesimal velocity $\underset{\sim}{\epsilon}$ may be written

$$L_{\underset{\sim}{\epsilon}} = 1 - i\underset{\sim}{\epsilon}\cdot\underset{\sim}{K} \quad , \qquad (2.F.11)$$

where

$$\begin{aligned} (K_k)^i{}_j &= (K_k)^0{}_0 = 0 \quad , \\ (K_k)^i{}_0 &= (K_k)^0{}_i = i\delta_{ik} \quad . \end{aligned} \qquad (2.F.12)$$

Note also that (2.F.11) may be generalized to finite boosts

$$L(p) = \exp\left\{-i\,\hat{p}\cdot\underset{\sim}{K}\,\sinh^{-1}\left(\frac{|p|}{m}\right)\right\} \qquad (2.F.13)$$

The generators $\underset{\sim}{J}$, $\underset{\sim}{K}$ obey the familiar commutation relations

$$[J_i, J_j] = i\,\epsilon_{ijk}\,J_k \quad , \qquad (2.F.14)$$

$$[J_i, K_j] = i\,\epsilon_{ijk}\,K_k \quad , \qquad (2.F.15)$$

$$[K_i, K_j] = -i\,\epsilon_{ijk}\,J_k \quad . \qquad (2.F.16)$$

The unitary matrix representation of the infinitesimal rotation (2.F.9) is

$$D^{(j)}[R_{\underset{\sim}{\epsilon}}] = 1 + i\underset{\sim}{\epsilon}\cdot\underset{\sim}{J}^{(j)} \qquad (2.F.17)$$

where $\underset{\sim}{J}^{(j)}$ forms the usual Hermitian $(2j+1)\times(2j+1)$ matrix representation of the Lie algebra (2.F.14) . In particular, an infinitesimal boost $L_{\underset{\sim}{\epsilon}}$ produces the infinitesimal Wigner rotation

$$L^{-1}(L_{\underset{\sim}{\epsilon}}\underset{\sim}{p})L_{\underset{\sim}{\epsilon}}L(\underset{\sim}{p}) = 1 + i\,\underset{\sim}{J}\cdot\frac{\underset{\sim}{\epsilon}\times\underset{\sim}{p}}{E+m} \qquad (2.F.18)$$

and the matrix representation may be obtained by replacing $\underset{\sim}{J}$ with $\underset{\sim}{J}^{(j)}$.

G. Soft Photon Vertices

As a preparation for our derivation of the. soft photon theorems, we shall next examine the structure of the vertex for emission of a single soft photon, defined by:

$$\langle \underset{\sim}{p}',\sigma'|\, J^\mu(0)\,|\underset{\sim}{p},\sigma\rangle \equiv \frac{\Gamma^\mu_{\sigma'\sigma}(\underset{\sim}{p}',\underset{\sim}{p})}{(2\pi)^3\sqrt{4E(\underset{\sim}{p}')E(\underset{\sim}{p})}} \qquad (2.G.1)$$

Here $|\underset{\sim}{p},\sigma\rangle$ is a one-particle state with momentum $\underset{\sim}{p}$ and spin z-component σ, and $E(\underset{\sim}{p}) \equiv \sqrt{\underset{\sim}{p}^2+m^2}$ We will be making use of five general properties of this vertex:

(a) Charge Non-renormalization: According to Eqs. (2.D.2) and (2.D.14), the vertex at $\underset{\sim}{p}'=\underset{\sim}{p}$ has the value

$$\Gamma^\mu_{\sigma'\sigma}(\underset{\sim}{p},\underset{\sim}{p}) = 2e\,p^\mu\,\delta_{\sigma'\sigma} \qquad (2.G.2)$$

(b) Current Conservation: The single-particle matrix element of $J^\mu(x)$ is proportional to a factor

$$\exp\{\,i\underset{\sim}{x}\cdot(\underset{\sim}{p}-\underset{\sim}{p}') - ix^0(E(\underset{\sim}{p})-E(\underset{\sim}{p}'))\}$$

and hence the conservation condition $\partial_\mu J^\mu=0$ implies that

$$(\underset{\sim}{p}-\underset{\sim}{p}')\cdot\underset{\sim}{\Gamma}(\underset{\sim}{p}',\underset{\sim}{p}) = \big(E(\underset{\sim}{p})-E(\underset{\sim}{p}')\big)\Gamma^0(\underset{\sim}{p}',\underset{\sim}{p}) \qquad (2.G.3)$$

(c) Lorentz Invariance: From Eqs. (2.F.3) and (2.F.8), it follows that Γ^μ obeys the Lorentz transformation rule

$$\Gamma^\mu(\underset{\sim}{p}',\underset{\sim}{p}) = D^{(j)-1}[L^{-1}(\Lambda\underset{\sim}{p}')\Lambda L(\underset{\sim}{p}')]\;\Gamma^\nu(\Lambda\underset{\sim}{p}',\Lambda\underset{\sim}{p})\,D^{(j)}[L^{-1}(\Lambda\underset{\sim}{p})\Lambda L(\underset{\sim}{p})]\,\Lambda_\nu{}^\mu$$
$$(2.G.4)$$

(d) Parity: The current J^i carries negative parity, while J^0 carries positive parity, so the vertex obeys the inversion rules

$$\Gamma^0(\underset{\sim}{p}', \underset{\sim}{p}) = \Gamma^0(-\underset{\sim}{p}', -\underset{\sim}{p}) \qquad (2.G.5)$$

$$\Gamma^i(\underset{\sim}{p}', \underset{\sim}{p}) = -\Gamma^i(-\underset{\sim}{p}', -\underset{\sim}{p}) \qquad (2.G.6)$$

(e) Reality: The current four-vector is an Hermitian operator, so Γ^μ obeys the rule

$$\Gamma^{\mu\dagger}(\underset{\sim}{p}', \underset{\sim}{p}) = \Gamma^\mu(\underset{\sim}{p}, \underset{\sim}{p}') \qquad (2.G.7)$$

the dagger here denoting a transposition of spin indices as well as a complex conjugation.

The aim of this section is to construct the first few terms of an expansion of $\Gamma^\mu(\underset{\sim}{p}', \underset{\sim}{p})$ in powers of $\underset{\sim}{p}'$ and $\underset{\sim}{p}$ which will satisfy the above five requirements. Our strategy will be first to construct an expansion for $\Gamma^\mu(\underset{\sim}{q}, \underset{\sim}{0})$ in powers of $\underset{\sim}{q}$ which will satisfy charge non-renormalization, current conservation, rotational invariance, and parity; then to boost the result to obtain a power series for $\Gamma^\mu(\underset{\sim}{p}', \underset{\sim}{p})$ with $\underset{\sim}{p} \neq \underset{\sim}{0}$; and then to impose reality.

According to (2.G.5) , $\Gamma^0(\underset{\sim}{q}, \underset{\sim}{0})$ is even in $\underset{\sim}{q}$, so it has the expansion, for $\underset{\sim}{q} \to \underset{\sim}{0}$,

$$\Gamma^0(\underset{\sim}{q}, \underset{\sim}{0}) \longrightarrow 2me\,1 + Q_{ij}\, q^i q^j + O(q^4), \qquad (2.G.8)$$

where Q_{ij} is a constant matrix. Rotational invariance requires, via the Wigner-Eckart theorem, that Q_{ij} be some linear combination of δ_{ij} and an anti-commutator of spin matrices; conventionally, this is written

$$Q_{ik} = e\left(\frac{1}{4m^2} - \frac{R^2}{6}\right)\delta_{ik} 1 + Q\left(\frac{1}{2}\{J_i^{(J)}, J_k^{(J)}\} - \frac{1}{3}J(J+1)\,\delta_{ik} 1\right) \qquad (2.G.9)$$

By studying the scattering of our particle in an homogeneous electric field, you can convince yourselves that R and Q are what are usually called the charge radius and electric quadrupole moment of the particle.

According to (2.G.3), the component of $\boldsymbol{\Gamma}$ along $\mathbf{q}$ is determined as

$$\mathbf{q}\cdot\boldsymbol{\Gamma}(\mathbf{q},\mathbf{0}) = [E(\mathbf{q})-E(\mathbf{0})]\,\Gamma^0(\mathbf{q},\mathbf{0}) \rightarrow e\mathbf{q}^2 1 + O(\mathbf{q}^4) \qquad (2.G.10)$$

But (2.G.6) requires that $\boldsymbol{\Gamma}(\mathbf{q},\mathbf{0})$ be odd in $\mathbf{q}$, so it has the power series expansion, for $\mathbf{q}\rightarrow\mathbf{0}$,

$$\boldsymbol{\Gamma}(\mathbf{q},\mathbf{0}) \longrightarrow e\mathbf{q}1 + 2im\boldsymbol{\mu}\times\mathbf{q} + O(\mathbf{q}^3) \qquad (2.G.11)$$

where $\boldsymbol{\mu}$ is some constant matrix. Rotational invariance requires, again via the Wigner-Eckart theorem, that $\boldsymbol{\mu}$ be proportional to the spin matrix $\mathbf{J}^{(J)}$:

$$\boldsymbol{\mu} = \mu\,\frac{\mathbf{J}^{(J)}}{J} \qquad (2.G.12)$$

By considering the scattering of our particle in an homogeneous magnetic field, you can convince yourselves that μ is what is usually called the total magnetic moment of the particle.

We now perform our boost. Setting $\Lambda = L^{-1}(\mathbf{p})$ in Eq. (2.G.4) gives

$$\Gamma^\mu(\mathbf{p}',\mathbf{p}) = D^{(J)}[L^{-1}(\mathbf{p}')L(\mathbf{p})L(\mathbf{q})]\,\Gamma^\nu(\mathbf{q},\mathbf{0})\,L^\mu{}_\nu(\mathbf{p}) \qquad (2.G.13)$$

where $\mathbf{q}$ is now a function of $\mathbf{p}$ and $\mathbf{p}'$:

$$q^\mu = L_\nu{}^\mu(\mathbf{p})\,p'^\nu \qquad (2.G.14)$$

For $\mathbf{p}$ and $\mathbf{p}'$ small, (2.G.14) gives

$$\mathbf{q} \rightarrow \mathbf{p}' - \mathbf{p} + \text{cubic} \qquad (2.G.15)$$

where "cubic" denotes terms of third order in $\mathbf{p}$ and/or $\mathbf{p}'$.

To calculate the Wigner rotation in (2.G.11), we use the formula (2.F.13), whose expansion reads

$$L(p) \to 1 - i p \cdot \frac{K}{m} - \frac{1}{2m^2} (p \cdot K)^2 + \text{cubic}$$

After straightforward though tedious calculation, using the commutation rule (2.F.16), we find

$$L^{-1}(p') \, L(p) \, L(q) \to 1 - \frac{i}{2m^2} (p' \times p) \cdot J + \text{cubic} \tag{2.G.16}$$

A direct calculation, using (2.F.5), (2.G.8), and (2.G.11), gives

$$L^{i}{}_{\nu}(p) \, \Gamma^{\nu}(q, 0) = e(q_i + 2p_i) 1 + \frac{2im\mu}{J} (J \times q)_i + \text{cubic} \tag{2.G.17}$$

$$L^{0}{}_{\nu}(p) \Gamma^{\nu}(q, 0) = \frac{e}{m} q \cdot p + \frac{2i\mu}{J} (J \times q) \cdot p + 2me\left(1 + \frac{p^2}{2m}\right) + + Q_{ij} q^i q^j + \text{cubic} \tag{2.G.18}$$

Putting together (2.G.13) - (2.G.18), we find the general vertex function:

$$\Gamma(p', p) \to e(p' + p) 1 + \left(\frac{2im\mu}{J}\right) J \times (p' - p) + \text{cubic} \tag{2.G.19}$$

$$\Gamma^0(p', p) \to 2me + \frac{e}{m} p \cdot p' + Q_{ij} (p' - p)_i (p' - p)_j + + i\left(\frac{2\mu}{J} - \frac{e}{m}\right)(p' \times p) \cdot J + \text{quartic} \tag{2.G.20}$$

(Here and below, we drop the superscript (J) on the spin matrix.) Finally, the reality property (2.G.7) merely requires that e, μ, R, and Q be real.

H. Elastic Photon Scattering: General Properties

The results derived above will be applied to the elastic scattering of a photon on a general single-particle target. From (2.B.3), it follows that the S-matrix here is

$$\langle \sigma' p'; \lambda' q' | S | \sigma p; \lambda q \rangle = -\int d^4x \, d^4y \, e_{\nu}^{*}(q' \lambda') e_{\mu}(q \lambda) \times$$
$$\times (2\pi)^{-3} (4 q'^0 q^0)^{-1/2} e^{-iq' \cdot y} e^{iq \cdot x} \left[\langle \sigma' p' | T\{ J^{\nu}(y), J^{\mu}(x) \} | \sigma p \rangle + \text{seagulls} \right]$$

Here σ and σ' are the initial and final target spin z-components; $\mathbf{p}$ and $\mathbf{p}'$ are the initial and final target momenta; λ and λ' are the initial and final photon helicities; $\mathbf{q}$ and $\mathbf{q}'$ are the initial and final photon momenta; $e_\mu(\mathbf{q}\,\lambda)$ is the polarization vector for a photon of momentum $\mathbf{q}$ and helicity λ, with

$$e^0(\mathbf{q},\lambda) = \mathbf{q}\cdot\underset{\sim}{e}(\mathbf{q},\lambda) = 0 \qquad (2.H.1)$$

and "seagulls" denotes terms in which the two photons interact simultaneously with the target, rather than separately with the two currents. Translational invariance allows us to replace the arguments of J^ν and J^μ with 0 and $x-y$ respectively, so

$$\langle\sigma'\mathbf{p}';\lambda'\mathbf{q}'|S|\sigma\mathbf{p};\lambda\mathbf{q}\rangle = -\frac{1}{(2\pi)^2}\cdot\frac{e_\nu^*(\mathbf{q}'\lambda')\,e_\mu(\mathbf{q}\lambda)}{\sqrt{16q'^0q^0E'E}}\times$$

$$\times\;\delta^4(p+q-p'-q')\,M^{\nu\mu}_{\sigma'\sigma}(\mathbf{q};\mathbf{p}',\mathbf{p}) \qquad (2.H.2)$$

where

$$M^{\nu\mu}_{\sigma'\sigma}(\mathbf{q};\mathbf{p}',\mathbf{p})\frac{1}{(2\pi)^3\sqrt{4E'E}} \equiv \int d^4x\,e^{iq\cdot x}\Big[\langle\sigma'\mathbf{p}'|T\{J^\nu(0)J^\mu(x)\}|\sigma\mathbf{p}\rangle +$$

$$+\text{seagulls}\Big] \qquad (2.H.3)$$

Although q^μ in (2.H.2) is on the forward light cone, with $q^0=|\mathbf{q}|$, Eq. (2.H.3) can be used to define M for arbitrary four-vectors q^μ.

The fundamental current algebra result, to be used in deriving our soft photon theorems, is the formula

$$q_\mu M^{\nu\mu}_{\sigma'\sigma}(q;\mathbf{p}',\mathbf{p}) = 0 \qquad (2.H.4)$$

One way to derive this result is to make use of the conservation of current,

$$\partial_\mu J^\mu(x) = 0 \qquad (2.H.5)$$

and assume that J^ν commutes with J^0 like a neutral field

$$\delta(x^0)\,[J^0(x),J^\nu(0)] = 0 \qquad (2.H.6)$$

This isn't actually true - Schwinger showed long ago that for ν spacelike, this commutator must involve derivatives of δ-functions.[6] However, experience with specific models has shown that the "seagull" terms in (2.H.3) always cancel the Schwinger terms. Thus, if we forget the seagulls, then (2.H.5) and (2.H.6) can be used to derive the formula

$$\partial_\mu T\{J^\nu(0), J^\mu(x)\} = 0 \quad ,$$

which yields (2.H.4).

A different approach, which goes more deeply to the fundamental basis of electromagnetic theory, is to impose on the S-matrix the requirement of Lorentz invariance.[7] If the polarization vector e_μ were a four-vector, then (2.H.1) would automatically be Lorentz invariant. However, e_μ is not a four-vector, as shown by Eq. (2.H.1). Instead, a Lorentz transformation Λ carries e^μ into

$$\Lambda^\mu{}_\nu e^\nu - \frac{q^\mu}{q^0} \Lambda^0{}_\nu e^\nu$$

In order that the second term not disturb the Lorentz invariance of the S-matrix, it is necessary that the second term should give no contribution, which leads to (2.H.4), at least for q^μ on the light cone.

Yet another approach is to imagine the target particle to be scattered by an external electromagnetic potential

$$a^\mu(x) = \frac{1}{\sqrt{(2\pi)^3 2q^0}} \left[e^\mu e^{iq\cdot x} + e^{\mu *} e^{-iq\cdot x} \right] +$$

$$+ \frac{1}{\sqrt{(2\pi)^3 2q'^0}} \left[e'^\mu e^{iq'\cdot x} + e'^{\mu *} e^{-iq'\cdot x} \right] .$$

The S-matrix, to first order in e^μ and $e'^{\nu *}$, is just (2.H.2). But this result must be invariant with respect to gauge transformations

$$a_\mu \rightarrow a_\mu + \partial_\mu \Lambda \quad ,$$

which leads again to (2.H.4).

Another general property of the scattering matrix element, which will be important below, is the crossing relation:

$$M^{\nu\mu}_{\sigma'\sigma}(q; p', p) = M^{\mu\nu}_{\sigma'\sigma}(p'-p-q; p', p) \quad . \qquad (2.H.7)$$

The time-ordered product term in (2.H.3) obviously satisfies this condition.

The final general property of $M^{\nu\mu}$ which we need is its pole structure. The time-ordered product in (2.H.3) can be evaluated by inserting a complete set of states between currents:

$$\int d^4x\, e^{iq\cdot x} \langle \sigma' p' | T\{J^\nu(0) J^\mu(x)\} | \sigma p \rangle =$$
$$= -i(2\pi)^3 \sum_n \Big\{ \langle \sigma' p' | J^\nu(0) | n \rangle \langle n | J^\mu(0) | \sigma p \rangle \frac{\delta^3(p_n - p - q)}{E_n - E - q^0 - i\epsilon} +$$
$$+ \langle \sigma' p' | J^\mu(0) | n \rangle \langle n | J^\nu(0) | \sigma p \rangle \frac{\delta^3(p_n - p' + q)}{E_n - E' + q^0 - i\epsilon}$$

Both terms have a pole at $q^\mu = 0$ when the intermediate state $|n\rangle$ is a single-particle state with the same mass m as the initial and final particles. Isolating this pole, and using (2.G.1) to write the matrix elements of the currents in terms of the covariant vertices Γ^μ, we have then

$$M^{\nu\mu}(q; p', p) = \frac{-i\Gamma^\nu(p', p+q)\Gamma^\mu(p+q, p)}{[E(p+q) - E(p) - q^0 - i\epsilon]\, 2E(p+q)} +$$

$$\frac{-i\Gamma^\mu(p', p'-q)\Gamma^\nu(p'-q, p)}{[E(p'-q) - E(p') + q^0 - i\epsilon]\, 2E(p'-q)} + N^{\nu\mu}(q; p', p) \qquad (2.H.8)$$

where $N^{\nu\mu}$ **consists of all the terms in** $M^{\nu\mu}$ **which are** **not** singular at $q^\mu = 0$ **, including the states** $|n\rangle$ **with** $m_n \neq m$ **and the "seagulls" in (2.H.3). (We use a matrix notation, with spin indices suppressed.)**

The function here of the conservation condition is to provide information about the unknown non-pole matrix element $N^{\nu\mu}$ The vertex Γ^{μ} is conserved, in the sense that

$$\mathbf{q}\cdot\mathbf{\Gamma}(\mathbf{p}+\mathbf{q},\mathbf{p}) = [E(\mathbf{p}+\mathbf{q})-E(\mathbf{p})]\,\Gamma^{0}(\mathbf{p}+\mathbf{q},\mathbf{p})$$

$$\mathbf{q}\cdot\mathbf{\Gamma}(\mathbf{p}',\mathbf{p}'-\mathbf{q}) = [E(\mathbf{p}')-E(\mathbf{p}'-\mathbf{q})]\,\Gamma^{0}(\mathbf{p}',\mathbf{p}'-\mathbf{q})$$

so (2.H.4) and (2.H.8) give

$$q_{\mu}N^{\nu\mu}(q;p',p) = i\,\frac{\Gamma^{\nu}(\mathbf{p}',\mathbf{p}+\mathbf{q})\,\Gamma^{0}(\mathbf{p}+\mathbf{q},\mathbf{p})}{2E(\mathbf{p}+\mathbf{q})} + {}$$

$$-i\,\frac{\Gamma^{0}(\mathbf{p}',\mathbf{p}'-\mathbf{q})\,\Gamma^{\nu}(\mathbf{p}'-\mathbf{q},\mathbf{p})}{2E(\mathbf{p}'-\mathbf{q})} \qquad (2.H.9)$$

Also, the pole terms in (2.H.8) satisfy the crossing relation (2.H.7) separately, and therefore so does $N^{\nu\mu}$:

$$N^{\nu\mu}(q;p',p) = N^{\mu\nu}(p'-p-q;p',p) \qquad (2.H.10)$$

Obviously (2.H.9) and (2.H.10) do not determine $N^{\nu\mu}$ uniquely, but they do provide just enough information to allow us to calculate $N^{\nu\mu}$ up to first order in the photon energy.

I. Elastic Photon Scattering: Low Energy Limits

We wish to determine the behaviour of the amplitude $M^{\nu\mu}$ as q^{μ} and q'^{ν} go to zero. This problem was originally solved for a spin ½ target independently by Low[8] and by Gell-Mann and Goldberger,[9] and since then generalized by many authors. The method I will follow is based on that of Low.

Let us work in the laboratory frame, with $\mathbf{p}=0$ The independent variables can then be taken as $\mathbf{q}$, $\mathbf{q}'$, and q^{0}, with $\mathbf{p}'=\mathbf{q}-\mathbf{q}'$, and

$$q^{0\prime} = q^{0}+m-E(\mathbf{q}-\mathbf{q}') = q^{0}-\frac{1}{2m}(\mathbf{q}-\mathbf{q}')^{2}+\cdots \qquad (2.I.1)$$

We can then expand $N^{\nu\mu}$, which by construction is analytic near $q^\mu = q'^\mu = 0$, in powers of $\mathbf{q}$, $\mathbf{q}'$, and q^0:

$$N^{\nu\mu} \to A^{\nu\mu} + B^{\nu\mu i} q_i + \bar{B}^{\nu\mu i} q'_i + C^{\nu\mu} q^0 + \cdots \tag{2.I.2}$$

the neglected terms being of second and higher order in $\mathbf{q}$, $\mathbf{q}'$, and q^0. According to (2.H.10), $N^{\nu\mu}$ must be unchanged when we interchange μ and ν, $\mathbf{q}$ and $-\mathbf{q}'$, and q^0 and $-q^{0\prime}$. But $q^{0\prime}$ equals q^0 within the accuracy of (2.I.2), so crossing requires

$$A^{\nu\mu} = A^{\mu\nu}, \quad \bar{B}^{\nu\mu i} = -B^{\mu\nu i}, \quad C^{\nu\mu} = -C^{\mu\nu} \tag{2.I.3}$$

Our strategy will be to determine the coefficients A, B, and C by using (2.I.2) in (2.H.9), and then to use the results in (2.H.8) to determine the low energy limit of $M^{\nu\mu}$.

First, let's check that the conservation condition (2.H.9) determines A, B, and C <u>uniquely</u>. Suppose we find two solutions of (2.H.8) of the form

$$N^{\nu\mu}_{1,2} \to A^{\nu\mu}_{1,2} + B^{\nu\mu i}_{1,2} q_i - B^{\mu\nu i}_{1,2} q'_i + C^{\nu\mu}_{1,2} q^0 + \cdots \tag{2.I.4}$$

The difference $\Delta N^{\nu\mu}$ has a similar expansion

$$\Delta N^{\nu\mu} \equiv N_1^{\nu\mu} - N_2^{\nu\mu} \to \Delta A^{\nu\mu} + \Delta B^{\nu\mu i} q_i - \Delta B^{\mu\nu i} q'_i + \Delta C^{\nu\mu} q^0 + \cdots \tag{2.I.5}$$

with

$$\Delta A^{\nu\mu} = \Delta A^{\mu\nu} \equiv A_1^{\nu\mu} - A_2^{\mu\nu}$$

$$\Delta B^{\nu\mu i} \equiv B_1^{\nu\mu i} - B_2^{\nu\mu i}$$

$$\Delta C^{\nu\mu} = -\Delta C^{\mu\nu} \equiv C_1^{\nu\mu} - C_2^{\nu\mu}$$

and it satisfies the homogeneous equation

$$0 = q_\mu \Delta N^{\nu\mu} \to \Delta A^{\nu i} q_i - \Delta A^{\nu 0} q^0 + \Delta B^{\nu j i} q_i q_j +$$

$$- \Delta B^{\nu 0 i} q_i q^0 - \Delta B^{j \nu i} q'_i q_j + \Delta B^{0 \nu i} q'^i q^0 +$$

$$+ \Delta C^{\nu i} q^0 q_i - \Delta C^{\nu 0} (q^0)^2 + \cdots$$

Even if we only impose this condition on the photon mass shell, where $(q^0)^2 = \mathbf{q}^2$, we can still conclude that

$$\Delta A^{\nu i} = 0 \qquad (2.I.6)$$

$$\Delta A^{\nu 0} = 0 \qquad (2.I.7)$$

$$\Delta B^{\nu j i} + \Delta B^{\nu i j} = 2\,\Delta C^{\nu 0}\,\delta_{ij} \qquad (2.I.8)$$

$$\Delta B^{\nu 0 i} = \Delta C^{\nu i} \qquad (2.I.9)$$

$$\Delta B^{j \nu i} = 0 \qquad (2.I.10)$$

$$\Delta B^{0 \nu i} = 0 \qquad (2.I.11)$$

From (2.I.6), (2.I.7), (2.I.10), and (2.I.11), it immediately follows that

$$\Delta A^{\nu\mu} = \Delta B^{\mu\nu i} = 0 \qquad (2.I.12)$$

so (2.I.8) and (2.I.9) give

$$\Delta C^{\nu\mu} = 0 \qquad (2.I.13)$$

Hence A, B, and C are unique. Note that we could not obtain a unique solution of the conservation equation if we carried the expansion of $N^{\nu\mu}$ to terms which are quadratic in q and q', because then we could include terms such as $\epsilon^{\nu\mu\rho\sigma} q'_\rho q_\sigma$, which are separately conserved and crossing-symmetric, and whose coefficients can therefore not be determined from these conditions.

Returning now to (2.H.9), we can use the low energy limits (2.G.19) and (2.G.20) for the vertex functions to calculate that

$$q_\mu N^{i\mu} \longrightarrow 2ie^2\,1\,q_i + \text{cubic} \qquad (2.I.14)$$

$$q_\mu N^{0\mu} \longrightarrow \frac{4e}{m}\left(\frac{\mu}{J} - \frac{e}{2m}\right)(\mathbf{q}' \times \mathbf{q}) \cdot \mathbf{J} + \text{cubic} \qquad (2.I.15)$$

where "cubic" denotes terms of third or higher order in $\mathbf{q}$, $\mathbf{q}'$, and q^0. Condition (2.I.15) can be satisfied with

$$N^{00} \longrightarrow \text{quadratic}$$

$$N^{0i} \longrightarrow 4e\left(\frac{\mu}{J} - \frac{e}{2m}\right)(\mathbf{J} \times \mathbf{q}')_i + \text{quadratic}$$

so crossing symmetry gives

$$N^{io} \longrightarrow -4e\left(\frac{\mu}{J}-\frac{e}{2m}\right)\left(\underset{\sim}{J}\times \underset{\sim}{q}\right)_i \ + \text{quadratic}$$

and (2.I.14) can then be satisfied with

$$N^{ji} \longrightarrow 2ie^2\delta_{ij}\,1 - 4e\left(\frac{\mu}{J}-\frac{e}{m}\right)\epsilon_{ijk}\,J_k\,q^0 + \text{quadratic}. \quad (2.\text{I}.16)$$

Having found one solution of (2.I.14) and (2.I.15), we are assured by our previous argument that this solution is unique.

For q^μ and q'^μ going to zero, with $\underset{\sim}{p}=0$, Eq. (2.H.8) gives

$$M^{ji} \rightarrow \frac{i}{2mq^0}\left[\Gamma^j_{(1)}(\underset{\sim}{q}-\underset{\sim}{q}',\underset{\sim}{q})\Gamma^i_{(1)}(\underset{\sim}{q},\underset{\sim}{0}) - \Gamma^i_{(1)}(\underset{\sim}{q}-\underset{\sim}{q}',-\underset{\sim}{q})\Gamma^j_{(1)}(-\underset{\sim}{q},\underset{\sim}{0})\right] + N^{ji}_{(1)} + \text{quadratic} \quad (2.\text{I}.17)$$

the subscript (1) indicating that we keep terms only up to first order in $\underset{\sim}{q}$, $\underset{\sim}{q}'$, and q^0 . To the accuracy of this result, the S-matrix is given by (2.H.2) as

$$S = -\frac{\delta^4(\;)}{16\pi^2 m q^0}\, e_j'^*\, e_i\, M^{ji}$$

and the usual photon <u>scattering amplitude</u> is defined by $-2\pi i q^0$ times the coefficient of the δ-function in S , or

$$f(\underset{\sim}{q}'\lambda',\underset{\sim}{q}\lambda) \longrightarrow \frac{i}{8\pi m}\, e_j'^*\, e_i\, M^{ji}(\underset{\sim}{q}',\underset{\sim}{q},q^0). \quad (2.\text{I}.18)$$

Inserting the limiting formulas (2.G.19) and (2.I.16) in (2.I.17), and recalling that $\underset{\sim}{e}'^*$ and $\underset{\sim}{e}$ are respectively orthogonal to $\underset{\sim}{q}'$ and $\underset{\sim}{q}$, we have then

$$f(\underset{\sim}{q}'\lambda',\underset{\sim}{q}\lambda) \longrightarrow -\frac{e^2}{4\pi m}\,\underset{\sim}{e}'^*\cdot\underset{\sim}{e}\,1 - \frac{ieq^0}{2\pi m}\left(\frac{\mu}{J}-\frac{e}{2m}\right)(\underset{\sim}{e}'^*\times\underset{\sim}{e})\cdot\underset{\sim}{J} +$$

$$-\frac{ie\mu}{4\pi m J q^0}(\underset{\sim}{e}'^*\cdot\underset{\sim}{q})(\underset{\sim}{e}\cdot(\underset{\sim}{J}\times\underset{\sim}{q})) + \frac{ie\mu}{4\pi m J q^0}(\underset{\sim}{e}\cdot\underset{\sim}{q}')(\underset{\sim}{e}'^*\cdot(\underset{\sim}{J}\times\underset{\sim}{q}')) +$$

$$-\frac{\mu^2}{4\pi J^2 q^0}\left[\underset{\sim}{e}'^*\cdot(\underset{\sim}{J}\times\underset{\sim}{q}'),\ \underset{\sim}{e}\cdot(\underset{\sim}{J}\times\underset{\sim}{q})\right] + \text{quadratic} \quad (2.\text{I}.19)$$

a place even in calculations involving hadrons. However, if we tried to use our Lagrangian to calculate the ω^2 or higher terms, we would have no reason to expect that the result calculated to zeroth order in the strong interaction coupling constants need have any connection with reality.

J. The Drell-Hearn Sum Rule

One other important property of the photon forward-scattering amplitudes, which we have not yet used, is their asymptotic behaviour at high energy. If this asymptotic behaviour is sufficiently good, then the amplitudes will satisfy unsubtracted dispersion relations, and by evaluating the dispersion integrals at zero photon energy we can derive useful sum rules.

To see how this works in detail, consider the two forward-scattering amplitudes

$$f_+(\omega^2) \equiv \tfrac{1}{2}\left[f(\omega,+1) + f(-\omega,+1)\right] \qquad (2.J.1)$$

$$f_-(\omega^2) \equiv \tfrac{1}{2\omega}\left[f(\omega,+1 - f(-\omega,+1)\right] \qquad (2.J.2)$$

The covariant amplitude $M^{\nu\mu}(q',q)$ is unchanged when we interchange μ with ν and q with $-q'$, so the forward scattering amplitude is unchanged if we replace ω by $-\omega$ and polarization vectors by their complex conjugates, which amounts to replacing λ by $-\lambda$:

$$f(\omega,\lambda) = f(-\omega,-\lambda) \qquad (2.J.3)$$

Hence our two amplitudes $f_\pm(\omega^2)$ may be written in terms of the physical forward scattering amplitudes $f(\omega,\lambda)$ for $\omega > 0$:

$$f_+(\omega^2) = \tfrac{1}{2}\left[f(\omega,+1) + f(\omega,-1)\right] \qquad (2.J.4)$$

$$f_-(\omega^2) = \tfrac{1}{2\omega}\left[f(\omega,+1) - f(\omega,-1)\right] \qquad (2.J.5)$$

The leading term in (2.I.19) is the famous Thomson scattering amplitude, and yields a total cross-section $\sigma_T = \frac{8\pi e^2}{3m^2}$ in the limit of zero photon energy. (Recall that $e^2 = \frac{4\pi}{137}$ in our notation.) The remaining terms provide a correction, valid to first order in the photon energy, for target particles of arbitrary spin.

We will chiefly be interested in the special case of forward photon scattering. Let us take $\mathbf{q} = \mathbf{q}'$ in the z-direction, so that

$$\mathbf{q} = \mathbf{q}' = (0,0,\omega) \quad , \quad q^0 \equiv \omega$$

The photon helicity is then conserved, so $\lambda = \lambda' = \pm 1$, and

$$\underset{\sim}{e} = \underset{\sim}{e}' = \frac{1}{\sqrt{2}}(1, i\lambda, 0)$$

A straightforward calculation then gives

$$f(\omega,\lambda) \longrightarrow -\frac{e^2}{4\pi m}1 + \frac{\lambda J_z \omega}{4\pi}\left(\frac{\mu}{J} - \frac{e}{m}\right)^2 + O(\omega^2) \qquad (2.I.20)$$

Note that the term of order ω receives contributions both from N and from the poles.

This has been an unpleasantly long calculation. However, once we know that we can get a definite answer in this way up to first order in the photon energy, we can use a trick to get the precise answer by much more familiar means. Note that (2.I.19) and (2.I.20) give results which are of second order in the coupling constants μ and e , but which do not involve the strong interaction coupling constants, except in so far as the strong interactions play a role in determining the physical values of μ and m . If we take any Lagrangian for the strong interactions, and put in photon couplings in a manner consistent with the requirements of gauge invariance of the second kind, we will necessarily get (2.I.19) and (2.I.20) to zeroth order in the strong interactions, provided we insert the physical values for the magnetic moment, charge, and mass of the hadron target. In this way, phenomenological Lagrangians find

The amplitudes $f(\omega,\lambda)$ are real analytic functions of ω, except for cuts on the real ω-axis, which for $\omega > 0$ are given by the optical theorem

$$\mathrm{Im}\, f(\omega,\lambda) = \frac{\omega}{4\pi}\,\sigma(\omega,\lambda) \qquad (2.J.6)$$

where σ is the total cross-section for a photon of helicity λ and energy ω on our target particle. It follows that $f_\pm(\omega^2)$ are real analytic functions of ω^2, except for cuts on the positive real ω^2 axis, with

$$\mathrm{Im}\, f_+(\omega^2) = \frac{\omega}{8\pi}\left[\sigma(\omega,+1) + \sigma(\omega,-1)\right] \qquad (2.J.7)$$

$$\mathrm{Im}\, f_-(\omega^2) = \frac{1}{8\pi}\left[\sigma(\omega,+1) - \sigma(\omega,-1)\right] \qquad (2.J.8)$$

it being understood that ω is here the positive square root of ω^2. If $f_+(\omega^2)$ and/or $f_-(\omega^2)$ vanish for $|\omega^2| \to \infty$, then they satisfy unsubtracted dispersion relations

$$f_\pm(\omega^2) = \frac{1}{\pi}\int_0^\infty \frac{\mathrm{Im}\, f_\pm(\omega'^2)}{\omega'^2 - \omega^2 - i\epsilon}\, d\omega'^2 \qquad (2.J.9)$$

Using (2.J.7) and (2.J.8) , these relations read

$$f_+(\omega^2) = \frac{1}{4\pi^2}\int_0^\infty \frac{\left[\sigma(\omega',+1) + \sigma(\omega',-1)\right]}{\omega'^2 - \omega^2 - i\epsilon}\,\omega'^2 d\omega' \qquad (2.J.10)$$

$$f_-(\omega^2) = \frac{1}{4\pi^2}\int_0^\infty \frac{\left[\sigma(\omega',+1) - \sigma(\omega',-1)\right]}{\omega'^2 - \omega^2 - i\epsilon}\,\omega' d\omega' \qquad (2.J.11)$$

Let me repeat, that these relations are respectively valid if and only if $f_+(\omega^2)$ and $f_-(\omega^2)$ vanish as $|\omega| \to \infty$.

Now let's apply our low energy theorems. According to (2.I.20) , we have

$$f_+(0,\lambda) = -\frac{e^2}{4\pi m} \qquad (2.J.12)$$

$$f_-(0,\lambda) = \frac{J_z}{4\pi}\left(\frac{\mu}{J} - \frac{e}{m}\right)^2 \qquad (2.J.13)$$

If $f_+(\omega^2)$ vanishes at infinity, then we can conclude from (2.J.10) and (2.J.12) that

$$-\frac{\pi e^2}{m} = \int_0^\infty [\sigma(\omega,+1) + \sigma(\omega,-1)]\, d\omega \qquad (2.J.14)$$

This is clearly impossible, because the cross sections are positive-definite. It is truly remarkable (and very unusual) that a low energy theorem can be used to put a lower bound on the asymptotic behavior of a scattering amplitude. On the other hand, $f_-(\omega^2)$ has a much better chance of vanishing at infinity, both because of the presence of a factor ω in the denominator of (2.J.2) , and because we might guess that the residues of the leading Regge trajectories are independent of helicity. If $f_-(\omega^2)$ does vanish at infinity, then we can conclude from (2.J.11) and (2.J.13) that

$$\pi J_z \left(\frac{\mu}{J} - \frac{e}{m}\right)^2 = \int_0^\infty \frac{[\sigma(\omega,+1) - \sigma(\omega,-1)]}{\omega}\, d\omega \qquad (2.J.15)$$

For instance, for a proton target we have

$$\mu = 2.79\left(\frac{e}{2m}\right), \quad J = 1/2, \quad \frac{e^2}{4\pi} = \alpha$$

so (2.J.15) **gives**

$$\int_0^\infty [\sigma_P(\omega) - \sigma_A(\omega)]\, \frac{d\omega}{\omega} = \frac{2\pi^2 \alpha}{m_p^2}(1.79)^2 = 205\,\mu b \quad (2.J.16)$$

where σ_P and σ_A are the cross sections for parallel $\left(J_z = \frac{\lambda}{2}\right)$ and antiparallel $\left(J_z = -\frac{\lambda}{2}\right)$ photon and proton spins. This is the Drell-Hearn sum rule. [10]

The physical significance of the general result (2.J.15) can best be seen by applying it to a charged particle that does not participate in the strong interactions. In this case the integral in (2.J.15) arises only from higher order weak and electromagnetic interactions, which we have been neglecting, so $\mu \simeq \frac{eJ}{m}$

The gyromagnetic ratio is defined as $\frac{\mu}{J(\frac{e}{2m})}$ so this just says that $g \simeq 2$.

This prediction, as you know, is in essential agreement with the g-values of the electron and muon,

$g_e = 2\times[1.001159557 \pm 0.000000030]$ and $g_\mu = 2\times[1.00116614 \pm 0.00000031]$,

the small corrections being explained by higher-order electromagnetic effects. A classical uniformly charged rotating body would have $g=1$, so it was regarded as a great triumph that Dirac's theory of the electron gave $g_e = 2$.

However, there is nothing in the Dirac theory that really requires a g-value of 2, as shown by the fact that a gauge invariant Pauli term (2.B.15) can be added to the Lagrangian with an arbitrary coefficient μ , which could alter g. It is sometimes said that a Pauli term does not appear in the electron Lagrangian because it represents a "non-minimal" coupling, which is not required by the gauge-invariant procedure of replacing all derivatives $\partial/\partial x^\mu$ by the gauge invariant derivatives D_μ , defined by Eq. (2.B.5). However, this is a pretty thin justification, because we can add to the electron kinematic Lagrangian a term $-i(\frac{\mu'}{e})\bar{\psi}\sigma^{\mu\nu}\partial_\mu\partial_\nu\psi$ which vanishes identically, but which generates the Pauli term (2.B.15) when ∂_μ is replaced with D_μ . Anyway, where is it written that an interaction which is not compulsory is forbidden? A much better reason that is often offered for the absence of a Pauli term is that it would be non-renormalizable. Renormalizability is a matter of asymptotic behaviour at infinite momentum, so it is not surprising that in quantum electrodynamics renormalizability would be achieved for just that zeroth order g-value, $g=2$, which is required by the condition that $f_-(\omega^2)$ vanish as $\omega \to \infty$ in forward Compton scattering.

These rather academic remarks would become much more interesting if the experimental physicists would at last find the spin-one charged "intermediate boson" which has been hypothesized as the agent of the weak interactions. As far as I know, the electrodynamics of charged particles of unit spin is unrenormalizable for any value of g. However, even if we can't calculate them, higher order electromagnetic (and weak) effects on photon-intermediate boson scattering amplitudes ought to be small, so if these amplitudes behave well at infinity, we must again have $g \simeq 2$. This conclusion is by no means universally accepted - indeed, in a recent issue of the Physical Review Letters, Hagen and Hurley present a formal argument that g should be $\frac{1}{J}$ for a particle of spin J.[11] One of the many reasons for hoping that an intermediate boson can be found is that the value of its magnetic moment would tell us a good deal about what principles really determine the electromagnetic properties of the elementary particles.

K. Other Soft Photon Theorems and Sum Rules

The problem we solved in the last three sections was complicated by the presence of two soft photons, one incoming and one outgoing. We can also derive useful results for the emission or absorption of one soft photon.[12]

Consider some process $i \to f$, where $|i\rangle$ and $|f\rangle$ are single-particle or multiparticle states which can have $p_f^\mu = p_i^\mu$. The matrix element for absorbing or emitting a soft photon of momentum q^μ in this process will have the form

$$M_{fi}^\mu(q) = P_{fi}^\mu(q) + N_{fi}^\mu(q) \qquad (2.K.1)$$

where $P_{fi}^\mu(q)$ contains all the terms which can become singular near $q^\mu = 0$ and $N_{fi}^\mu(q)$ contains everything else. The important thing about $P^{\mu}_{fi}(q)$ is that it is explicitly known in terms of the matrix element M_{fi} for the process $i \to f$, because the only terms in M^{μ}_{fi} which can have singularities near $q^{\mu}=0$ are those arising from diagrams in which the soft photon is emitted or absorbed from an external line of the original process. In contrast, $N^{\mu}_{fi}(q)$ is not known, but it is analytic near $q^{\mu}=0$, so it can be determined at $q^{\mu}=0$ from the conservation condition

$$0 = q_{\mu} M^{\mu}_{fi}(q) = q_{\mu} P^{\mu}_{fi}(q) + q_{\mu} N^{\mu}_{fi}(q) \qquad (2.K.2)$$

as in Section 2.I.

The differences between the present problem and the photon scattering problem discussed in Sections 2.H-J, arise from the different structure of the pole term P^{μ}_{fi}. The contribution from any one pole diagram, say one in which a soft photon is emitted from the n-th particle in the final or initial states, is proportional to a factor

$$[(p_n \pm q)^2 + m_n^2]^{-1} = [\pm 2 p_n \cdot q]^{-1}$$
$$(+ \text{ for } n \text{ in } |f\rangle \text{ ; } - \text{ for } n \text{ in } |i\rangle)$$

from the external line propagator; a factor Γ_n from the photon-hadron vertex; and a factor M_{fi}, with p_n replaced with $p_n + q$. Thus the pole part of M^{μ} may be expanded in terms $P^{(n)\mu}$ of order n in q^{μ}:

$$P^{\mu}_{fi}(q) = P^{(-1)\mu}_{fi}(q) + P^{(0)\mu}_{fi}(q) + O(q) \qquad (2.K.3)$$

Using (2.G2) to determine the photon-hadron vertices at $q^{\mu}=0$, we can easily calculate the term of order -1 in q^{μ} as

$$P^{(-1)\mu}_{fi}(q) = \Delta \left\{ \sum_n \frac{e_n p_n^{\mu}}{p_n \cdot q} \right\} M_{fi} \qquad (2.K.4)$$

where Δ denotes the difference between the final and initial states. (This is discussed in detail in my 1964 Brandeis lectures.) The term $P^{(0)}$ of zeroth order in q^{μ} is more complicated (it is not a constant) but can be explicitly calculated in terms of particle charges e_n and magnetic moments μ_n, **and the matrix element M_{fi} and its first derivatives with respect to external particle momenta. (In our previous problem of elastic photon scattering, the pole terms in M^{ji} started with** order $+1$ in q^μ , because there both hadrons had momenta of order q^μ or q'^μ , so that both hadron-photon vertices started with terms of first order in these momenta.)

The pole term of order -1 in q^μ is separately conserved, because charge is conserved:

$$q_\mu P_{fi}^{(-1)\mu}(q) = \Delta\left\{\sum_n e_n\right\} M_{fi} = 0 \qquad (2.K.5)$$

The pole term of order 0 in q^μ is not generally separately conserved, but $q_\mu P^{(0)\mu}(q)$ is analytic near $q^\mu = 0$, for the same reasons as in Section 2.H. The non-pole term $N^\mu(q)$ is of course analytic near $q^\mu = 0$, so its value at $q^\mu = 0$ can be determined from the conservation condition (2.K. 2), which gives

$$N_{fi}^\mu(0) = -\left[\frac{\partial}{\partial q_\mu}\left\{q_\nu P_{fi}^{(0)\nu}(q)\right\}\right]_{q=0} . \qquad (2.K.6)$$

The soft-photon matrix element can thus be calculated up to order 0 in q^μ :

$$M_{fi}^\mu(q) = P_{fi}^{(-1)\mu}(q) - \left[q_\nu \frac{\partial}{\partial q_\mu} P_{fi}^{(0)\nu}(q)\right]_{q=0} + O(q) \qquad (2.K.7)$$

The only hard part is to calculate the pole terms to order 0 in q^μ .

One important case where this calculation goes through in a rather elegant way is in the problem of inelastic photon scattering, where the initial target particle α is converted into a final particle β of different mass. For definiteness, let's suppose that the scattering is exothermic, with $m_\alpha > m_\beta$

We can think of the scattering process then as a "hard photon" decay, $\alpha \to \beta + \gamma$, with a soft photon being absorbed during the decay process. Let us agree to fix the initial and final target momenta p_α and p_β, and set the final photon momentum q' equal to $p_\alpha - p_\beta + q$, so that q' varies slightly as $q \to 0$. Both q' and q must stay on the light cone, so

$$0 = q'^2 = (p_\alpha - p_\beta + q)^2 = (p_\alpha - p_\beta)^2 + 2q\cdot(p_\alpha - p_\beta) \ ,$$

and therefore, since $p_\alpha - p_\beta$ is fixed,

$$(p_\alpha - p_\beta)^2 \qquad q\cdot(p_\alpha - p_\beta) = 0 \qquad (2.K.8)$$

The incoming photon polarization vector e_μ is orthogonal to q^μ, and since (2.K.8) gives $(q-q')^2 = 0$ e_μ must also be orthogonal to q'^μ, and hence

$$e\cdot(p_\alpha - p_\beta) = e\cdot(q' - q) = 0 \qquad (2.K.9)$$

Also $e_\alpha = e_\beta$ and $e_\gamma = 0$, so the leading pole term (2.K.4) here vanishes

$$P^{(-1)\mu}_{\beta\gamma,\alpha}(q) = 0$$

The matrix element is then given by the zeroth order term in (2.K.7). The momentum dependence of the hard-photon decay matrix element $M_{\beta\gamma,\alpha}$ is entirely determined by Lorentz invariance, and the derivatives of this matrix element with respect to its external momenta can therefore be calculated by performing infinitesimal rotations on its external spins. Specializing to the case where both the soft incoming and hard outgoing photons travel in the $+3$ -direction with equal helicities $\lambda' = \lambda$, and the initial and final targets α, β move in the -3 -direction with equal spin z-components $\sigma' = \sigma$, we find for the inelastic amplitude at zero incoming photon frequency

$$M^{ji}_{\beta\alpha}(0) = \frac{1}{2}\epsilon_{ij3}\Big\{-\Big(\frac{\mu}{J}-\frac{e}{m}\Big)_\alpha\Big(\Gamma^1_{\beta\alpha}J_1+\Gamma^2_{\beta\alpha}J_2\Big) + \\ +\Big(\frac{\mu}{J}-\frac{e}{m}\Big)_\beta\Big(J_1\Gamma^1_{\beta\alpha}+J_2\Gamma^2_{\beta\alpha}\Big)\Big\} \qquad (2.K.10)$$

with Γ^μ a covariant vertex defined just as in (2.G.1) . A sum rule can be derived from this result, but the imaginary part of the relevant photon scattering amplitude cannot here be expressed in terms of observable cross sections.

More low energy theorems and sum rules can be derived if we make use of the fact that the electromagnetic current of the hadrons consists of an isoscalar plus an isovector part. For instance, consider the forward elastic scattering of an isovector photon, with initial isovector index a and final isovector index b . (Here a and b run over the values 1, 2, 3.) By the same method used in Sections 2.I. and 2.H., we can prove a low energy theorem for this forward scattering amplitude

$$f_{ba}(\omega,\lambda',\lambda) = -\frac{1}{8\pi m}\{t_a,t_b\}\delta_{\lambda'\lambda} + \\ +\frac{i\lambda}{8\pi m}\epsilon_{abc}t_c g_v J_z\delta_{\lambda'\lambda} + \\ +\frac{i\omega}{2\pi}\epsilon_{abc}t_c\Big\{\delta_{\lambda'\lambda}\Big(-\frac{1}{6}R_v^2+\frac{J(J+1)}{24m^2}g_v^2\Big) + \\ + e_j'^{*}e_i\Big[\frac{1}{2}\{J_i,J_j\}-\frac{1}{3}\delta_{ij}J(J+1)\Big]\Big[Q_v+\frac{\lambda\lambda' g_v^2}{8m^2}\Big]\Big\} + \\ -\frac{\omega}{32\pi m^2}\delta_{\lambda'\lambda}\lambda\{t_a,t_b\}J_z(g_v-2)^2+O(\omega^2) \qquad (2.K.11)$$

where g_v , R_v^2 , and Q_v are the isovector parts of the gyromagnetic ratio, charge radius-squared, and quadrupole moment of the target particle, with a factor e extracted in defining the isovector current. The whole amplitude can be divided into four parts, according as it is even or odd under a sign change of ω and an interchange of a and b, and a sum rule can be derived from (2.K.11) for each of these parts which satisfies an unsubtracted dispersion relation. The sum rules are listed below:

photon frequency	photon isovector indices	sum rule
even	even	none
odd	even	isovector Drell-Hearn
even	odd	Beg(?) [13]
odd	odd	Cabibbo-Radicati [14]

The even-even part cannot satisfy an unsubtracted dispersion relation, for the same reason as in the last section, and the even-odd part probably doesn't either.

Finally, one can also derive low energy theorems for the leading part of an amplitude involving the emission and or absorption of an arbitrary number of soft photons. This leading part arises solely from diagrams in which the photons are absorbed or emitted from external lines. If N soft photons with momenta $q_1 \cdots q_N$ and polarization vectors $e_1 \cdots e_N$ are emitted from a particular outgoing charged-particle line of momentum p^μ, the contribution to the over-all matrix element is proportionsl to a factor

$$e^N (e_1 \cdot p)(e_2 \cdot p) \cdots (e_N \cdot p) \left\{ \frac{1}{(p \cdot q_1)(p \cdot [q_1 + q_2]) \cdots (p \cdot [q_1 + \cdots + q_N])} + \text{permutations} \right\} \qquad (2.K.12)$$

Amazingly, when we sum over all permutations of $1, 2, \cdots, N$, we find that the whole product factorizes into

$$\left[e \, \frac{p \cdot e_1}{p \cdot q_1} \right] \left[e \, \frac{p \cdot e_2}{p \cdot q_2} \right] \cdots \left[e \, \frac{p \cdot e_N}{p \cdot q_N} \right] \qquad (2.K.13)$$

As a result, the matrix element for emitting an arbitrary number of soft photons is to get a product of factors like that appearing on the right-hand-side of (2.K.4), one factor for each soft photon. It is this factorization that allows the elimination of infrared divergences, a la Bloch-Nordsieck, to all orders in e . It is also this same factorization, although in a slightly different context, that allows the summation of infinite series of diagrams in the eikonal approximation.

L. Algebraization

The whole point of this chapter has been to emphasize how different are the consequences of a dynamic symmetry like gauge invariance of the second kind from those of an algebraic symmetry like gauge invariance of the first kind. And yet there is a sense in which a dynamic symmetry, together with assumptions about high energy behaviour, can lead to results of a distinctly algebraic flavor.

Let us consider the contribution of a single very narrow resonance to the dispersion integral for the forward elastic scattering amplitude $f_{(-)}(\omega^2)$, with a target particle α at rest. By the same reasoning that led to (2.H.8) , **a discrete state** n of mass m_n **will lead to a pole in** $f(\omega,\lambda)$ **of the form**

$$f(\omega,\lambda) \rightarrow \left(\frac{i}{8\pi m_\alpha}\right) \frac{-i\,(\underset{\sim}{e}^*_\lambda \cdot \underset{\sim}{\Gamma}_{\alpha n})(\underset{\sim}{e}_\lambda \cdot \underset{\sim}{\Gamma}_{n\alpha})}{[\sqrt{m_n^2+\omega^2} - m_\alpha - \omega]\, 2\sqrt{m_n^2+\omega_n^2}} \qquad (2.L.1)$$

$$\text{for } \omega \rightarrow \omega_n$$

where ω_n is the root of the equation

$$\sqrt{m_n^2+\omega_n^2} = m_\alpha + \omega_n \qquad (2.L.2)$$

and $\underset{\sim}{\Gamma}_{\alpha\beta}$ is a covariant vertex defined by

$$\langle \underset{\sim}{p}'\sigma'\beta \,|\, \underset{\sim}{J}(0) \,|\, \underset{\sim}{p}\sigma\alpha \rangle = \frac{\underset{\sim}{\Gamma}_{\beta\alpha}}{(2\pi)^3\sqrt{4E_\alpha E_\beta}} \qquad (2.L.3)$$

where $\underset{\sim}{p}'$ and $\underset{\sim}{p}$ are in the -3 direction, and $(p'-p)$ is in the light cone. (In this case the components of $\underset{\sim}{\Gamma}$ transverse to the direction of $\underset{\sim}{p}$ are invariant under Lorentz boosts along the 3-direction, so no momentum arguments need be written for $\underset{\sim}{\Gamma}$.) The pole energy is, explicitly,

$$\omega_n = \frac{m_n^2 - m_\alpha^2}{2m_\alpha} , \qquad (2.L.4)$$

and (2.L.1) can thus be written

$$f(\omega,\lambda) \to -\frac{1}{32\pi m_\alpha^2} \cdot \frac{(\Gamma^x_{\alpha n} - i\lambda\Gamma^y_{\alpha n})(\Gamma^x_{n\alpha} + i\lambda\Gamma^y_{n\alpha})}{\omega - \omega_n} \qquad (2.L.5)$$

(Spin sums are implicit here.)

The amplitude $f_{(-)}(\omega^2)$ which satisfies an unsubtracted dispersion relation therefore has the pole, given by (2.L.5) and (2.J.5) :

$$f_{(-)}(\omega^2) \to -\frac{1}{32\pi m_\alpha^2} \cdot \frac{1}{\omega^2 - \omega_n^2} \cdot \{(\Gamma^x_{\alpha n} - i\Gamma^y_{\alpha n})(\Gamma^x_{n\alpha} + i\Gamma^y_{n\alpha}) + - (\Gamma^x_{\alpha n} + i\Gamma^y_{\alpha n})(\Gamma^x_{n\alpha} - i\Gamma^y_{n\alpha})\} \qquad (2.L.6)$$

Now let us suppose that the whole absorptive part of $f_{(-)}(\omega^2)$ can be well approximated by a sum, possibly infinite, of discrete contributions from narrow resonances. (Such assumptions underlie all the work on Veneziano models, discussed here by Mandelstam.) Since $f_{(-)}(\omega^2)$ is presumed to satisfy an unsubtracted dispersion relation, it is then given by a sum of terms like (2.L.6) , and in particular

$$f_{(-)}(0) = \frac{1}{4\pi} \sum_{n \neq \alpha} \{(\kappa^+)_{\alpha n} (\kappa)_{n\alpha} - (\kappa)_{\alpha n} (\kappa^+)_{n\alpha}\} , \qquad (2.L.7)$$

where κ is the matrix

$$\kappa_{\beta\alpha} \equiv \frac{\Gamma^x_{\beta\alpha} + i\Gamma^y_{\beta\alpha}}{i\sqrt{2}\,(m_\beta^2 - m_\alpha^2)} \quad \text{for } m_\beta \neq m_\alpha \qquad (2.L.8)$$

(Recall that $\underset{\sim}{\Gamma}$ is Hermitian.) However, the low energy theorem (2.J.13) may be written

$$f_{(-)}(0) = -\frac{1}{4\pi} \{ (\kappa^+)_{\alpha\alpha} (\kappa)_{\alpha\alpha} - (\kappa)_{\alpha\alpha} (\kappa^+)_{\alpha\alpha} \} , \qquad (2.L.9)$$

where now

$$\kappa_{\alpha\alpha} \equiv \frac{1}{\sqrt{2}} (J_x + iJ_y) \left(\frac{\mu}{J} - \frac{e}{m}\right)_\alpha \qquad (2.L.10)$$

Hence the low energy theorem, plus the narrow-resonance-saturated unsubtracted dispersion relation for $f_{(-)}(\omega^2)$, leads to the compact formula

$$[\mathcal{K}, \mathcal{K}^{+}]_{\alpha\alpha} = 0 \tag{2.L.11}$$

Even better, the same approach, when applied to the inelastic forward scattering low energy theorem (2.K. 10), leads to the result

$$[\mathcal{K}, \mathcal{K}^{+}]_{\beta\alpha} = 0 \quad \text{for } m_\beta \neq m_\alpha \tag{2.L.12}$$

Both the elastic result (2.L. 11) and the inelastic result (2.L. 12) can thus be put together as a single statement

$$[\mathcal{K}, \mathcal{K}^{+}] = 0 \tag{2.L.13}$$

Thus, out of a dynamical symmetry, we extract a Lie algebra $\{\mathcal{K}, \mathcal{K}^{+}\}$, albeit one of a very simple kind. The Lie algebra connects states of different mass, so it does not lead immediately to any very simple predictions about actual physical particles, but it puts constraints on anomalous magnetic moments and photon transition amplitudes of a useful, if complicated, sort.

A great deal of work remains to be done in deriving various sum rules and algebraizing them. For instance, what are the sum rules for inelastic isovector photon scattering, or for photon scattering on a multiparticle state, and how do they look when saturated with narrow resonances?

I have not discussed a different, and often more direct, method for deriving sum rules with algebraic interpretations, that of saturating current commutation relations at infinite momentum. This omission is for two reasons: I do not understand in detail the relation between the validity conditions for the infinite momentum method and the assumptions about the asymptotic behaviour of physical scattering amplitudes used in deriving sum rules from low energy theorems; and I do not see how the infinite momentum method can be extended to processes more complicated than simple scattering.

III. LOCAL SYMMETRIES

A. General Realizations

To start in a very general way, let us now consider an abstract Lie group G , which transforms a set of fields ψ_n into functions of the fields:

$$\psi_n(x) \longrightarrow f_n(\psi(x); g) \quad \text{for } g \in G \qquad (3.A.1)$$

The difference between such groups and the older dynamic symmetries, gauge invariance of the second kind and general covariance, is that f_n is now not allowed to depend explicitly on position in space-time. In order that (3.A.1) be consistent with the structure of the group, it is necessary that we get the same result when we apply g_2 and then apply g_1 as when we apply the group element $g_1 g_2$:

$$f_n(\psi; g_1 g_2) = f_n(f(\psi; g_2); g_1) \qquad (3.A.2)$$

For a Lie group, we can restrict our attention to a neighborhood of the identity, where $g = 1 + i\,\epsilon_\alpha \Gamma_\alpha$ (3.A.3) with Γ_α a set of group generators and ϵ_α a set of infinitesimal parameters. The group structure is then embodied in a set of commutation relations

$$[\Gamma_\alpha , \Gamma_\beta] = i\, C_{\alpha\beta\gamma} \Gamma_\gamma \tag{3.A.4}$$

with $C_{\alpha\beta\gamma}$ an array of constants, called the structure constants of the group. The Γ_α are said to span an abstract Lie algebra. The change in any field ψ_n under the infinitesimal transformation (3.A.2) is assumed to be linear in the ϵ_α

$$f_n(\psi ; 1 + i\epsilon_\alpha \Gamma_\alpha) = \psi_n + i\, f_{n\alpha}(\psi)\, \epsilon_\alpha \quad ; \tag{3.A.5}$$

the functions $f_{n\alpha}(\psi)$ characterizing the action of the Lie algebra on ψ

For a given group or algebra, the functions $f_{n\alpha}$ cannot be chosen in a perfectly arbitrary way, but must obey certain differential equations. To derive these equations, consider two arbitrary elements of the group g_1, g_2, and apply (3.A.2) to the group element $g_1 g_2 g_1^{-1}$:

$$f_n(\psi ; g_1 g_2 g_1^{-1}) = f_n(f(\psi ; g_2 g_1^{-1}) ; g_1) =$$
$$= f_n(f(f(\psi ; g_1^{-1}) ; g_2) ; g_1) \tag{3.A.6}$$

Now let g_2 approach the identity

$$g_2 \longrightarrow 1 + i\epsilon_\alpha \Gamma_\alpha \tag{3.A.7}$$

so that

$$g_1 g_2 g_1^{-1} \longrightarrow 1 + i\epsilon_\alpha M_{\alpha\beta}(g_1) \Gamma_\beta \tag{3.A.8}$$

where M is a matrix defined by

$$M_{\alpha\beta}(g)\, \Gamma_\beta \equiv g\, \Gamma_\alpha\, g^{-1} \tag{3.A.9}$$

In this limit, Eq. (3.A.6) reads

$$f_{n\beta}(\psi)\,M_{\alpha\beta}(g_1) = f_{n,m}(f(\psi;g_1^{-1});g_1)f_{m\alpha}(f(\psi;g_1^{-1})), \tag{3.A.10}$$

where

$$f_{n,m}(\psi;g) \equiv \frac{\partial f_n(\psi;g)}{\partial \psi_m} . \tag{3.A.11}$$

Now let g_1 approach the identity :

$$g_1 \longrightarrow 1 + i\,\epsilon'_\gamma \Gamma_\gamma , \tag{3.A.12}$$

so that (3.A.9) and (3.A.4) give

$$M_{\alpha\beta}(g_1) \longrightarrow C_{\alpha\gamma\beta}\,\epsilon'_\gamma . \tag{3.A.13}$$

In this limit, Eq. (3.A.10) reads

$$f_{n\beta}(\psi)\,C_{\alpha\gamma\beta} = -i\,\frac{\partial f_{n\alpha}(\psi)}{\partial \psi_m}\,f_{m\gamma}(\psi) + i\,\frac{\partial f_{n\gamma}(\psi)}{\partial \psi_m}\,f_{m\alpha}(\psi) . \tag{3.A.14}$$

Any set of functions $f_{n\alpha}(\psi)$ which satisfy these equations will be said to provide a realization of the Lie algebra. A representation of the algebra is just a linear realization, with

$$f_{n\alpha}(\psi) = (\tau_\alpha)_{nm}\,\psi_m \tag{3.A.15}$$

In this case the fundamental equations (3.A.14) read

$$i\,C_{\alpha\gamma\beta}\,\tau_\beta = [\tau_\alpha, \tau_\gamma] \tag{3.A.16}$$

which just says that the matrices τ_α form a representation of the Lie algebra Γ_α .

A familiar feature of Lie algebra representation theory is that two representations τ_α and τ'_α are equivalent if there is a non-singular matrix M such that $\tau'_\alpha = M\tau_\alpha M^{-1}$. Similarly, we say that two realizations $f_{n\alpha}(\psi)$ and $f'_{n\alpha}(\psi')$ are equivalent if we can go from one to the other by a functional transformation $\psi \longrightarrow \psi'(\psi)$. In this case, the the transformation rule for ψ' under an infinitesimal group transformation $1 + i\epsilon_\alpha \Gamma_\alpha$ will be

$$\psi'_n \longrightarrow \psi'_n\left(\psi_m + i\epsilon_\alpha f_{m\alpha}(\psi)\right) = \psi'_n(\psi) + i\epsilon_\alpha \frac{\partial \psi'_n(\psi)}{\partial \psi_m} f_{m\alpha}(\psi)$$

and hence

$$f'_{n\alpha}\left(\psi'(\psi)\right) = \frac{\partial \psi'_n(\psi)}{\partial \psi_m} f_{m\alpha}(\psi) \qquad (3.A.17)$$

Two realizations $f'(\psi')$ and $f(\psi)$ are thus equivalent if and only if it is possible to construct a function $\psi'(\psi)$ satisfying (3.A.17). The reader may check that if $f_{n\alpha}(\psi)$ satisfies the fundamental differential equations (3.A.14), then so does the function $f'_{n\alpha}(\psi')$ defined by (3.A.17).

B. Currents

Symmetries are generally made useful by exploiting the properties of the currents they generate. By analogy with the electromagnetic current, these are here defined by

$$J_\alpha{}^\mu \equiv -i \frac{\partial \mathcal{L}}{\partial(\partial_\mu \psi_n)} f_{n\alpha}(\psi) \qquad (3.B.1)$$

(The repeated index n implies a sum over field types.) By use of the Euler-Lagrange equations, we can write the divergence of the current as

$$\partial_\mu J_\alpha{}^\mu = -i \frac{\partial \mathcal{L}}{\partial \psi_n} f_{n\alpha}(\psi) - i \frac{\partial \mathcal{L}}{\partial(\partial_\mu \psi_n)} \partial_\mu f_{n\alpha}(\psi)$$

But then $-\epsilon_\alpha \partial_\mu J_\alpha{}^\mu$ is just the change in $\mathcal{L}$ under the infinitesimal transformation $1 + i\epsilon_\alpha \Gamma_\alpha$ and so if $\mathcal{L}$ is invariant under this Lie algebra, we have

$$\partial_\mu J_\alpha{}^\mu = 0 \qquad (3.B.2)$$

In addition to this conservation property, the currents also satisfy important commutation rules. Note first that

$$J_\alpha{}^0 = -i\,\pi_n f_{n\alpha}(\psi) \;, \qquad (3.B.3)$$

where π_n is the canonical conjugate to ψ_n :

$$\pi_n \equiv \frac{\partial \mathcal{L}}{\partial(\partial_0 \psi_n)} \qquad (3.B.4)$$

At equal times, the quantities π_n and ψ_n obey canonical commutation (or anti-commutation) rules

$$\left[\pi_n(\underset{\sim}{x},t),\, \psi_m(\underset{\sim}{y},t)\right]_\pm = -i\,\delta^3(\underset{\sim}{x}-\underset{\sim}{y})\,\delta_{nm}$$

$$\left[\pi_n(\underset{\sim}{x},t),\, \pi_m(\underset{\sim}{y},t)\right]_\pm = \left[\psi_n(\underset{\sim}{x},t),\, \psi_m(\underset{\sim}{y},t)\right]_\pm = 0$$

$$(3.B.5)$$

so (3. B. 3) gives the commutation (not anti-commutation) rules

$$\left[J_\alpha^0(\underset{\sim}{x},t),\, \psi_n(\underset{\sim}{y},t)\right] = -\,\delta^3(\underset{\sim}{x}-\underset{\sim}{y})\, f_{n\alpha}(\psi(\underset{\sim}{y},t)) \qquad \leftarrow (3.B.6)$$

$$(3.B.7)$$

$$\left[J_\alpha^0(\underset{\sim}{x},t),\, f_{n\beta}(\psi(\underset{\sim}{y},t))\right] = -\,\delta^3(\underset{\sim}{x}-\underset{\sim}{y})\, \frac{\partial f_{n\beta}(\psi(\underset{\sim}{y},t))}{\partial \psi_m(\underset{\sim}{y},t)}\, f_{m\alpha}(\psi(\underset{\sim}{y},t))$$

$$\left[J_\alpha^0(\underset{\sim}{x},t),\, \pi_n(\underset{\sim}{y},t)\right] = \delta^3(\underset{\sim}{x}-\underset{\sim}{y})\, \pi_m(\underset{\sim}{y},t)\, \frac{\partial f_{m\alpha}(\psi(\underset{\sim}{y},t))}{\partial \psi_n(\underset{\sim}{y},t)} \qquad (3.B.8)$$

From (3. B. 7) and (3. B. 8) we find

$$\left[J_\alpha^0(\underset{\sim}{x},t),\, J_\beta^0(\underset{\sim}{y},t)\right] = i\,\delta^3(\underset{\sim}{x}-\underset{\sim}{y})\, \pi_n(\underset{\sim}{y},t)\, \frac{\partial f_{n\beta}(\psi(\underset{\sim}{y},t))}{\partial \psi_m(\underset{\sim}{y},t)}\, f_{m\alpha}(\psi(\underset{\sim}{y},t)) \;+$$

$$-\,i\,\delta^3(\underset{\sim}{x}-\underset{\sim}{y})\, \pi_m(\underset{\sim}{y},t)\, \frac{\partial f_{m\alpha}(\psi(\underset{\sim}{y},t))}{\partial \psi_n(\underset{\sim}{y},t)}\, f_{n\beta}(\psi(\underset{\sim}{y},t))$$

The indices n and m may be interchanged in the last term, and the result evaluated with the aid of the fundamental differential equations (3. A. 14) . We find then

$$\left[J_\alpha^0(\underset{\sim}{x},t)\,,\,J_\beta^0(\underset{\sim}{y},t)\right] = i\,C_{\alpha\beta\gamma}\,J_\gamma^0(\underset{\sim}{y},t)\,\delta^3(\underset{\sim}{x}-\underset{\sim}{y}) \tag{3.B.9}$$

That is, the "densities" J_α^0 form an algebra which mimics the abstract Lie algebra of the underlying symmetry group. This is a familiar enough result for theories in which the fields provide representations of the symmetry group, but I wanted to make it clear that this result also holds when the fields provide more general, non-linear, realizations of the group.

C. Goldstone Bosons

Normally it is a quick and easy step from the current properties discussed in the last section to the algebraic properties of the S-matrix. Since J_α^μ is conserved, we can construct "charges"

$$Q_\alpha = \int d^3x\; J_\alpha^0(\underset{\sim}{x},t) \tag{3.C.1}$$

which are time-independent and hence commute with the Hamiltonian ,

$$\left[Q_\alpha\,,H\right] = 0\ , \tag{3.C.2}$$

and, it follows, also with the S-matrix

$$\left[Q_\alpha\,,S\right] = 0\,. \tag{3.C.3}$$

These charges satisfy commutation relations given by (3. B. 9) :

$$\left[Q_\alpha\,,Q_\beta\right] = i\,C_{\alpha\beta\gamma}\,Q_\gamma\ . \tag{3.C.4}$$

Taken together, (3. C. 2) -- (3. C. 4) would impose a multiplet structure on the spectrum of physical states, and would provide "Clebsch-Gordan" rules for ratios of matrix elements for transitions among these states.

It is nevertheless possible to have a symmetry of the Lagrangian which is not visible upon inspection of the Rosenfeld wallet-card table of elementary particle states. One way is to have massless bosons in the theory, which produce long-range forces which prevent the convergence of the integral (3.C.1). Another possibility, which was suggested by Nambu, Bogoliubov, and others on the basis of an analogy with the BCS theory of superconductivity, is to have a degenerate vacuum, which would infect all multiparticle states built upon this vacuum with a hidden degeneracy.[15] Examples of this latter sort were examined by Nambu, Goldstone, and others around 1960, and turned out always to involve massless bosons, leading us back to the first alternative. Goldstone conjectured that this would always be so, that to each Lie generator Γ_α which was a good symmetry of the Lagrangian but not of physical Hilbert space there would have to correspond a massless spin-zero boson.[16] This conjecture was subsequently proved in 1962 by Goldstone, Salam, and myself.[17] I will give a generalized version of one of the proofs we offered then.

Consider a general "Green's function", defined as a vacuum expectation value of a time-ordered product of an arbitrary number of fields,

$$\langle T\{\psi_n(x),\ \psi_m(y),\quad\}\rangle_0 \tag{3.C.5}$$

Suppose for a moment that the charge (3.C.1) exists, and annihilates the vacuum:

$$Q_\alpha|0\rangle = 0$$

According to (3.B.6), we have

$$\left[Q_\alpha, \psi_n(x)\right] = - f_{n\alpha}(\psi(x))$$

so by commuting Q_α past all the ψ's in (3.C.5), we would find

$$0 = \delta_\alpha \langle T\{\psi_n(x), \psi_m(y), \cdots\}\rangle_0$$

where

$$\delta_\alpha \langle T\{\psi_n(x), \psi_m(y), \cdots\}\rangle_0 \equiv \langle T\{f_{n\alpha}(\psi(x)), \psi_m(y), \cdots\}\rangle_0 +$$
$$+ \langle T\{\psi_n(x), f_{m\alpha}(y), \cdots\}\rangle_0 + \cdots$$

Now suppose that the symmetry is broken, either because Q_α doesn't exist or because it doesn't annihilate the vacuum, in such a way that

$$\delta_\alpha \langle T\{\psi_n(x), \psi_m(y), \cdots\}\rangle_0 \neq 0 . \qquad (3.C.6)$$

To prove the necessity of massless bosons, we use (3.B.2) and (3.B.6), which give

$$\frac{\partial}{\partial z^\mu} \langle T\{J^\mu_\alpha(z), \psi_n(x), \psi_m(y), \cdots\}\rangle_0 =$$
$$= - \delta^4(z-x) \langle T\{f_{n\alpha}(\psi(x)), \psi_m(y), \cdots\}\rangle_0 +$$
$$- \delta^4(z-y) \langle T\{\psi_n(x), f_{m\alpha}(y), \cdots\}\rangle_0 +$$
$$- \cdots$$

Multiplying with $e^{-iq\cdot z}$, integrating over z and letting $q\to 0$, then gives

$$\lim_{q\to 0} iq_\mu \int d^4z\, e^{-iq\cdot z} \langle T\{ J_\alpha^\mu(z), \psi_n(x), \psi_m(y), \cdots\}\rangle_0 =$$

$$= - \delta_\alpha \langle T\{ \psi_n(x), \psi_m(y), \cdots\}\rangle_0 \qquad (3.C.7)$$

Clearly (3.C.6) is impossible unless the Fourier transform on the left-hand side of (3.C.7) has a singularity at $q_\mu \to 0$. Such a singularity can arise from a zero-mass spin zero particle which couples to the current J_α^μ In this case, (3.C.7) can be satisfied with

$$\int d^4z\, e^{-iq\cdot z} \langle T\{ J_\alpha^\mu(z), \psi_n(x), \psi_m(y), \cdots\}\rangle_0 \to \frac{iq^\mu}{q^2} \delta_\alpha \langle T\{\psi_n(x)\psi_m(y)\cdots\}\rangle_0 , \qquad (3.C.8)$$

the pole at $q^2 = 0$ coming from the massless boson propagator. However, if there were no massless particles in the theory, there would be no way that this Fourier transform could acquire a singularity at $q^\mu = 0$, and (3.C.7) would have to vanish .

The original proof of Goldstone, Salam, and myself **followed the same lines,** but made use of the single-field **case of Eq. (3.C.7) :**

$$\lim_{q\to 0} iq_\mu \int d^4z\, e^{-iq\cdot z} \langle T\{ J_\alpha^\mu(z), \psi_n(x)\}\rangle_0 \to$$

$$\to - \delta_\alpha \langle \psi_n(x)\rangle_0 \equiv - \langle f_{n\alpha}(\psi(x))\rangle_0$$

which requires a massless boson whenever the vacuum expectation values of the $f_{n\alpha}(\psi)$ do not vanish. The present proof has the advantage that massless bosons are required whenever any Green's function violates a symmetry of the Lagrangian.

Eq. (3.C.8) not only tells us when a theory needs massless bosons; it also tells us how strongly these bosons are coupled to various sets of fields. Note the reciprocity embodied in (3.C.8) : for processes in which a Goldstone boson is not strongly emitted or absorbed, the corresponding symmetry plays the role of an approximate algebraic symmetry, which requires $\delta_\alpha \langle\cdots\rangle$ to be small, while for processes in which a Goldstone boson is strongly coupled, the corresponding symmetry plays a purely dynamical role, dictating the matrix elements for the Goldstone boson emission and absorption processes.

D. Non-linear Realizations for Goldstone Bosons

In the general case, a group G of local symmetries may have a subgroup H of algebraic symmetries, like isospin or hypercharge, which are not associated with Goldstone bosons. The Lie algebra $\{\Gamma_\alpha\}$ then breaks up into a set of generators T_i for the algebraic subgroup H, and a set of generators X_a, each of which is associated with a Goldstone boson. It is natural to look for a realization in which the Goldstone boson fields $\xi_a(x)$ transform into functions of themselves, which are required to be linear for the "ordinary" symmetries T_i, but may be nonlinear for the algebraic symmetries X_a .

A general algorithm for constructing such realizations is provided by the work of Callan, Coleman, Wess, and Zumino.[18,20] Consider the product of a general group element g with the group element $\exp[i\xi_a X_a]$. Since this product must belong to the group, it can be expressed in the form

$$g \exp[i\xi_a X_a] = \exp[i f_a(\xi;g) X_a] \exp[i \mu_i(\xi;g) T_i] \tag{3.D.1}$$

The CCWZ realization is defined by setting the transformed Goldstone boson field equal to f_a:

$$\xi'_a(x) \equiv f_a(\xi(x);g) \tag{3.D.2}$$

Note that this is a realization, because

$$g_1 g_2 \exp[i\xi_a X_a] = g_1 \exp[i f_a(\xi;g_2) X_a] \exp[i\mu_i(\xi;g_2)T_i] =$$

$$= \exp[i f_a(f(\xi;g_2);g_1) X_a] \exp[i\mu_i(f(\xi;g_2);g_1)T_i] \exp[i\mu_i(\xi;g_2)T_i] \, . \tag{3.D.3}$$

Since H is a subgroup, the product of the last two factors can be expressed as a single exponential

$$\exp[i\mu_i(\xi;g_1 g_2)T_i] \equiv \exp[i\mu_i(f(\xi;g_2);g_1)T_i] \exp[i\mu_i(\xi;g_2)T_i] \tag{3.D.4}$$

and (3.D.3) is thus of the same form as (3.D.1) , with

$$f_a(\xi;g_1 g_2) = f_a(f(\xi;g_2);g_1) \tag{3.D.5}$$

as required for a realization of G . Also note that if g

belongs to the algebraic subgroup H, then

$$g \exp[i\xi_a X_a] = \exp[i\xi_a h_{ab} X_b]\, g \ , \qquad (3.D.6)$$

where h_{ab} is the representation of H afforded by the generators X_a:

$$g X_a g^{-1} = h_{ab}(g) X_b \quad \text{for } g \in H \qquad (3.D.7)$$

But any element $g \in H$ can be expressed in the form

$$g \equiv \exp[i u_i(g) T_i] \quad \text{for } g \in H \ , \qquad (3.D.8)$$

so (3.D.6) has the same form as (3.D.1), with

$$u_i(\xi; g) = u_i(g) \qquad (3.D.9)$$

and with ξ_a transforming linearly

$$f_b(\xi; g) = h_{ab}(g)\xi_a \quad \text{for } g \in H \qquad (3.D.10)$$

as we wished.

We see that Eq. (3.D.1) does in general provide us with a realization of the group G, in which the Goldstone boson fields transform linearly under the subgroup H of "ordinary" symmetries. It is natural to ask whether, in this sense, this realization is unique. Uniqueness here must of course be understood to mean, not that there is no other such realization, but that all such realizations are equivalent, any one being transformed into any other by a suitable functional transformation of the Goldstone boson fields ξ_a. The uniqueness proof for the special case of chiral SU(2) × SU(2) was given by myself by direct solution of the differential equation (3.A.14).[19] Subsequently

Coleman, Wess, and Zumino gave a much more elegant proof for the general case.[20]

Now that we have defined a realization $f_a(\xi;g)$ for the Goldstone boson fields ξ_a, we can use it to construct a large variety of other realizations. Consider a field ψ which transforms under the algebraic symmetries T_i according to some finite matrix representation D of the group H:

$$\psi'(x) = D(g)\,\psi(x) \quad \text{for } g \in H \tag{3.D.11}$$

We will define its transformation under general elements of the larger group G by

$$\psi'(x) \equiv D\left(\exp[i\mu_i(\xi;g)\,T_i]\right)\psi(x) \;, \tag{3.D.12}$$

where μ_i is the function of the Goldstone boson field ξ_a and of g, defined by Eq. (3.D.1). Note from (3.D.8) and (3.D.9) that (3.D.12) reduces to (3.D.11) in the case where $g \in H$. Note also that (3.D.12) does provide us with a realization of G, because if we first act on ψ and ξ_a with g_2 and then with g_1, we find that ψ is transformed into

$$\psi'' = D\left(\exp[i\mu_i(f(\xi;g_2);g_1)\,T_i]\right) \times$$

$$\times\, D\left(\exp[i\mu_i(\xi;g_2)\,T_i]\right)\psi \;.$$

Since D is a representation of H, we can now use (3.D.4), which gives

$$\psi'' = D\left(\exp[i\mu_i(\xi;g_1g_2)\,T_i]\right)\psi$$

as required for a realization of G.

Given a set of fields ξ_a and ψ, how are we to construct a Lagrangian which will be invariant under G? If the Lagrangian were to involve only the fields ψ, and no derivatives, the answer would be simple. We would simply put together any function of the ψ fields which is invariant under the "ordinary" symmetries H, and then (3.D.12) would ensure that it is also invariant under the full group G. However, we want to build up our Lagrangian not only of ψ but also of $\partial_\mu \psi$ and, at least, $\partial_\mu \xi_a$. The solution is to be found by defining <u>covariant derivatives</u>, in much the same way as in general relativity or electrodynamics.

To this end, we shall first define functions $D_{ab}(\xi)$ and $E_{ai}(\xi)$ in terms of variational derivatives of the exponential $\exp[i\xi_a X_a]$. For an infinitesimal variation $\delta\xi_a$ the group element $\exp[-i\xi_a X_a]\exp[i(\xi_a+\delta\xi_a)X_a]$ is close to unity, and so it can be written as

$$\exp[-i\xi_a X_a]\exp[i(\xi_a+\delta\xi_a)X_a] = 1 + i\,\delta\xi_a[D_{ab}(\xi)X_a + E_{ai}(\xi)T_i]. \tag{3.D.13}$$

The covariant derivative of the Goldstone boson field ξ_a is defined by

$$D_\mu \xi_a \equiv \partial_\mu \xi_b D_{ba}(\xi) \tag{3.D.14}$$

and the covariant derivative of a general hadron field ψ is defined by

$$D_\mu \psi \equiv \partial_\mu \psi + i\,\partial_\mu \xi_a E_{aj}(\xi) t_j \psi \quad , \tag{3.D.15}$$

where t_j is the matrix representing the generator T_j in the matrix representation $D(g)$ of H associated with the field ψ. These definitions are chosen so that $D_\mu \xi_a$ and $D_\mu \psi$ will transform just like ψ. Since $h_{ab}(g)$ is just the representation of H appropriate to the Goldstone boson field, this means that

$$D_\mu \xi_a' = D_\mu \xi_b \, h_{ba}\left(\exp[i u_i(\xi; g) T_i]\right) \qquad (3.D.16)$$

and

$$D_\mu \psi' = D\left(\exp[i u_i(\xi; g) T_i]\right) D_\mu \psi \qquad (3.D.17)$$

with ξ' and ψ' given by (3.D.2) and (3.D.11).

In order to verify the desired transformation rules (3.D.16) and (3.D.17), we first differentiate Eq. (3.D.1):

$$g \exp[i\xi_a X_a] \, i\partial_\mu \xi_a \{D_{ab}(\xi) X_b + E_{ai}(\xi) T_i\} =$$

$$= \exp[i\xi_a' X_a] \, i\partial_\mu \xi_a' \{D_{ab}(\xi') X_b + E_{ai}(\xi') T_i\} \exp[i u_i T_i] +$$

$$+ \exp[i\xi_a' X_a] \, \partial_\mu \exp[i u_i T_i]$$

Multiplying on the left and right with $\exp[-i\xi_a' X_a]$ and $\exp[-i u_i T_i]$ respectively gives

$$\partial_\mu \xi_a' \{D_{ab}(\xi') X_b + E_{ai}(\xi') T_i\} =$$

$$= \exp[i u_i T_i] \, \partial_\mu \xi_a \{D_{ab}(\xi) X_b + E_{ai}(\xi) T_i\} \exp[-i u_i T_i] +$$

$$- \left[\partial_\mu \exp[i u_i T_i]\right] \exp[-i u_i T_i] \qquad (3.D.18)$$

Using (3.D.7) and equating coefficients of X_b gives (3.D.16) The rest of (3.D.18) relates elements of the algebra of H :

$$\partial_\mu \xi'_a E_{ai}(\xi') T_i = \{ \exp[iu_i T_i] \partial_\mu \xi_a E_{ai}(\xi) T_i + + i \partial_\mu \exp[iu_i T_i] \} \exp[-iu_i T_i] \tag{3.D.19}$$

The derivative of (3.D.12) gives

$$\partial_\mu \psi' = D(\exp[iu_i T_i]) \partial_\mu \psi + + \partial_\mu D(\exp[iu_i T_i]) \psi \tag{3.D.20}$$

and putting (3.D.19) and (3.D.20) together gives (3.D.17) .

Since ψ, $D_\mu \psi$ and $D_\mu \xi_a$ all have the same transformation rule (3.D.12) , any Lagrangian constructed from just these ingredients will automatically be invariant under the full group G if it is invariant under the subgroup H .

4. CHIRALITY

A. The Pion as a Goldstone Boson

Are there any exact or approximate symmetries of the strong interaction which are manifested dynamically rather than algebraically ? According to the results of the last section, this question may be approached by asking whether there are any massless or nearly massless hadrons which might play the role of a Goldstone boson. Of course, there are no massless hadrons, but there is one spin-zero meson, the pion, whose mass (or rather, squared mass, which is what really matters) is very much smaller than that of any other hadron.

Historically, the idea that the pion might be a Goldstone boson[21] arose from the effort to understand a result of weak interaction theory, the Goldberger-Treiman relation, to which we shall come later on. In line with the task of re-writing history, I want to explore the idea that the pion is a Goldstone boson corresponding to an approximate dynamical symmetry of the strong interactions from the point of view of the strong interactions alone, and put off the application of this idea to the weak interactions until Section 4C.

First, let us adopt the fiction that the pion mass is zero, and that the corresponding dynamical symmetry is exact. That is, there must be a conserved current to which the pion can couple, so that a pion pole can occur in matrix elements like (3.C.8). Since the pion has isospin unity and negative parity, this current must be an isovector axial-vector $A_a^\mu(x)$, where a is a Cartesian isospin index running over the values 1, 2, 3, and

$$\partial_\mu A_a^\mu = 0 \qquad (4.A.1)$$

In addition, these currents must form part of a Lie algebra, in the sense of Eq. (3.B.9), so that

$$\left[A_a^0(\underset{\sim}{x},t),\, A_b^0(\underset{\sim}{y},t)\right] = i\, C_{abn}\, V_n^0(\underset{\sim}{x},t)\,\delta^3(\underset{\sim}{x}-\underset{\sim}{y})$$

where C_{abn} are a set of structure constants, and V_n^μ are a set of conserved currents. In addition, C_{abn} is antisymmetric in a and b, so it must be proportional to the totally antisymmetric tensor ϵ_{abc}

$$C_{abn} = \epsilon_{abc} C_{cn}$$

and isospin conservation then tells us that the current $C_{cn} V_n^\mu$

must be an isovector, which we shall call $2\eta V_c^\mu$. The current commutation relation then reads

$$[A_a^0(\underset{\sim}{x},t), A_b^0(\underset{\sim}{y},t)] = 2i\eta\,\epsilon_{abc} V_c^0(\underset{\sim}{x},t)\,\delta^3(\underset{\sim}{x}-\underset{\sim}{y}) \qquad (4.A.2)$$

with V_c^μ unknown, except that it is some vector isovector current which is conserved,

$$\partial_\mu V_c^\mu = 0 \qquad (4.A.3)$$

Now, what kind of symmetry does the current V_c^μ represent? It might be a dynamical symmetry, but then there would have to be an approximately massless Goldstone boson of isospin unity and spin-parity 0^+. A glance at the wallet card shows that the lightest particle with these quantum numbers is the δ-resonance, with mass 960 MeV, far too high to qualify as any sort of Goldstone boson. We are only left with the possibilities that V_c^μ might vanish, or might be the current of an ordinary algebraic symmetry, with which we would presumably be already familiar. The only algebraic symmetry of the strong interactions whose generators are isovectors is isotopic spin itself, so V_c^μ must be proportional to the current generated by isospin transformations. We still have a constant η at our disposal in (4.A.2), so there is no loss of generality in fixing V_c^μ to be twice the isospin current, with commutation relations given according to (3.B.9) by the structure of the isospin algebra

$$[V_a^0(\underset{\sim}{x},t), V_b^0(\underset{\sim}{y},t)] = 2i\,\epsilon_{abc} V_c^0(\underset{\sim}{x},t)\,\delta^3(\underset{\sim}{x}-\underset{\sim}{y}) \qquad (4.A.4)$$

(The factor 2 is inserted because these currents were first discussed in connection with nucleon β-decay, where it is traditional to write the vector current as $\bar{N}\gamma^{\mu}\underset{\sim}{\tau}N$, with $\underset{\sim}{\tau}$, the Pauli matrix, equal to twice the isospin matrix for $T = 1/2$.) . Further, (3.B.6) shows that the commutator of V_a^0 with any field is given by multiplying the field with $-2\delta^3(\ )$ times the appropriate isospin matrix, so that

$$\left[V_a^0(\underset{\sim}{x},t),\ A_b^0(\underset{\sim}{y},t)\right] = 2i\,\epsilon_{abc}\,A_c^0(\underset{\sim}{x},t)\,\delta^3(\underset{\sim}{x}-\underset{\sim}{y}) \qquad (4.A.5)$$

Thus the densities V_a^0 and A_a^0 span a complete Lie algebra, with just one degree of freedom, represented by the constant η. The currents are Hermitian, so η is real, and without changing the Hermiticity of the currents we can simplify the commutation relations further by absorbing into A^{μ} a factor $\frac{1}{\sqrt{\eta}}$ if $\eta > 0$ or a factor $\frac{1}{\sqrt{-\eta}}$ if $\eta < 0$. With axial currents normalized in this "standard" manner, the commutation relations are given by (4.A.2) , (4.A.4) , and (4.A.5) , with

$$\eta = \begin{cases} +1 \\ 0 \\ -1 \end{cases} \qquad (4.A.6)$$

These three values of η correspond to three different Lie algebras: for $\eta = +1$, we have the compact algebra $SU(2) \times SU(2)$, which is isomorphic to the four-dimensional rotation algebra SO(4) ; for $\eta = 0$ we have a non-semi-simple algebra, with the A_a^0 forming an invariant Abelian sub-algebra; and for $\eta = -1$, we have the non-compact algebra SO(3, 1) . For brevity I will refer to these three possibilities as <u>compact chirality</u> $(\eta = +1)$, <u>Abelian chirality</u>

$(\eta=0)$, and non-compact chirality $(\eta=-1)$. The choice among them must be made on the basis of empirical evidence, by comparing soft-pion theorems with experiment.

It is already apparent from our proof of Goldstone's theorem that the soft pion theorems will involve as a fundamental constant the strength of the pion coupling to the axial current. Lorentz invariance and isospin conservation require that this coupling take the form

$$\langle 0| A_a^{\mu}(x)|\pi_b, q\rangle = \frac{\delta_{ab}\, i q^{\mu} e^{iq\cdot x} F_{\pi}}{(2\pi)^{3/2}\sqrt{2q^0}} \qquad (4.A.7)$$

where F_π is a real number with the dimensions of a mass. For $\eta=\pm 1$ the normalization of the currents is fixed by Eq. (4.A.2), so F_π is a well-defined fundamental constant. For $\eta=0$ the question of the normalization of A_a^{μ} is still open, and can be adjusted to make F_π whatever we like. It should be noted that (4.A.7) is consistent with the notion of a conserved axial-vector current, because it gives $\langle 0|\partial_\mu A^\mu|\pi\rangle$ proportional to q^2, which vanishes for massless pions.

Now let us return to the real world, in which pions have a small mass. We can still suppose that there are vector and axial-vector currents which obey the commutation relations (4.A.2), (4.A.4), and (4.A.5), with V_c^{μ} the conserved current of isotopic spin. Also, (4.A.7) will still be valid, since it follows purely from Lorentz and isospin invariance. However, the axial current can now no longer be conserved (unless $F_\pi=0$, in which case A^μ would have nothing to do with pions) because (4.A.7) now gives

$$\langle 0|\partial_\mu A_a^{\mu}(x)|\pi_b, q\rangle = \frac{\delta_{ab}\, m_\pi^2 F_\pi e^{iq\cdot x}}{(2\pi)^{3/2}\sqrt{2q^0}} \qquad (4.A.8)$$

The best we can do is to assume that $\partial_\mu A_a{}^\mu$ is a small operator, proportional to a factor m_π^2 This assumption is usually implemented by defining a field π_a:

$$\partial_\mu A_a^\mu \equiv F_\pi m_\pi^2 \pi_a \qquad (4.A.9)$$

which is normalized properly to serve as a "renormalized pion field", i.e.,

$$\langle 0 | \pi_a(x) | \pi_b, q \rangle = \frac{\delta_{ab} e^{iq\cdot x}}{(2\pi)^{3/2}\sqrt{2q^0}} \qquad (4.A.10)$$

If we think of π_a as behaving something like a canonical field, then matrix elements of $\partial_\mu A_a^\mu$ will be small, of order m_π^2, unless some singularity, such as a one-pion pole, intervenes to make them large. The assumption that $\partial_\mu A_a^\mu$ may be neglected except for pion-pole contributions is known as the assumption of a partially conserved axial-vector current.[22,23]

B. One Soft Pion

As a first application of dynamical chirality, let us calculate the matrix element for emission of a single soft pion in an arbitrary multiparticle reaction $i \to f$. We will first consider the case of zero pion mass, where exact results can be obtained, and will then extend these results to real pions.

Consider the matrix element

$$\langle f | A_a^\mu(x) | i \rangle \equiv e^{iq\cdot x} M_a^\mu \quad , \qquad (4.B.1)$$

where $q = p_i - p_f$. We can split M_a^μ into a pion pole term, and a term N_a^μ which is defined not to contain a pole at $q^2 = 0$:

$$M_a^\mu \equiv \frac{-iF_\pi q^\mu}{q^2} M_a + N_a^\mu \qquad (4.B.2)$$

There are different ways of defining M_a away from $q^2=0$, and for different definitions of M_a we get different no-pole terms $N_a{}^\mu$, but the important point is that M_a gives the correct physical one-pion matrix element at $q^2=0$, while $N_a{}^\mu$ has no one-pion pole.

Since the axial current is presumed here to be conserved for $m_\pi = 0$, it follows from (4.B.2) that

$$0 = q_\mu M_a{}^\mu = -i F_\pi M_a + q_\mu N_a{}^\mu$$

so

$$M_a = \frac{-i}{F_\pi} q_\mu N_a{}^\mu \qquad (4.B.3)$$

We don't know $N_a{}^\mu$ but we can evaluate it in the limit $q \to 0$. In this limit $N_a{}^\mu$ has poles arising from graphs in which the axial vector current hooks on to the external lines of the "core" process $i \to f$. The propagator of an outgoing or ingoing line, which loses a momentum q^μ to the axial current, will have a denominator

$$\left[(p \pm q)^2 + m^2\right]^{-1} \longrightarrow \pm\left[2\, p \cdot q\right]^{-1} \qquad (4.B.4)$$

(compare Section 2K) so these poles will yield a contribution to (4.B.3) which is of order zero in q^μ as $q^\mu \to 0$, which can be evaluated exactly in terms of the axial-vector coupling constants and the matrix element for the process $i \to f$. For nucleon processes, (4.B.3) just says that soft pions are emitted from the external lines of the process $i \to f$ with an <u>effective</u> pion-nucleon coupling

$$\mathcal{L}_{EFF} = i g_A F_\pi^{-1} \bar{N} \gamma_5 \gamma^\mu \underset{\sim}{\tau} N \cdot \partial_\mu \underset{\sim}{\pi} , \qquad (4.B.5)$$

where g_A is the axial-vector coupling constant of the nucleon, defined by the formula

$$\langle N' | \underset{\sim}{A}^\mu | N \rangle \to (2\pi)^{-3} (4E'E)^{-1/2} g_A \bar{u}_N' \gamma_5 \underset{\sim}{\tau} [-i\gamma^\mu + 2m_N q^\mu / q^2] u_N \qquad (4.B.6)$$

In processes like nucleon-nucleon scattering, where the nucleons are on the mass shell, (4B.5) is equivalent to the more familiar pseudoscalar interaction

$$\mathcal{L}'_{EFF} = i G \bar{N} \gamma_5 \underset{\sim}{\tau} N \cdot \underset{\sim}{\pi} \qquad (4.B.7)$$

with an effective pseudo-scalar coupling constant given by

$$\frac{G}{2m_N} = \frac{g_A}{F_\pi} \qquad (4.B.8)$$

This last result is not particularly useful unless we find some independent method of calculating or measuring g_A. However, (4B.3) always allows us to determine M_a for $q \to 0$ up to the unknown parameter g_A. In particular, (4B.3) tells us that M_a vanishes as $q^\mu \to 0$ for processes which do not have poles like (4B.4). This is known as the Adler condition.[24]

The calculation is a little different for real pions. Separating the pion-pole and no-pion-pole terms in (4B.1) now gives

$$M_a^\mu \equiv \frac{-i F_\pi q^\mu}{q^2 + m_\pi^2} M_a + N_a^\mu \qquad (4.B.9)$$

We can now define the one-pion matrix element off the pion mass shell by using the pion field defined by Eq. (4A.9),

$$\langle f | \pi_a(x) | i \rangle \equiv \frac{-M_a e^{iq\cdot x}}{q^2 + m_\pi^2} \qquad (4.B.10)$$

If we now take the divergence of (4B.1) and use (4B.9), (4B.10), and (4A.9), we find

$$\frac{F_\pi q^2 M_a}{q^2 + m_\pi^2} + i q_\mu N_a^\mu = - \frac{F_\pi m_\pi^2 M_a}{q^2 + m_\pi^2}$$

which again yields (4.B.4). However, this result is now a mere tautology, following directly from identities and definitions. The physical input comes when we assume that N_a^μ is well approximated by pole terms, in which the axial-vector current hooks on to external lines, for momentum transfers q_μ of order m_π. This assumption is valid if (4.B.9) does not introduce any terms of order $\frac{1}{m_\pi^2}$ into N_a^μ, which in turn is valid if (4.B.10) does not yield a matrix element which varies rapidly away from the pion mass shell.

C. Chirality and Weak Interactions

The weak interaction responsible for nucleon beta-decay has been known, since the 1957 work of Feynman, Gell-Mann,[25] Marshak, and Sudarshan,[26] to be mediated by vector and axial-vector currents. Further, in order to explain the equality of the vector matrix elements in O^{14} decay and μ decay, it was suggested by Feynman, Gell-Mann, Gerstein, and Zeldovich that the vector current of the weak interactions is proportional to the conserved current of isospin

$$V^\mu_{a,WK} = C_V V^\mu_a \,, \tag{4.C.1}$$

where $C_V = 1.02 \cdot 10^{-5}\, m_p^{-2}$ is the Fermi beta-decay coupling constant. It is natural then to assume, following Gell-Mann,[27] that the axial weak current is proportional to the axial current of an $SU(2) \times SU(2)$ algebra:

$$A^\mu_{a,WK} = C_{A0} A^\mu_a \tag{4.C.2}$$

We have put a subscript "o" on C_A because the observed axial-vector coupling constant C_A of neutron beta decay will be renormalized by the strong interactions (unlike the vector constant), so that

$$C_A = g_A C_{A0} \tag{4.C.3}$$

where g_A is the strong-interaction parameter introduced in (4.B.6). Experimentally $C_A / C_V \simeq 1.2$, so (4.C.3) gives

$$g_A \frac{C_{A0}}{C_V} \simeq 1.2 \quad . \tag{4.C.4}$$

Also, (4.C.2) and (4.A.7) give a matrix element for pion decay proportional to

$$\langle 0 | A^{\mu}_{a,WK}(0) | \pi_b \rangle = \frac{i \delta_{ab} C_{A0} F_\pi q^\mu}{(2\pi)^{3/2} \sqrt{2q^0}} \tag{4.C.5}$$

and the experimental pion lifetime then yields

$$C_{A0} F_\pi \simeq 190 \; C_V \text{ Mev} \tag{4.C.6}$$

From (4.C.4) and (4.C.5) we have then

$$\frac{g_A}{F_\pi} \simeq \frac{1.2}{190 \text{ Mev}} \tag{4.C.7}$$

and (4.B.8) then gives a pseudo-scalar pion-nucleon coupling constant

$$G \simeq \frac{2\,(938 \text{ Mev})\,(1.2)}{(190 \text{ Mev})} \simeq 12 \tag{4.C.8}$$

roughly in agreement with the coupling-constant measured in dispersion-theory analyses of pion-nucleon scattering. This result is essentially the same as the formula for the pion decay rate, derived by quite a different method, of Goldberger and Treiman.[28] The fair success of this result may be taken as evidence that the weak axial-vector current is proportional to the "partially conserved" current A^{μ}_{a}, but it does not tell us what the proportionality constant C_{A0} is, and it does not distinguish among the three possible kinds of chirality. In order to settle these questions, we must turn to problems involving at least two soft pions.

D. Two Soft Pions - Pion Scattering Lengths

The problem of two soft pions, and in particular, the problem of pion scattering near threshold, plays a fundamental role in revealing the nature of the chiral symmetry group, in verifying the universality of weak couplings, and in determining the properties of pion transition amplitudes which are needed for the study of many soft pions. In the early days of current algebra (say, 1965) it was generally thought that useful soft pion results could only be obtained at the unphysical point of zero pion energy and momentum, and that these soft pion results would have to be converted into sum rules before using PCAC to get into the real world of finite pion mass. Even after the derivation of an approximate theorem for the physical pion scattering lengths a certain amount of confusion remained concerning the validity of the approximations employed. Personally, I was never happy about my own 1966 papers in Physical Review Letters,[29] partly because the brevity required by that journal did not allow a full discussion of the problem. The following is a longer, and I think better, derivation that was in my desk at that time, somewhat upgraded by adopting some subsequent suggestions of Fubini and Furlan.[30]

The first step in dealing with the problem of two soft pions is to decompose the time ordered product of two axial currents into terms in which either, or neither, or both currents hook on to single-pion lines. In analogy with (4.B.9), we can define a Green's function $T\{A_b^\mu, \pi_a\}_N$, in which the axial current is not allowed to couple to a single pion, by the formula

$$T\{A_b^\mu(x), \pi_a(y)\} \equiv T\{A_b^\mu(x), \pi_a(y)\}_N + F_\pi \frac{\partial}{\partial x_\mu} T\{\pi_b(x), \pi_a(y)\} \tag{4.D.1}$$

and then define a Green's function $T\{A_b^\mu, A_a^\nu\}_N$ in which neither current is allowed to couple to a single pion, by the formula

$$T\{A_b^\mu(x), A_a^\nu(y)\} \equiv T\{A_b^\mu(x), A_a^\nu(y)\}_N + F_\pi \frac{\partial}{\partial x_\mu} T\{\pi_b(x), A_a^\nu(y)\}_N + F_\pi \frac{\partial}{\partial y_\nu} T\{A_b^\mu(x), \pi_a(y)\}_N + F_\pi^2 \frac{\partial}{\partial x_\mu}\frac{\partial}{\partial y_\nu} T\{\pi_b(x), \pi_a(y)\} . \tag{4.D.2}$$

Using (4.D.1) in (4 D.2) gives

$$T\{A_b^\mu(x), A_a^\nu(y)\} = T\{A_b^\mu(x), A_a^\nu(y)\}_N + F_\pi \frac{\partial}{\partial x_\mu} T\{\pi_b(x), A_a^\nu(y)\} + F_\pi \frac{\partial}{\partial y_\nu} T\{A_b^\mu(x), \pi_a(y)\} + - F_\pi^2 \frac{\partial}{\partial x_\mu}\frac{\partial}{\partial y_\nu} T\{\pi_b(x), \pi_a(y)\} \tag{4.D.3}$$

This is only a definition of $T\{A_b^\mu(x), A_a^\nu(y)\}_N$, but we are eventually going to make qualitative assumptions about its behaviour, which are valid only if (4.D.1) and (4.D.2), with π_a defined by (4.A.9), really do a good job of isolating the single and double one-pion-pole terms in $T\{A_b^\mu, A_a^\nu\}$

The next step is to take the divergence of Eq (4.D.3) with respect to both x^μ and y^ν :

$$\frac{\partial}{\partial x^\mu}\frac{\partial}{\partial y^\nu} T\{A_b^\mu(x), A_a^\nu(y)\} = \frac{\partial}{\partial x^\mu}\frac{\partial}{\partial y^\nu} T\{A_b^\mu(x), A_a^\nu(y)\}_N +$$
$$+ F_\pi \Box_x^2 \frac{\partial}{\partial y^\nu} T\{\pi_b(x), A_a^\nu(y)\} + F_\pi \Box_y^2 \frac{\partial}{\partial x^\mu} T\{A_b^\mu(x), \pi_a(y)\} +$$
$$- F_\pi^2 \Box_x^2 \Box_y^2 T\{\pi_b(x), \pi_a(y)\}. \qquad (4.D.4)$$

ThePCAC definition (4.A.9) yields

$$\frac{\partial}{\partial x^\mu} T\{A_b^\mu(x), A_a^\nu(y)\} = m_\pi^2 F_\pi T\{\pi_b(x), A_a^\nu(y)\} +$$
$$+ \delta(x^0-y^0)\,[A_b^0(x), A_a^\nu(y)],$$

$$\frac{\partial}{\partial x^\mu} T\{A_b^\mu(x), \pi_a(y)\} = m_\pi^2 F_\pi T\{\pi_b(x), \pi_a(y)\} +$$
$$+ \delta(x^0-y^0)\,[A_b^0(x), \pi_a(y)]$$

and a similar expression with a, ν, y interchanged with b, μ, x

Inserting these formulas in (4.D.4.) gives

$$m_\pi^4 F_\pi^2 T\{\pi_b(x), \pi_a(y)\} + m_\pi^2 F_\pi \delta(x^0-y^0)[A_a^0(y), \pi_b(x)] +$$
$$+ \frac{\partial}{\partial y^\nu} \delta(x^0-y^0)[A_b^0(x), A_a^\nu(y)] = \frac{\partial}{\partial x^\mu}\frac{\partial}{\partial y^\nu} T\{A_b^\mu(x), A_a^\nu(y)\}_N +$$
$$+ F_\pi^2 m_\pi^2 \Box_x^2 T\{\pi_b(x), \pi_a(y)\} + F_\pi \Box_x^2 \delta(x^0-y^0)[A_a^0(y), \pi_b(x)] +$$
$$+ F_\pi^2 m_\pi^2 \Box_y^2 T\{\pi_b(x), \pi_a(y)\} + F_\pi \Box_y^2 \delta(x^0-y^0)[A_b^0(x), \pi_a(y)] +$$
$$- F_\pi^2 \Box_x^2 \Box_y^2 T\{\pi_b(x), \pi_a(y)\},$$

or collecting terms,

$$(\Box_x^2 - m_\pi^2)(\Box_y^2 - m_\pi^2) T\{\pi_b(x), \pi_a(y)\} = F_\pi^{-2} \frac{\partial}{\partial x^\mu}\frac{\partial}{\partial y^\nu} T\{A_b^\mu(x), A_a^\nu(y)\}_N +$$
$$+ F_\pi^{-1} \Box_y^2 \delta(x^0-y^0)[A_b^0(x), \pi_a(y)] + F_\pi^{-1}(\Box_x^2 - m_\pi^2)\delta(x^0-y^0)[A_a^0(y), \pi_b(x)] +$$
$$- F_\pi^{-2} \frac{\partial}{\partial y^\nu} \delta(x^0-y^0)[A_b^0(x), A_a^\nu(y)] \qquad (4.D.5)$$

This is the fundamental identity on which are based all applications of current algebra to problems involving two soft pions.

Let us now specialize to the problem of pion scattering, in which a target particle α is converted to a final particle β which may or may not have the same mass. According to the LSZ formalism, the S-matrix here is

$$S_{\beta b,\alpha a} = -(2\pi)^{-3}(2q^0 2q^{0\prime})^{-1/2}\int d^4x\, d^4y\, e^{-iq'\cdot x}\, e^{+iq\cdot y} \times$$

$$\times(\Box_x^2 - m_\pi^2)(\Box_y^2 - m_\pi^2)\langle\beta|T\{\pi_b(x),\pi_a(y)\}|\alpha\rangle\,, \tag{4.D.6}$$

where q' and q are the final and initial pion four-momenta, with

$$q'^2 = q^2 = -m_\pi^2 \,. \tag{4.D.7}$$

Inserting (4.D.5) gives then

$$S_{\beta b,\alpha a} = -(2\pi)^{-3}(2q^0 2q^{0\prime})^{-1/2}\int d^4x\, d^4y\, e^{-iq'\cdot x}\, e^{+iq\cdot y} \times$$

$$\times\langle\beta|\{F_\pi^{-2} q'_\mu q_\nu T\{A_b^\mu(x), A_a^\nu(y)\}_N + F_\pi^{-1} m_\pi^2 \delta(x^0-y^0)[A_b^0(x),\pi_a(y)] +$$

$$+ i q_\nu F_\pi^{-2}\delta(x^0-y^0)[A_b^0(x), A_a^\nu(y)]\}|\alpha\rangle\,.$$

Translational invariance allows us to shift the arguments of $A_b^\mu(x)$ and $A_a^\nu(y)$ by an amount $-x$, with a resulting extra factor $\exp\{ix\cdot(p_\alpha - p_\beta)\}$. The x integral can then be done explicitly, and we find

$$S_{\beta b,\alpha a} = \frac{-i(2\pi)^4 M_{\beta b,\alpha a}\,\delta^4(p_\alpha + q - p_\beta - q')}{(2\pi)^6\,[2q^0\,2q^{0\prime}\,2p_\alpha^0\,2p_\beta^0]^{1/2}}\,, \tag{4.D.8}$$

where M, the usual Feynman amplitude, is given (with $z = y - x$) by

$$\frac{M_{\beta b,\alpha a}}{(2\pi)^3[2p_\alpha^0\,2p_\beta^0]^{1/2}} = -i\int d^4z\, e^{iq\cdot z}\langle\beta|\{F_\pi^{-2} q'_\mu q_\nu T\{A_b^\mu(0), A_a^\nu(z)\}_N +$$

$$+ F_\pi^{-1} m_\pi^2\,\delta(z^0)[A_b^0(0), \pi_a(z)] + iq_\nu F_\pi^{-2}\,\delta(z^0)[A_b^0(0), A_a^\nu(z)]\}|\alpha\rangle \tag{4.D.9}$$

In general, this formula does not tell us very much about the pion scattering amplitude. However, there is one physical point in elastic pion scattering where this formula can be used to estimate the amplitude. This distinctive point is the threshold. As particularly emphasized by Fubini and Furlan, it is very convenient to estimate the threshold amplitude by going to the laboratory reference frame, where

$$p_\alpha = p_\beta = (0,0,0,m)\,,\quad q = q' = (0,0,0,m_\pi)$$

The S-matrix here is conventionally written in terms of a scattering length $a_{\beta b,\alpha a}$, defined by

$$S_{\beta b,\alpha a} = \frac{i\,a_{\beta b,\alpha a}\,\delta^4(\;)}{2\pi\left(\frac{m\,m_\pi}{m + m_\pi}\right)}\,, \tag{4.D.10}$$

and (4.D.9) gives

$$a_{\beta b,\alpha a} = - \frac{M_{\beta b,\alpha a}(\text{threshold})}{8\pi(m+m_\pi)} =$$

$$= \frac{2i\pi^2 m}{m+m_\pi} \int d^4z\, e^{-im_\pi z^0} \langle\beta| \Big\{ F_\pi^{-2} m_\pi^2\, T\{A_b^0(0), A_a^0(z)\}_N +$$

$$+ F_\pi^{-1} m_\pi^2\, \delta(z^0)\left[A_b^0(0), \pi_a(z)\right] - i F_\pi^{-2} m_\pi\, \delta(z^0)\left[A_b^0(0), A_a^0(z)\right]\Big\}|\alpha\rangle. \tag{4.D.11}$$

Up to now, everything has been exact. The essential approximation now is to treat m_π as a very small number, and assume that the m_π factors displayed explicitly in (4.D.11) give the correct order in m_π of the three terms. This would not be true, for instance, if it happened that the first term had singularities near the physical threshold. We have presumably eliminated the one-pion pole terms correctly (if π_a is a "good" pion field) but small denominators could still enter if there were to exist low-lying states γ, for which $\langle\beta|A_b^0(0)|\gamma\rangle\langle\gamma|A_a^0(0)|\alpha\rangle \neq 0$ and/or $\langle\beta|A_a^0(0)|\gamma\rangle\langle\gamma|A_b^0(0)|\alpha\rangle \neq 0$ (with α, β, γ at rest) and $|m_\gamma - m_\alpha| \lesssim m_\pi$. Such states would have to have the same spin and opposite parity as α and β, so they would form something like a chiral multiplet, which should be visible on the wallet card tables of elementary particles. Needless to say, such chiral multiplets do not seem to exist: there is no $\tfrac{1}{2}^-, T=\tfrac{1}{2}$ state near the nucleon mass; there is no $0^+, T=1$ state near the pion mass; and so on. With this proviso, that there are no chiral multiplets, we can then neglect the first two terms in (4.D.11). The last term may be evaluated with the aid of the current (or rather, density) commutation relation (4.A.2), and we find

$$a_{\beta b,\alpha a} \simeq \frac{4i\pi^2 m m_\pi F_\pi^{-2}\eta}{m+m_\pi} \epsilon_{bac} \langle\beta| V_c^0(0) |\alpha\rangle. \quad (4.D.12)$$

But by the same argument by which we showed that the strong interactions do not renormalize electric charges, the matrix element of V_c^0 is just

$$\langle\beta| V_c^0(0) |\alpha\rangle = \frac{2\,(t_c)_{\beta\alpha}}{(2\pi)^3} \quad (4.D.13)$$

Also, the isotopic spin matrix of the pion is just

$$\left(t_c^\pi\right)_{ba} = -i\,\epsilon_{bac} , \quad (4.D.14)$$

and, by the same trick used to derive the Lande g-factor,

$$\underset{\sim}{t}^\pi \cdot \underset{\sim}{t} = \tfrac{1}{2}\left[T(T+1) - t(t+1) - 2\right] \quad (4.D.15)$$

where t is the target isospin, T is the total isospin of the scattering state, and 2 is $(\underset{\sim}{t}^\pi)^2$. Using (4.D.13), (4.D.14), and (4.D.15) in (4.D.12), we arrive at a universal formula for the pion scattering length

$$a_T \simeq \frac{-L\eta}{1+\frac{m_\pi}{m}}\left[T(T+1) - t(t+1) - 2\right] , \quad (4.D.16)$$

where

$$L \equiv \frac{m_\pi}{2\pi F_\pi^2} \quad (4.D.17)$$

This is "universal", except that it doesn't hold if the target mass is comparable to the pion mass, as in pion-pion scattering, or if there are states of the same spin, opposite parity, and approximately equal mass as the target, as in the case of pion-nucleus scattering.

In using (4.D.16), we must recall that $\eta = +1, 0, \text{or} -1$, according as the underlying symmetry is compact, Abelian, or non-compact chirality. Also, if we assume that the unrenormalized weak axial-vector coupling constant C_{A0} is equal to the vector coupling constant C_V, then according to Eq. (4.C.6), the observed pion lifetime gives

$$F_\pi \simeq 190 \text{ Mev}, \tag{4.D.18}$$

and hence

$$L \simeq 0.086\, m_\pi^{-1} \tag{4.D.19}$$

Confrontation of (4.D.16) with a single empirical pion scattering length can in principle settle the question of the nature of the chiral symmetry group for which the pion serves as Goldstone boson, and tell us whether or not there is a universal strength for semi-leptonic $\Delta S = 0$ weak interactions, with $C_{A0} = C_V$.

By far the best known pion scattering lengths are those for π-N scattering. Here the target isospin t is $1/2$, so there are two scattering states, with $T = 3/2$ and $T = 1/2$. According to Eqs. (4.D.16) and (4.D.19), the scattering lengths in these two states should be

$$a_{3/2} \simeq \frac{-L\eta}{1 + \frac{m_\pi}{m_N}} \simeq -0.075\, \eta\, m_\pi^{-1}, \tag{4.D.20}$$

$$a_{1/2} \simeq \frac{2L\eta}{1 + \frac{m_\pi}{m_N}} \simeq +0.15\, \eta\, m_\pi^{-1}, \tag{4.D.21}$$

a result obtained more-or-less simultaneously by Tomozawa[31]; Raman and Sudarshan;[32] Balachandran, Gundzik, and Nicodemi; and myself.[29] Experimentally, we have (these are 1963 values)

$$a_{3/2} = (-0.088 \pm 0.004)\, m_\pi^{-1} \quad , \quad a_{1/2} = (0.171 \pm 0.005)\, m_\pi^{-1}$$

There thus seems little doubt that $\eta = +1$, so that the dynamic symmetry of the strong interactions is compact chirality, or $SU(2) \times SU(2)$. We can also conclude that C_{A0} is at least approximately equal to C_V.

The results (4.D.20) and (4.D.21) can be put into better perspective if we write them in terms of the t -channel isotopic-spin amplitudes (now setting $\eta = +1$)

$$\bar{a}_0 \equiv 2a_{3/2} + a_{1/2} \simeq 0 \qquad (4.D.22)$$

$$\bar{a}_1 \equiv a_{1/2} - a_{3/2} \simeq \frac{3L}{1 + \frac{m_\pi}{m_N}} \qquad (4.D.23)$$

The result (4.D.22), which really means that $\bar{a}_0$ is less than $\bar{a}_1$ by a factor of order $\frac{m_\pi}{m_N}$ was known to be true experimentally and was regarded as an interesting mystery before the advent of current algebra. The old pseudoscalar coupling theory in Born approximation gives about the right value for $\bar{a}_1$ but grossly over-estimates $\bar{a}_0$, while the old pseudovector coupling theory in Born approximat ion gives a small value for $\bar{a}_0$ but grossly under-estimates $\bar{a}_1$. (This last remark actually follows from (4.D.16), because the pseudo-vector coupling theory obeys an Abelian chiral symmetry, with $\eta = 0$.) It is only with the aid of current algebra, based on compact chirality, that we are able to understand the values of both $\bar{a}_0$ and $\bar{a}_1$.

It should also be mentioned that the correction terms in

(4.D.23) are expected to be smaller than L by a factor of order $\left(\frac{m_\pi}{m_N}\right)^2$ rather than $\frac{m_\pi}{m_N}$, because $\bar{a}_1$ receives contributions only from the part of (4.D.11) that is antisymmetric in a and b, and crossing symmetry allows only odd powers of m_π in such terms.

E. The Adler - Weisberger Sum Rule

Ten years before current algebra came into flower, Goldberger, Miyazawa, and Oehme wrote down a sum rule for the pion-nucleon scattering lengths:[33]

$$\frac{1}{3}(a_{1/2}-a_{3/2})\left(1+\frac{m_\pi}{m_N}\right) = \frac{G^2 m_\pi}{8\pi m_N^2}\left(1-\frac{m_\pi^2}{4m_N^2}\right)^{-1} + $$
$$+\frac{m_\pi}{4\pi^2}\int_{m_\pi}^{\infty}\frac{d\nu}{\sqrt{\nu^2-m_\pi^2}}\left[\sigma_{\pi^-p}(\nu)-\sigma_{\pi^+p}(\nu)\right] \qquad (4.E.1)$$

where $\sigma_{\pi p}(\nu)$ are the total pion-proton scattering cross sections at laboratory energy ν. The derivation, which you can read in Goldberger's 1960 Les Houches lecture notes,[34] has nothing to do with chirality; you simply form the odd part of the pion-proton forward scattering amplitude

$$M^{(-)}(\nu^2) \equiv \frac{1}{2\nu}\left[M_{\pi^-p}(\nu)-M_{\pi^-p}(-\nu)\right] =$$
$$= \frac{1}{2\nu}\left[M_{\pi^-p}(\nu)-M_{\pi^+p}(\nu)\right], \qquad (4.E.2)$$

assume that $M^{(-)}(\nu^2)$ satisfies on unsubtracted dispersion relation, with a pole residue proportional to G^2 and a cut discontinuity proportional to $\sigma_{\pi^-p}(\nu)-\sigma_{\pi^+p}(\nu)$, and then evaluate the dispersion relation at threshold. The only possible way that (4.E.1) could fail is if $M^{(-)}$ did not satisfy an unsubtracted dispersion relation. However, Regge pole theory suggests that the part of the pion-nucleon forward scattering amplitude with $T=1$ in the t-channel should behave like $\nu^{\alpha_\rho(0)}$ as $\nu\to\infty$, where $\alpha_\rho(0)\simeq 1/2$ is the ρ-trajectory intercept. If this is true (and experiment seems to bear it out) then we would expect that, for $\nu \to \infty$,

$$M^{(-)}(\nu^2) \sim \nu^{\alpha_\rho(0)-1} \to 0 \quad , \qquad (4.E.3)$$

so that $M^{(-)}(\nu^2)$ does satisfy an unsubtracted dispersion relation.

The GMO sum rule becomes relevant to current algebra when we insert on the left-hand-side the soft-pion prediction (4.D.23) for $a_{1/2} - a_{3/2}$. Using (4D.17) to express L in terms of G and g_A, we find

$$\frac{1}{g_A^2} = \frac{1}{1 - \frac{m_\pi^2}{4m_N^2}} + \frac{2m_N^2}{\pi G^2} \int_{m_\pi}^{\infty} \frac{d\nu}{\sqrt{\nu^2 - m_\pi^2}} \left[\sigma_{\pi^- p}(\nu) - \sigma_{\pi^+ p}(\nu) \right]. \qquad (4.E.4)$$

This may be compared with the celebrated sum rule of Adler and Weisberger,[35] which in one version reads

$$\frac{1}{g_A^2} = 1 + \frac{2m_N^2}{\pi G^2} \int_{m_\pi}^{\infty} \frac{d\nu \sqrt{\nu^2 - m_\pi^2}}{\nu^2} \left[\sigma_{\pi^- p}(\nu) - \sigma_{\pi^+ p}(\nu) \right], \qquad (4.E.5)$$

and yields a value $g_A \simeq 1.2$ which with (4.C.4) means again that $C_V \simeq C_{A0}$. This latter sum rule was obtained by first deriving an exact soft pion theorem at the unphysical point $q^\mu = q'^\mu = 0$, then using an unsubtracted dispersion relation to express the result in terms of cross-sections for massless pions, and then using PCAC to extrapolate to the pion mass shell. (Adler also presented a derivation based on the saturation of current commutation relations at infinite momentum.) It was only later that other theorists realized that an approximate soft pion theorem could be obtained at the physical threshold, which in conjunction with the old G-M-O sum rule, would lead to (4.E.4). Obviously (4.E.4) and (4.E.5) agree up to terms of order m_π^2 (provided that the integrals do not receive important contributions from near threshold) and I have never been able to convince myself that there is any theoretical reason to expect that (4.E.4) or (4.E.5) should be more accurate. It certainly would have been more economical if the soft-pion theorem (4.E.1) had been derived first, and then put together with dispersion theory to derive Eq (4.E.4), but that is not the way it happened.

F. The χ - Matrix

We are going to generalize the Adler - Weisberger sum rule in the next section,[36] but first we need a general notation for axial-vector coupling constants and pion transition amplitudes. From now on we will work in the idealized case of zero pion mass and conserved axial-vector currents, so as not to confuse the drift of our argumen t with the different, and rather technical, problem of how to interpret PCAC in the real world.

Let α and β be particles with momenta along the $+3$ or -3 direction, and assume that

$$\pm(p_\alpha^\mu - p_\beta^\mu) = (0, 0, \omega, \omega), \tag{4.F.1}$$

as would be the case for a transition $\alpha \to \beta + \pi$ or $\beta \to \alpha + \pi$ with π a massless pion having energy ω and momentum in the $+3$ direction. Eliminating ω gives then

$$p_\alpha^0 - p_\alpha^3 = p_\beta^0 - p_\beta^3 . \tag{4.F.2}$$

Since α and β are on the mass shell, both p_α^μ and p_β^μ are entirely determined once we specify p_α^3 or p_β^3 .

The χ -matrix may now be defined by

$$\langle\beta| A_{aN}^0(0) - A_{aN}^3(0) |\alpha\rangle \equiv \frac{4(p_\alpha^0 - p_\alpha^3)(\chi_a)_{\beta\alpha}}{(2\pi)^3\sqrt{4p_\alpha^0 p_\beta^0}} , \tag{4.F.3}$$

the label N denoting the part of the axial current excluding the one-pion pole. The properties of the $\underset{\sim}{\chi}$ -matrix are determined by those strong interaction symmetries which leave the kinematic condition (4.F.1) unchanged, except perhaps for a change in ω . These are:

(1) Invariance with respect to Lorentz boosts along the 3 -direction tells us that χ_a is a fixed numerical matrix, independent of the particular value chosen for p_α^3 or p_β^3 .

(2) Invariance with respect to rotations around the 3 -axis tells us that χ_a conserves helicity

$$(\chi_a)_{\beta\alpha} = 0 \quad \text{unless} \quad \lambda_\beta = \lambda_\alpha \, . \tag{4.F.4}$$

(3) Invariance with respect to a combined space inversion and 180° rotation around the 2 -axis gives

$$(\chi_a(\lambda))_{\beta\alpha} = (\chi_a(-\lambda))_{\beta\alpha} \, \Pi_\beta \Pi_\alpha (-1)^{j_\beta - j_\alpha + 1} \, , \tag{4.F.5}$$

where Π and j are the particles' parity and spin. In particular, for zero helicity we have

$$(\chi_a)_{\beta\alpha} = 0 \quad \text{unless} \quad \Pi_\beta \Pi_\alpha (-1)^{j_\beta - j_\alpha + 1} = +1 \, . \tag{4.F.6}$$

(4) Since A_a^μ is an isovector, isospin conservation gives

$$[t_a, \chi_b] = i \epsilon_{abc} \chi_c \quad , \tag{4.F.7}$$

where $\underset{\sim}{t}$ is the isospin matrix, which obeys the commutation relations

$$[t_a, t_b] = i \epsilon_{abc} t_c \tag{4.F.8}$$

(5) For non-strange mesons, G -parity conservation gives

$$(\chi_a)_{\beta\alpha} = 0 \quad \text{unless} \quad G_\beta = -G_\alpha \tag{4.F.9}$$

In addition, the axial current is Hermitian so (4.F.2) and (4.F.3) give χ_a Hermitian

$$(\chi_a)_{\beta\alpha}^{*} = (\chi_a)_{\alpha\beta} \tag{4.F.10}$$

As an illustration, the $\underset{\sim}{\chi}$ -matrix between nucleon states can be calculated by comparing the results of the usual Dirac formalism with (4.F.3). This gives

$$(\chi_a)_{NN} = \pm g_A t_a \quad \text{for } \lambda = \pm \tfrac{1}{2}\,, \qquad (4.F.11)$$

where

$$g_A \simeq 1.2$$

The connection between the $\underset{\sim}{\chi}$ -matrix elements and pion decay amplitudes is provided by the conservation of the axial vector current, as in the derivation of the Goldberger-Treiman relation. Adding the one-pion pole contribution to $\underset{\sim}{A}^{\mu}_{N}$ gives, for $m_\alpha > m_\beta$

$$\langle\beta|A^{\mu}_{a}|\alpha\rangle = \langle\beta|A^{\mu}_{aN}|\alpha\rangle + \frac{M(\alpha\to\beta+\pi_a)F_\pi}{(2\pi)^3\sqrt{4p^0_\alpha p^0_\beta}}\cdot\frac{(p_\alpha{}^{\mu}-p_\beta{}^{\mu})}{(p_\alpha-p_\beta)^2}$$

where M is the Feynman amplitude for the pion decay $\alpha\to\beta+\pi$. Axial current conservation says that this must vanish when multiplied with $q_\mu = (p_\alpha - p_\beta)_\mu$, so

$$\frac{F_\pi M(\alpha\to\beta+\pi_a)}{(2\pi)^3\sqrt{4p^0_\alpha p^0_\beta}} = -(p_\alpha-p_\beta)_\mu \langle\beta|A^{\mu}_{aN}|\alpha\rangle\,. \qquad (4.F.12)$$

If we now adopt the co-linear kinematics used to define the $\underset{\sim}{\chi}$-matrix, the right-hand side becomes

$$(p^3_\alpha - p^3_\beta)\langle\beta|[A^0_{aN}-A^3_{aN}]|\alpha\rangle$$

and can be evaluating using

$$(p^3_\alpha - p^3_\beta)(p^0_\alpha - p^3_\alpha) = \tfrac{1}{2}(p^0_\alpha - p^0_\beta)(p^0_\alpha + p^0_\beta) +$$

$$-\tfrac{1}{2}(p^3_\alpha - p^3_\beta)(p^3_\alpha + p^3_\beta) = \tfrac{1}{2}(m^2_\alpha - m^2_\beta)$$

Hence (4.F.12) yields

$$\frac{M(\alpha\to\beta+\pi_a)}{(2\pi)^3\sqrt{4p^0_\alpha p^0_\beta}} = 2F_\pi^{-1}(m^2_\alpha - m^2_\beta)(\chi_a)_{\beta\alpha}\,. \qquad (4.F.13)$$

(We could have used this result to define the $\underset{\sim}{X}$ -matrix, except for the case $m_\alpha = m_\beta$, where (4.F.3) is needed.) In particular, (4.F.13) gives the pion decay matrix element for α at rest, and the rest-frame decay rate for the process $\alpha \to \beta + \pi_a$ is

$$\Gamma(\alpha \to \beta + \pi_a) = \frac{(m_\alpha^2 - m_\beta^2)^3}{4\pi F_\pi^2 m_\alpha^3 (2j_\alpha + 1)} \left|(X_a)_{\beta\alpha}\right|^2 . \qquad (4.F.14)$$

G. Algebraization

There is an Adler - Weisberger relation for every stable hadron, but none of them is as useful as the π-N sum rule, because the only convenient targets are composed of nucleons. However, suppose we assume, as done in the Veneziano model, that the widths of all hadron resonances may consistently be neglected, and that these narrow hadron resonances completely dominate all dispersion relations. Then we have an infinity of Adler - Weisberger sum rules for general reactions $\pi + \alpha \to \pi + \beta$, where α and β may or may not have the same mass, and each such sum rule may be expressed as a quadratic relation among the pion decay amplitudes for these various resonances. How can we make sense of all these relations?

We shall work in the limit of zero pion mass and conserved axial currents from the beginning. According to Eq.(4.D.9) , the Feynman amplitude for the reaction $\pi + \alpha \to \pi + \beta$, is given, for $m_\pi = 0$, by

$$\frac{M_{\beta b,\alpha a}}{(2\pi)^3\sqrt{2p_\alpha^0 2p_\beta^0}} = -i\int d^4z\, e^{iq\cdot z}\langle\beta|\{F_\pi^{-2} q'_\mu q_\nu T\{A_b^\mu(0), A_a^\nu(z)\}_N +$$

$$+ iq_\nu F_\pi^{-2}\delta(z^0)[A_b^0(0), A_a^\nu(z)]\}|\alpha\rangle . \quad (4.G.1)$$

We shall be studying forward scattering, and it is convenient to adopt a laboratory frame of reference, with

$$\begin{aligned} p_\alpha &= (0,0,0,m_\alpha) & p_\beta &= (0,0;-p,\sqrt{p^2+m_\beta^2}) \\ q &= (0,0,\omega,\omega) & q' &= (0,0,\omega',\omega') \end{aligned} \quad (4.G.2)$$

Energy and momentum conservation give, respectively,

$$m_\alpha + \omega = \sqrt{p^2+m_\beta^2} + \omega'$$

$$\omega = \omega' - p$$

and eliminating $\omega'-\omega$, we have then

$$m_\alpha = \sqrt{p^2+m_\beta^2} + p$$

whose solution is

$$p = \frac{m_\alpha^2 - m_\beta^2}{2m_\alpha} \quad (4.G.3)$$

The energies of the particles in the final state are then

$$E_\beta = \sqrt{p^2+m_\beta^2} = \frac{m_\alpha^2+m_\beta^2}{2m_\alpha} , \quad (4.G.4)$$

$$\omega' = \omega + \frac{m_\alpha^2 - m_\beta^2}{2m_\alpha} . \quad (4.G.5)$$

The only independent variable left here is ω, and (4.G.1) **may now be written**

$$M_{\beta b,\alpha a}(\omega) = (2\pi)^3\sqrt{4E_\beta m_\alpha}\int d^4z\, e^{iq\cdot z} \times$$

$$\times\{-iF_\pi^{-2}\omega\omega'\langle\beta|T\{A_b^{0-3}(0), A_a^{0-3}(z)\}_N|\alpha\rangle + \quad (4.G.6)$$

$$-\omega F_\pi^{-2}\delta(z^0)\langle\beta|[A_b^0(0), A_a^{0-3}(z)]|\alpha\rangle\}$$

where we have used the convenient abbreviation

$$A_a^{0-3} \equiv A_a^0 - A_a^3 \ .$$

Both in order to derive our low energy theorems and to write down dispersion relations, we need to know the pole structure of $M(\omega)$. Suppose that there is a single particle state γ, with internal quantum numbers such that $\gamma \to \beta + \pi_b$ and $\gamma \to \alpha + \pi_a$ are both allowed transitions. Then (4.F.6) will have a pole, at which

$$M_{\beta b, \alpha a}(\omega) \longrightarrow -i\,(2\pi)^3 F_\pi^{-2} \sqrt{4E_\beta m_\alpha}\ \omega\omega' \times \tag{4.G.7}$$

$$\times \int d^4z\, \theta(-z^0) \int d^3p''\, e^{iq\cdot z} \langle \beta | A_{bN}^{0-3}(0) | \mathbf{p}'', \gamma \rangle \langle \mathbf{p}'', \gamma | A_{aN}^{0-3}(0) | \alpha \rangle\, e^{i(p_\alpha - p_\gamma)\cdot z} .$$

The integral over $\mathbf{z}$ immediately gives $(2\pi)^3$, times a δ-function which fixes $\mathbf{p}''$ to be

$$\mathbf{p}_\gamma = \mathbf{p}_\alpha + \mathbf{q} = (0,0,\omega) \ .$$

(We shall simply drop the $\mathbf{p}''$ integration and the momentum label in $|\gamma\rangle$.) The integral over z^0 now gives

$$\int_{-\infty}^{0} dz^0 \exp\{i z^0 (p_\gamma^0 - p_\alpha^0 - q^0)\} = \frac{-i}{p_\alpha^0 + q^0 - p_\gamma^0 - i\epsilon} =$$

$$= \frac{i}{m_\alpha + \omega - (m_\gamma^2 + \omega^2)^{1/2} + i\epsilon} \ .$$

This has a pole at $\omega = \frac{m_\gamma^2 - m_\alpha^2}{2m_\alpha}$, and behaves near the pole like

$$\frac{2iE_\gamma}{2m_\alpha\omega - m_\gamma^2 + m_\alpha^2 + i\epsilon} ,$$

where $E_\gamma = \frac{m_\gamma^2 + m_\alpha^2}{2m_\alpha}$ is the energy of particle γ.

Finally, we note that

$$p_\alpha^0 - p_\alpha^3 = p_\beta^0 - p_\beta^3 = p_\gamma^0 - p_\gamma^3 = m_\alpha$$

so (4.F.3) gives

$$\langle \beta | A_{bN}^{0-3}(0) | \gamma \rangle = \frac{4m_\alpha}{(2\pi)^3 \sqrt{4E_\beta E_\gamma}} (\chi_b)_{\beta\gamma} \ ,$$

$$\langle \gamma | A_{aN}^{0-3}(0) | \alpha \rangle = \frac{4m_\alpha}{(2\pi)^3 \sqrt{4E_\gamma m_\alpha}} (\chi_a)_{\gamma\alpha} \ .$$

Putting all this back into (4.G.7) we find that all factors $2E$ and $(2\pi)^3$ cancel, and near the pole

$$M_{\beta b,\alpha a}(\omega) \longrightarrow \frac{16 m_\alpha^2 F_\pi^{-2}\,\omega\omega'(\chi_b)_{\beta\gamma}(\chi_a)_{\gamma a}}{2m_\alpha\omega - m_\gamma^2 + m_\alpha^2 + i\epsilon} \qquad (4.G.8)$$

The whole amplitude must be invariant under the interchange of a with b and ω with $-\omega'$, so the $\theta(+z^0)$ term in the time-ordered product must yield a pole at $-2m_\alpha\omega' = m_\gamma^2 - m_\alpha^2$, or in other words, at $2m_\alpha\omega = m_\beta^2 - m_\gamma^2$, with

$$M_{\beta b,\alpha a}(\omega) \longrightarrow \frac{16 m_\alpha^2 F_\pi^{-2}\,\omega\omega'(\chi_a)_{\beta\gamma}(\chi_b)_{\gamma\alpha}}{-2m_\alpha\omega - m_\gamma^2 + m_\beta^2 + i\epsilon} \qquad (4.G.9)$$

Let us now write down our general low energy theorems. For $m_\alpha = m_\beta$ we have $\omega' = \omega$, so in the limit $\omega \to 0$ the first term in (4.G.6) is of order ω^2, except for pole terms with $m_\gamma = m_\alpha$ The density commutation relations give

$$\delta(z^0)\langle\beta|[A_b^0(0), A_a^0(z)]|\alpha\rangle = \delta^4(z)\langle\beta|\{-2i\epsilon_{abc}V_c^0(0)\}|\alpha\rangle =$$

$$= -4i\,\epsilon_{abc}\,\delta^4(z)\,(t_c)_{\beta\alpha}\,(2\pi)^{-3}, \qquad (4.G.10)$$

while parity kills the contribution from $[A_b^0, A_a^3]$. Using (4.G.8), (4.G.9), and (4.G.10) in (4.G.6) gives, for $m_\alpha = m_\beta$ and $\omega \to 0$,

$$M_{\beta b,\alpha a}(\omega) \longrightarrow 8m_\alpha F_\pi^{-2}\omega\Big[\sum_{\gamma: m_\gamma = m_\alpha}\{(\chi_b)_{\beta\gamma}(\chi_a)_{\gamma\alpha} + $$
$$ - (\chi_a)_{\beta\gamma}(\chi_b)_{\gamma\alpha}\} + i\,\epsilon_{abc}(t_c)_{\beta\alpha}\Big] \qquad (4.G.11)$$

For $m_\alpha \neq m_\beta$, we have ω' non-zero in the limit $\omega \to 0$, so both terms in (4.G.6) vanish like ω as $\omega \to 0$, except for those pole terms where $m_\gamma = m_\alpha$ or $m_\gamma = m_\beta$. Inspection of (4.G.8) and (4.G.9) shows that, for $\omega \to 0$ with $m_\alpha \neq m_\beta$,

$$M_{\beta b,\alpha a}(\omega) \to 4(m_\alpha^2 - m_\beta^2)F_\pi^{-2}\Big\{\sum_{\gamma: m_\gamma = m_\alpha}(\chi_b)_{\beta\gamma}(\chi_a)_{\gamma\alpha} + $$
$$ - \sum_{\gamma: m_\gamma = m_\beta}(\chi_a)_{\beta\gamma}(\chi_b)_{\gamma\alpha}\Big\} \qquad (4.G.12)$$

These results can be expressed more elegantly as formulas for the even and odd parts of the scattering amplitude at $\omega = 0$. Define

$$M^{(+)}_{\beta b,\alpha a}(\omega) = \frac{1}{2}\left[M_{\beta b,\alpha a}(\omega) + M_{\beta a,\alpha b}(\omega)\right] \qquad (4.G.13)$$

$$M^{(-)}_{\beta b,\alpha a}(\omega) = \frac{1}{\omega+\omega'}\left[M_{\beta b,\alpha a}(\omega) - M_{\beta a,\alpha b}(\omega)\right]. \qquad (4.G.14)$$

Crossing symmetry says that

$$M_{\beta b,\alpha a}(\omega) = M_{\beta a,\alpha b}(-\omega') \qquad (4.G.15)$$

so we can also write

$$M^{(+)}_{\beta b,\alpha a}(\omega) = \frac{1}{2}\left[M_{\beta b,\alpha a}(\omega) + M_{\beta b,\alpha a}(-\omega')\right], \qquad (4.G.16)$$

$$M^{(-)}_{\beta b,\alpha a}(\omega) = \frac{1}{\omega+\omega'}\left[M_{\beta b,\alpha a}(\omega) - M_{\beta b,\alpha a}(-\omega')\right], \qquad (4.G.17)$$

and therefore

$$M^{(\pm)}_{\beta b,\alpha a}(\omega) = M^{(\pm)}_{\beta b,\alpha a}(-\omega') \qquad (4.G.18)$$

The low energy theorems (4.G.11) and (4.G.12) now read

$$M^{(+)}_{\beta b,\alpha a}(0) = 2F_\pi^{-2}\sum_\gamma{}'(2m_\gamma^2 - m_\alpha^2 - m_\beta^2)\left\{(X_b)_{\beta\gamma}(X_a)_{\gamma\alpha} + (X_a)_{\beta\gamma}(X_b)_{\gamma\alpha}\right\} \qquad (4.G.19)$$

$$M^{(-)}_{\beta b,\alpha a}(0) = 8m_\alpha F_\pi^{-2}\left[\sum_\gamma{}'\left\{(X_b)_{\beta\gamma}(X_a)_{\gamma\alpha} - (X_a)_{\beta\gamma}(X_b)_{\gamma\alpha}\right\} + i\epsilon_{abc}(t_c)_{\beta\alpha}\right] \qquad (4.G.20)$$

both for $m_\alpha = m_\beta$ and for $m_\alpha \neq m_\beta$. (Recall that $(t_c)_{\beta\alpha}$ vanishes if $m_\beta \neq m_\alpha$.) The prime on the sums over γ indicate that we are to sum only over states γ with $m_\gamma = m_\alpha$ or $m_\gamma = m_\beta$

As discussed in Sec. 4E, we expect $M^{(-)}(\omega)$ to vanish as $\omega \to \infty$ something like $\frac{1}{\sqrt{\omega}}$, so that it should satisfy an unsubtracted dispersion relation. This dispersion relation, when saturated with single particle states, just amounts to a sum of pole terms, each defined to vanish as $|\omega| \to \infty$. Using (4.G.8), (4.G.9) and (4.G.5), we have then (dropping the $i\epsilon$)

$$M^{(-)}_{\beta b,\alpha a}(\omega) = 8m_\alpha F_\pi^{-2} \sum_\gamma \frac{(m_\gamma^2-m_\alpha^2)(m_\gamma^2-m_\beta^2)}{2m_\gamma^2-m_\alpha^2-m_\beta^2} \times$$

$$\times \left\{ \frac{1}{2m_\alpha\omega - m_\gamma^2+m_\alpha^2} + \frac{1}{-2m_\alpha\omega - m_\gamma^2+m_\beta^2} \right\} \times$$

$$\times \left\{ (\chi_b)_{\beta\gamma}(\chi_a)_{\gamma\alpha} - (\chi_a)_{\beta\gamma}(\chi_b)_{\gamma\alpha} \right\} \qquad (4.G.21)$$

The factor $(m_\gamma^2-m_\alpha^2)(m_\gamma^2-m_\beta^2)$ kills the contribution from any terms with $m_\gamma = m_\alpha$ or $m_\gamma = m_\beta$. Combining denominators gives then

$$M^{(-)}_{\beta b,\alpha a}(0) = -8m_\alpha F_\pi^{-2} \sum_\gamma{}'' \{(\chi_b)_{\beta\gamma}(\chi_a)_{\gamma\alpha} +$$

$$- (\chi_a)_{\beta\gamma}(\chi_b)_{\gamma\alpha}\} , \qquad (4.G.22)$$

the double prime indicating that terms with $m_\gamma = m_\alpha$ or $m_\gamma = m_\beta$ are to be left out in the sum. The sum of Σ' and Σ'' is just the sum over all γ, so (4.G.20) and (4.G.22) give

$$\sum_\gamma \{ (\chi_a)_{\beta\gamma}(\chi_b)_{\gamma\alpha} - (\chi_b)_{\beta\gamma}(\chi_a)_{\gamma\alpha}\} = i\epsilon_{abc}(t_c)_{\beta\alpha}$$

or, more simply,

$$[\chi_a, \chi_b] = i\epsilon_{abc} t_c \qquad (4.G.23)$$

Together with (4.F.7) and (4.F.8), this says that the 6 matrices χ_a and χ_b form a representation of the compact chiral algebra, SU(2) × SU(2) .

The result (4.G.23) may be regarded as analogous with the algebraic version (2.L.13) of the Drell-Hearn Compton scattering sum rule. In both cases, the algebraic interpretation of the sum rule may be obtained most easily by the infinite momentum method, but the method used here seems to be indispensable if we want to avoid all assumptions about high-momentum behaviour which do not follow clearly from the asymptotic behaviour of physical scattering amplitudes. If we were to assume that the $\underset{\sim}{\chi}$ -matrix connects nucleon states only with other nucleon states, then (4.G.23) and (4.G.11) would require the axial-vector coupling constant g_A to be ± 1, a result analogous to the statement that if one-proton states saturate the dispersion relation for proton-photon forward scattering, then the proton must have gyromagnetic ratio 2. Actually, higher states like the 3-3 resonances do contribute strongly in both pion-nucleon and photon-nucleon scattering, which is why g_A is not unity and g_p is not 2.

The Pomeranchuk and P' trajectories prevent $M^{(+)}(\omega)$ from vanishing as $\omega \to \infty$, so we cannot derive any general sum rule from the low energy theorem (4.G.19) for $M^{(+)}(0)$. However, we can separate out the part of the scattering amplitude with $T=2$ in the t-channel:

$$M^{(2)}_{\beta b,\alpha a}(\omega) = M^{(+)}_{\beta b,\alpha a}(\omega) - \tfrac{1}{3}\delta_{ba} M^{(+)}_{\beta c,\alpha c}(\omega) . \qquad (4.G.24)$$

Fubini and others have suggested that the absence of known trajectories with $T=2$ requires that $M^{(2)}(\omega)$ vanishes as $\omega \to \infty$, so that $M^{(2)}(\omega)$ obeys an unsubtracted dispersion relation.[37] With (4.G.8) and (4.G.9), the pole-saturated form of this dispersion relation reads

$$M^{(2)}_{\beta b,\alpha a}(\omega) = 2F_\pi^{-2}\sum_\gamma (m_\gamma^2 - m_\alpha^2)(m_\gamma^2 - m_\beta^2) \times$$
$$\times \left\{ \frac{1}{2m_\alpha\omega - m_\gamma^2 + m_\alpha^2} + \frac{1}{-2m_\alpha\omega - m_\gamma^2 + m_\beta^2} \right\} \times \qquad (4.G.25)$$
$$\times \left\{ (\chi_b)_{\beta\gamma}(\chi_a)_{\gamma\alpha} + (\chi_a)_{\beta\gamma}(\chi_b)_{\gamma\alpha} - \tfrac{2}{3}\delta_{ba}(\chi_c)_{\beta\gamma}(\chi_c)_{\gamma\alpha} \right\}$$

and therefore

$$M^{(2)}_{\beta b,\alpha a}(0) = -2F_\pi^{-2}\sum_\gamma{}'' (2m_\gamma^2 - m_\alpha^2 - m_\beta^2) \times$$
$$\times \left\{ (\chi_b)_{\beta\gamma}(\chi_a)_{\gamma\alpha} + (\chi_a)_{\beta\gamma}(\chi_b)_{\gamma\alpha} - \tfrac{2}{3}\delta_{ba}(\chi_c)_{\beta\gamma}(\chi_c)_{\gamma\alpha} \right\}. \qquad (4.G.26)$$

The low energy theorem (4.G.19) gives

$$M^{(2)}_{\beta b,\alpha a}(0) = 2F_\pi^{-2}\sum_\gamma{}' (2m_\gamma^2 - m_\alpha^2 - m_\beta^2)\times \tag{4.G.27}$$

$$\times\{(\chi_b)_{\beta\gamma}(\chi_a)_{\gamma\alpha} + (\chi_a)_{\beta\gamma}(\chi_b)_{\gamma\alpha} - \tfrac{2}{3}\delta_{ab}(\chi_c)_{\beta\gamma}(\chi_c)_{\gamma\alpha}\}.$$

Putting together (4.G.26) and (4.G.27), we have then

$$\sum_\gamma (2m_\gamma^2 - m_\alpha^2 - m_\beta^2)\{(\chi_b)_{\beta\gamma}(\chi_a)_{\gamma\alpha} + (\chi_a)_{\beta\gamma}(\chi_b)_{\gamma\alpha} +$$

$$- \tfrac{2}{3}\delta_{ab}(\chi_c)_{\beta\gamma}(\chi_c)_{\gamma\alpha}\} = 0$$

or, more simply

$$[\chi_a, [\chi_b, m^2]] = \tfrac{1}{3}\delta_{ab}[\chi_c, [\chi_c, m^2]], \tag{4.G.28}$$

where m^2 is the mass matrix

$$(m^2)_{\beta\alpha} = \delta_{\beta\alpha} m_\alpha^2 \tag{4.G.29}$$

This result can also be derived by the infinite momentum method, but apparently only by starting with $m_\pi \neq 0$, and then passing to the limit $m_\pi = 0$. (Special cases of (4.G.23) and (4.G.28) were first derived by Gilman and Harari,[38] but without using the χ-matrix. The algebraic significance of their results was thus somewhat obscure.)

The commutation relation (4.G.28) has a very appealing geometric significance.

Define

$$[\chi_b, m^2] \equiv i\, m_b^2 . \tag{4.G.30}$$

Then (4.G.28) may be written

$$[\chi_a, m_b^2] = -i m_4^2 \delta_{ab} . \tag{4.G.31}$$

The Jacobi identity for χ_a, χ_b and m_c^2 reads :

$$[\chi_a, [\chi_b, m_c^2]] - [\chi_b, [\chi_a, m_c^2]] = [[\chi_a, \chi_b], m_c^2] =$$

$$= i\epsilon_{abd}[t_d, m_c^2] = -\epsilon_{abd}\epsilon_{dce} m_e^2 = \delta_{bc} m_a^2 - \delta_{ac} m_b^2 .$$

Inserting (4.G.31) gives then

$$[\chi_a, m_4^2] = i\, m_a^2 \tag{4.G.32}$$

Taken together, (4.G.31) and (4.G.32) imply that m_4^2 and

m_b^2 form the four components of a chiral four-vector, the representation $(\frac{1}{2},\frac{1}{2})$ of SU (2) x SU (2) . Subtracting (4.G.32) from (4.G.30) gives

$$[\chi_b , m^2 - m_4^2] = 0$$

or in other words

$$m^2 = m_4^2 + m_0^2 , \tag{4.G.33}$$

where m_0^2 is a chiral scalar

$$[\chi_b , m_0^2] = 0$$

Thus the mass matrix is the sum of a chiral scalar and the fourth component of a chiral four-vector.

If there were no P or P' trajectories, we could derive a sum rule for $M_{\beta c,\alpha c}(\omega)$, which would yield the result that m_4^2 in Eq (4.G.31) vanishes. It would then follow that m^2 commutes with the $\underset{\sim}{\chi}$-matrix, so particles of a given mass would have to form complete chiral multiplets. It is the trajectories with the quantum numbers of the vacuum that keep chirality from showing up as an obvious symmetry of the spectrum of hadron states.

H. Chiral Dynamics

The use of relations like (4.B.3) and (4.D.5) allows us to express the matrix element for the emission of n soft pions as the product of a factor F_π^{-n} times known matrix elements. The same must be true in any chiral invariant Lagrangian theory. If the Lagrangian is such that F_π^{-1} appears as a coupling constant, then the correct current algebra result must be obtained in n^{th} order perturbation theory, except that we have to put in the correct axial couplings χ_a and, in problems more complicated than pion scattering, the correct matrix element for the process sans soft pions.[39] Thus phenomenological Lagrangians can be used in chiral theories in just about the same way as in electrodynamics.

We can construct a suitable chiral-invariant Lagrangian by the method described in Section 3D. The group G here is $SU(2) \times SU(2)$, and its algebraic subgroup H is the isospin group $SU(2)$. Any Lagrangian will be chiral-invariant if it is isospin invariant, and constructed only out of various fields ψ, their covariant derivatives

$$D_\mu \psi = \partial_\mu \psi + 2i F_\pi^{-1} \partial_\mu \pi_a E_{ab}(\underset{\sim}{\pi}) t_b \psi, \tag{4.H.1}$$

and the covariant derivative of the pion field

$$D_\mu \pi_a \equiv \partial_\mu \pi_b D_{ab}(\underset{\sim}{\pi}) \tag{4.H.2}$$

We have here replaced the Goldstone boson field of Section 3D with

$$\xi_a \equiv \frac{2\pi_a}{F_\pi}, \tag{4.H.3}$$

where π_a is a conventionally normalized pion field. (This normalization will have to be checked later on.) The functions D_{ab} and E_{ab} are now defined by

$$\exp\left[-2i F_\pi^{-1} \pi_a X_a\right] \exp\left[2i F_\pi^{-1} (\pi_a + \delta\pi_a) X_a\right] =$$

$$= 1 + 2i F_\pi^{-1} \delta\pi_a \left[D_{ab}(\underset{\sim}{\pi}) X_b + E_{ab}(\underset{\sim}{\pi}) T_b\right], \tag{4.H.4}$$

where X_a and T_a are the generators of the abstract Lie algebra of $SU(2) \times SU(2)$, with

$$[X_a, X_b] = i\epsilon_{abc} T_c \;;\; [T_a, T_b] = i\epsilon_{abc} T_c \;;\; [T_a, X_b] = i\epsilon_{abc} X_c .$$

(4.H.5) (4.H.6) (4.H.7)

The functions D_{ab} and E_{ab} can be calculated directly from (4.H.4), using for $\underset{\sim}{X}$ and $\underset{\sim}{T}$ any finite matrix representation of the commutation relations (4.H.5) - (4.H.7). In this way, we find

$$D_{ab}(\underset{\sim}{\pi}) = \delta_{ab}\left(\frac{\sin\xi}{\xi}\right) + \frac{\pi_a \pi_b}{\underset{\sim}{\pi}^2}\left(1 - \frac{\sin\xi}{\xi}\right), \quad (4.H.8)$$

$$E_{ab}(\underset{\sim}{\pi}) = F_\pi \left(\frac{\sin^2\frac{\xi}{2}}{\underset{\sim}{\pi}^2}\right) \epsilon_{abc}\, \pi_c , \quad (4.H.9)$$

where $\xi \equiv 2(\underset{\sim}{\pi}^2)^{1/2} F_\pi^{-1}$ This is not the same realization that I used to employ, but it turns out in the next section to have some remarkable advantages.

The lowest order in F_π^{-1} in which n soft pions can be emitted or absorbed, is F_π^{-n}. Hence the chiral-invariant Lagrangian is to be used only in lowest order, keeping only graphs without loops, in which trees of soft pions are attached to external lines of hadron processes.

In carrying out such calculations, we are not really interested in the most general possible chiral invariant Lagrangian, because the results can only be believed to the lowest order in the soft pion four-momenta. For instance, the most general purely pionic chiral invariant Lagrangian would contain terms of the form

$$\mathcal{L}_\pi = -\frac{Z}{2}(D_\mu \underset{\sim}{\pi})\cdot(D^\mu \underset{\sim}{\pi}) - \lambda\left[(D_\mu \underset{\sim}{\pi})\cdot(D^\mu \underset{\sim}{\pi})\right]^2$$

If only the first term is kept, then the matrix element for any process involving only soft pions will be of second order in pion momenta, while a tree graph of order n in λ will be of order $2+2n$ in the pion momenta. The leading term in the soft pion limit will thus be generated by the first term alone, which for a conventionally normalized pion field, is

$$\mathcal{L}_{\pi} = -\tfrac{1}{2}(\mathcal{D}_{\mu}\underset{\sim}{\pi})\cdot(\mathcal{D}^{\mu}\underset{\sim}{\pi})$$

Similarly, the leading part of the soft pion interaction with other hadrons, generated by the second term in (4.H.1), is linear in $E_{ab}\partial_{\mu}\pi_a$:

$$2iF_{\pi}^{-1}\frac{\partial\mathcal{L}_N}{\partial(\partial_{\mu}\psi)}\partial_{\mu}\pi_a E_{ab}(\underset{\sim}{\pi})t_b\psi = -F_{\pi}^{-1}\partial_{\mu}\pi_a E_{ab}(\underset{\sim}{\pi})V^{\mu}_{b\,N},$$

where $\mathcal{L}_N$ and $V^{\mu}_{b\,N}$ are respectively the non-pionic parts of the Lagrangian and the isospin current. This term, taken to second order in the pion field, gives a "universal" pion scattering length in agreement with the results of Section 4D. This verifies that (4.H.3) does give a correct field normalization.

Finally, it is possible to have chiral-invariant interactions between hadrons and soft pions which are of first order in pion field derivatives, of the form

$$-F_{\pi}^{-1}\mathcal{D}_{\mu}\underset{\sim}{\pi}\cdot\underset{\sim}{A}^{\mu}_{N} = -F_{\pi}^{-1}\partial_{\mu}\pi_a\mathcal{D}_{ab}(\underset{\sim}{\pi})A^{\mu}_{b\,N}$$

Comparison with the results of Section 4B for a single soft pion shows that $A^{\mu}_{b\,N}$ is the non-pionic part of the axial-vector current. This conclusion, and also the verification of (4.H.3), can also be obtained by direct calculation of the axial-vector current.

The full effective Lagrangian for the interaction of soft pions with other hadrons is now given by

$$\mathcal{L}_{EFF} = -\tfrac{1}{2}\mathcal{D}_{\mu}\underset{\sim}{\pi}\cdot\mathcal{D}^{\mu}\underset{\sim}{\pi} - F_{\pi}^{-1}\partial_{\mu}\pi_a\left\{E_{ab}(\underset{\sim}{\pi})V^{\mu}_{b_N} + \mathcal{D}_{ab}(\underset{\sim}{\pi})A^{\mu}_{b_N}\right\} \tag{4.H.10}$$

This Lagrangian will give results which satisfy the requirements of current algebra, or in other words, of dynamic chirality, for any choice of the operator A_{bN}^{μ}, including the possibility that A_{bN}^{μ} vanishes. Where then do the restrictions on $\underset{\sim}{\chi}$ derived in the last section come from? The answer can be seen by using (4.H.10) to calculate the amplitude for forward scattering of a pion on arbitrary hadron targets. The EV term gives a scattering amplitude $M^{(-)}(\omega)$ which approaches a constant as $\omega \to \infty$, and in order to cancel this unphysical behaviour, we must put in a DA coupling, with a coupling matrix $\underset{\sim}{\chi}$ which satisfies the commutation relation (4.G.23). This was the route I personally took in arriving at these commutation relations, but the derivation given in the last section seems much more convincing.

I. Summing Soft Pions

Having derived a Lagrangian which reproduces the results of current algebra, it is a natural next step to ask whether this Lagrangian can be used to carry out realistic calculations beyond the lowest order of perturbation theory. As an example, consider the exchange of an arbitrary number of soft pions from the external lines of the hard hadron reaction $i \to f$.[40]

In the limit of large hadron momentum, the matrix elements of the currents in (4.D.11) approach the limits

$$(2\pi)^3 (4p'^0 p^0)^{1/2} \langle p\beta | \underset{\sim}{A}^{\mu} | p\alpha \rangle \longrightarrow 4p^{\mu} \underset{\sim}{\chi}_{\beta\alpha} , \qquad (4.I.1)$$

$$(2\pi)^3 (4p'^0 p^0)^{1/2} \langle p\beta | V^{\mu} | p\alpha \rangle \longrightarrow 4p^{\mu} \underset{\sim}{t}_{\beta\alpha} , \qquad (4.I.2)$$

where $\underset{\sim}{\chi}$ is the matrix introduced in Section F. The numerators of the external line propagators just become projection operators which, sandwiched between the V and A vertices, yield these mass-shell matrix elements. The denominators of the propagators have the familiar eikonal form $\frac{-i}{2p\cdot q}$, which, when Fourier transformed, becomes

$$-i\,\Delta_{EIK}(x_2-x_1;p) \equiv -i\int \frac{d^4q}{(2\pi)^4}\,\frac{1}{2p\cdot q-i\epsilon}\,\exp[i(p+q)\cdot(x_2-x_1)]$$

$$\longrightarrow \frac{1}{2|p|}\,\delta^3(\underset{\sim}{\mu}_2-\underset{\sim}{\mu}_1)\,\theta(v_2-v_1)\,, \qquad (4.I.3)$$

where $\underset{\sim}{\mu}$ and v are the auxiliary coordinates

$$\underset{\sim}{\mu} \equiv \underset{\sim}{x} - \hat{p}\,t \quad , \qquad v \equiv \tfrac{1}{2}(\hat{p}\cdot\underset{\sim}{x} + t)\,. \quad (4.I.4)$$

In an external pion field $\underset{\sim}{\pi}(x)$ (ignoring pion-pion interactions for the moment) the effect of sandwiching a string of vertices between hard hadron propagators is to replace the final and initial state "wave functions" $e^{ip\cdot x}$ and $e^{-ip\cdot x}$ by $e^{ip\cdot x}F(x;p)$ and $e^{-ip\cdot x}I(x;p)$ respectively, where

$$F = 1 + \sum_{n=1}^{\infty} i^n \int_v^{\infty} dv_n \int_{v_n}^{\infty} dv_{n-1} \cdots \int_{v_2}^{\infty} dv_1\, \Gamma(\underset{\sim}{\mu},v_1)\cdots\Gamma(\underset{\sim}{\mu},v_n)\,, \quad (4.I.5)$$

$$I = 1 + \sum_{n=1}^{\infty} i^n \int_{-\infty}^{v} dv_1 \int_{-\infty}^{v_1} dv_2 \cdots \int_{-\infty}^{v_{n-1}} dv_n\, \Gamma(\underset{\sim}{\mu},v_1)\cdots\Gamma(\underset{\sim}{\mu},v_n)\,, \quad (4.I.6)$$

where Γ is the matrix

$$\Gamma(x) = \frac{-2}{F_\pi}\,[\,D_{ab}(\pi)\,\chi_b + E_{ab}(\pi)\,t_b\,]\,\frac{\partial \pi_a}{\partial v} \qquad (4.I.7)$$

This looks complicated, but recalling the definition (4.H.4) of D and E we see that if χ and t obey the same commutation relations (4.H.5) (4.H.7) as X and T then

$$\Gamma = 2i\,\exp\{-2iF_\pi^{-1}\pi_a\chi_a\}\,\frac{\partial}{\partial v}\,\exp\{2iF_\pi^{-1}\pi_a\chi_a\}$$

$$(4.I.8)$$

The sums (4.I.5) and (4.I.6) are now easily carried out, and we find

$$F = \exp\{2i F_\pi^{-1} \underset{\sim}{\chi} \cdot \underset{\sim}{\pi}\} \quad , \tag{4.I.9}$$

$$I = \exp\{-2i F_\pi^{-1} \underset{\sim}{\chi} \cdot \underset{\sim}{\pi}\} \tag{4.I.10}$$

The over-all effect of an external pion field on a hard hadron reaction matrix element is then that we are to make the replacement

$$(2\pi)^4 \delta^4(p_f - p_i) \longrightarrow \int d^4x \, e^{-i(p_i - p_f)\cdot x} \exp\{2i F_\pi^{-1} \Delta \underset{\sim}{\chi} \cdot \underset{\sim}{\pi}(x)\}, \tag{4.I.11}$$

where $\Delta \underset{\sim}{\chi}$ is the sum of the $\underset{\sim}{\chi}$ -matrices in the final state, acting from the left in the final particle indices, minus the sum of all $\underset{\sim}{\chi}$ -matrices in the initial state, acting from the right on the initial particle indices.

It remains to tie together the external line insertions described by (4.I.11) with soft pion propagators. Inspired by current algebra, we will include pion-pion interactions only to the extent of summing over purely pionic trees. The summation technique has been described by Nambu;[41] we must first solve the nonlinear pion equation with an external source $\underset{\sim}{\eta}$:

$$\Box^2 \underset{\sim}{\pi} = -\underset{\sim}{\eta} - \underset{\sim}{J}(\underset{\sim}{\pi}) \quad , \tag{4.I.12}$$

where $\underset{\sim}{J}$ is the pionic current

$$\underset{\sim}{J} \equiv \frac{\partial \mathcal{L}'}{\partial \underset{\sim}{\pi}} - \partial_\mu \frac{\partial \mathcal{L}'}{\partial(\partial_\mu \underset{\sim}{\pi})} \quad , \tag{4.I.13}$$

with $\mathcal{L}'$ the pion self interaction

$$\mathcal{L}' - \tfrac{1}{2} \partial_\mu \underset{\sim}{\pi} \, \partial^\mu \underset{\sim}{\pi} \equiv -\tfrac{1}{2} D_\mu \underset{\sim}{\pi} \, D^\mu \underset{\sim}{\pi} \tag{4.I.14}$$

The sum of all pion trees for the vacuum-vacuum transition in an external pion field η is then

$$\Pi[\eta] = \exp\Big\{ \frac{i}{2}\int d^4x\, d^4y\, \eta(x)\cdot\eta(y)\,\Delta_\pi(x-y) + i\int d^4x\, \mathcal{L}' + -\frac{1}{2}\int d^4x\, d^4y\, \underset{\sim}{J}(x)\cdot\underset{\sim}{J}(y)\,\Delta_\pi(x-y)\Big\}, \qquad (4.I.15)$$

where Δ_π is the usual pion propagator

$$\Delta_\pi(x) \equiv \int \frac{d^4q}{(2\pi)^4}\, e^{iq\cdot x}\, \frac{1}{(q^2-i\epsilon)} \quad . \qquad (4.I.16)$$

By expanding the exponential (4.I.11) we see that the effect of tying together the pion insertions in external lines with pion trees is to make the replacement

$$(2\pi)^4\,\delta^4(p_f-p_i) \longrightarrow \int d^4x \sum_{n=0}^{\infty} \frac{1}{n!}\left(\frac{2}{F_\pi}\right)^n \Delta\chi_{a_1}\cdots\Delta\chi_{a_n}\times$$

$$\times\frac{\delta^n}{\delta\eta_{a_1}(x)\cdots\delta\eta_{a_n}(x)}\,\Pi[\eta]\Big|_{\eta=0}\, e^{-i(p_i-p_f)\cdot x} \qquad (4.I.17)$$

Translational invariance requires that the variational derivative here is x-independent, so we can replace $\frac{\delta}{\delta\eta(x)}$ by $\frac{\delta}{\delta\eta(0)}$. The x-integral then just gives a factor $(2\pi)^4\,\delta^4(\ )$, so the effect of pion exchange is just to insert into the S-matrix an additional factor

$$\Pi\left[\frac{2}{F_\pi}\,\Delta\underset{\sim}{\chi}\,\delta^4(x)\right] \qquad (4.I.18)$$

However, direct calculation shows that for a source $\eta(x) = \underset{\sim}{e}\,\delta^4(x)$, the solution of (4.I.12) is

$$\underset{\sim}{\pi}(x) = \underset{\sim}{e}\,\Delta_\pi(x)$$

because for this function, $\underset{\sim}{J} = 0$. Also $\mathcal{L}' = 0$, so (4.I.15)

gives
$$\Pi\,[\underset{\sim}{e}\,\delta^4(x)] = \exp\left\{\frac{i}{2}\,\underset{\sim}{e}^2\,\Delta_\pi(0)\right\}$$

It is necessary to employ a cutoff at momentum Λ to distinguish soft from hard pions. Summing only over soft pions in (4.I.16) gives
$$\Delta_\pi(0) = \frac{i\Lambda^2}{8\pi^2}$$

Hence the factor (4.I.18) is just (ERRATUM: See note [42])
$$\exp\left\{\frac{-\Lambda^2}{4\pi^2 F_\pi^2}\,(\Delta\underset{\sim}{\chi})^2\right\}, \tag{4.I.19}$$

and the whole effect of the exchange of arbitrary soft pion trees is just to multiply the uncorrected S-matrix by this factor.

It proves very convenient at this point to decompose the uncorrected S-matrix S_0 into terms S_{0N} which, with respect to multiple commutation with the matrix $\Delta\underset{\sim}{\chi}$ behave like the representation $(N/2, N/2)$ of the algebra SU(2) × SU(2), or in other words, like the $44\cdots4$ component of a traceless symmetric tensor $S_{0N}^{\alpha\beta\cdots}$ of rank N. We then have
$$\Delta\chi_a\,\Delta\chi_b\,S_{0N}^{4\cdots4} = -iN\,\Delta\chi_a\,S_{0N}^{b4\cdots4} =$$
$$= -iN\left[-i(N-1)\,S_{0N}^{ba4\cdots4} + i\,\delta_{ab}\,S_{0N}^{44\cdots4}\right]$$

and since S_{0N} is defined to be traceless,
$$(\Delta\underset{\sim}{\chi})^2\,S_{0N} = -iN\left[i(N-1)S_{0N} + 3iS_{0N}\right] = N(N+2)\,S_{0N}$$

Thus the effect of soft pion exchange is to suppress the part of the S-matrix which belongs to a representation $(N/2, N/2)$ by a factor
$$\exp\left\{\frac{-\Lambda^2}{4\pi^2 F_\pi^2}\,N(N+2)\right\} \tag{4.I.20}$$

The physical origin of this suppression is clear - the emission of soft pions takes place more strongly in the part of the S-matrix with N large, and not at all in the part with $N=0$, so for large N the S-matrix for processes in which soft pions are not emitted must be suppressed in order to avoid violating unitarity. We therefore expect that at high energy, where our approximations may be valid, the S-matrix is dominated by terms which conserve algebraic chirality $(N=0)$ or belong to low representations $(N \simeq 1)$. Of course, most S-matrix elements are already suppressed by unitarity at high energy, and we don't want to count this suppression twice, so our remarks apply to the unsuppressed S-matrix elements, and most notably, to forward elastic scattering. The conjecture which comes out of all this is then that the residue of the Pomeranchukon may belong to a low representation of the chiral algebra $\underset{\sim}{\chi}, \underset{\sim}{t}$. Unfortunately, to test this idea we have to know what $\underset{\sim}{\chi}$ is. A good deal of analysis of pion and photon decay matrix elements must be done before these results can be tested.

The relation between algebraic and dynamic symmetries has turned out to be far more complicated than we might have expected. Dynamic symmetries like gauge invariance of the second kind and chirality are directly manifested in low energy theorems for bosons like photons or pions. When put together with unsubtracted, single-particle-saturated dispersion relations, these low energy theorems yield algebraic relations for matrices like $\mathcal{K}$ and $\underset{\sim}{\chi}$. These relations are still far short of providing a full-fledged algebraic symmetry like isospin conservation, because $\mathcal{K}$ and $\underset{\sim}{\chi}$ do not commute with the mass matrix (a failure which arises for $\underset{\sim}{\chi}$ from the presence of the vacuum trajectories P, P') or with the S-matrix. However, the strong emission of soft pions in processes which violate algebraic chirality may lead to a suppression of the matrix elements for such processes at high energy, in which case chirality would play the role of an ordinary algebraic symmetry in the high energy limit.

REFERENCES

[1]. H. Lehmann, K. Symanzik, and W. Zimmermann, Nuovo Cimento 1, 205 (1955).

[2]. J. C. Ward, Phys. Rev. 78, 1824 (1950).

[3]. Y. Takahashi, Nuovo Cimento 6, 370 (1957).

[4]. E. P. Wigner, Ann. Math 40, 149 (1939).

[5]. S. Weinberg, Phys. Rev. 133, B1318 (1964).

[6]. J. Schwinger, Phys. Rev. Letters 3, 296 (1959).

[7]. S. Weinberg, Lectures on Particles and Field Theory, (Prentice-Hall, 1965), p. 476.

[8]. F. E. Low, Phys. Rev. 96, 1428 (1954).

[9]. M. Gell-Mann and M. L. Goldberger, Phys. Rev. 96, 1433 (1954).

[10]. S. D. Drell and A. Hearn, Phys. Rev. Letters 16, 908 (1966).

[11]. C. R. Hagen and W. J. Hurley, Phys. Rev. Letters 24, 1381 (1970).

[12]. F. E. Low, Phys. Rev. 110, 974 (1958).

[13]. M. B. Beg, Phys. Rev. 150, 1276 (1966).

[14]. N. Cabibbo and L. Radicati, Phys. Lett. 19, 697 (1966).

[15]. Y. Nambu and G. Jona-Lasinio, Phys. Rev. 122, 345 (1960).

[16]. J. Goldstone, Nuovo Cimento 19, 154 (1961).

[17]. J. Goldstone, A. Salam, and S. Weinberg, Phys. Rev. 127, 965 (1962).

[18]. C. Callan,. S. Coleman, J. Wess, and B. Zumino, Phys. Rev. 177, 2247 (1968).

[19]. S. Weinberg, Phys. Rev. 166, 1568 (1968).

[20]. S. Coleman, J. Wess, and B. Zumino, Phys. Rev. 177, 2239 (1968)

[21]. Y. Nambu, Phys. Rev. Letters 4, 380 (1960).

[22]. M. Gell-Mann and M. Lévy, Nuovo Cimento 16, 705 (1960).

[23]. J. Bernstein, S. Fubini, M. Gell-Mann, and W. Thirring, Nuovo Cimento 17, 757 (1960).

[24]. S. Adler, Phys. Rev. 137, B1022 (1965); 139, B1638 (1965).

[25]. R. Feynman and M. Gell-Mann, Phys. Rev. 109, 193 (1958).

[26]. R. E. Marshak and E. C. G. Sudarshan, Phys. Rev. 109, 1860 (1958).

[27]. M. Gell-Mann, Physics 1, 63 (1964).

[28]. M. Goldberger and S. B. Treiman, Phys. Rev. 110, 1178 (1958).

[29]. S. Weinberg, Phys Rev. Letters 17, 336 (1966); Phys. Rev. Letters 16, 879 (1966).

[30]. S. Fubini and G. Furlan, Ann. Phys. (U.S.A.) 48, 1322 (1968).

[31]. Y. Tomozawa, Nuovo Cimento 46A, 707 (1966).

[32]. K. Raman and E. C. G. Sudarshan, Phys. Lett. 21, 450 (1966).

[33]. M. L. Goldberger, H. Miyazawa, R. Oehme, Phys. Rev. 99, 986 (1955).

[34]. M. L. Goldberger, Dispersion Relations and Elementary Particles (Wiley, 1960), p. 146.

[35] S. Adler, Phys. Rev. Letters 14, 1051 (1965); Phys. Rev. 140, B736 (1965); W. Weisberger, Phys. Rev. Letters 14, 1047 (1965); Phys. Rev. 143, 1302 (1966).

[36] S. Weinberg, Phys. Rev. 177, 2604 (1969).

[37] R. deAlfaro, S. Fubini, G. Rosetti, and G. Furlan, Phys. Rev. Letters 21, 576 (1968).

[38] F. J. Gilman and H. Harari, Phys. Rev. 165, 1803 (1968).

[39] S. Weinberg, Phys. Rev. Letters 18, 188 (1967).

[40] S. Weinberg, Phys. Rev. (15 Aug. 1970).

[41] Y. Nambu, Phys. Letters 26B, 626 (1968).

[42] Added note: A serious algebraic error in the steps leading from (4.I.18) to (4.I.19) has been pointed out to me by Lowell Brown. According to (4.I.17), the coefficient of $\Delta\chi_{a_1} \cdots \Delta\chi_{a_n}$ in a power series expansion of $\Pi\left[\frac{2}{F_\pi}\Delta\underset{\sim}{\chi}\,\delta^4(x)\right]$ must be completely symmetrical in $a_1 \cdots a_n$, so it should be evaluated by expanding $\Pi\left[\underset{\sim}{e}\,\delta^4(x)\right]$ in a power series expansion in $\underset{\sim}{e}$, then symmetrizing the coefficient of $e_{a_1} \cdots e_{a_n}$ in the indices $a_1 \cdots a_n$ and only then replacing $\underset{\sim}{e}$ by $\frac{2}{F_\pi}\Delta\underset{\sim}{\chi}$. Thus, in place of (4.I.19), we should have a matrix

$$M = \sum_{n=0}^{\infty} \frac{(-1)^n}{n!} \lambda^{2n}\, \delta(a_1 \cdots a_{2n})\, \Delta\chi_{a_1} \cdots \Delta\chi_{a_{2n}}$$

where $\lambda \equiv \frac{\Lambda}{2\pi F_\pi}$, and δ is the product of Kronecker deltas, averaged over permutations of the indices. That is,

$$\delta(ab) = \delta_{ab},$$

$$\delta(abcd) = \tfrac{1}{3}\left(\delta_{ab}\delta_{cd} + \delta_{ac}\delta_{bd} + \delta_{ad}\delta_{bc}\right),$$

and so on. To evaluate M it is easiest to use a trick, and write

$$\delta(a_1 \cdots a_{2n}) = \frac{2n+1}{4\pi} \int d\Omega \, \mu_{a_1} \cdots \mu_{a_{2n}}$$

where $\underset{\sim}{\mu}$ is a unit vector and $d\Omega$ is the element of solid angle in the direction of $\underset{\sim}{\mu}$ Then we have

$$M = \frac{1}{4\pi} \int d\Omega \left[1 - 2\lambda^2 (\underset{\sim}{\mu} \cdot \Delta \underset{\sim}{\chi})^2 \right] \exp\left(-\lambda^2 (\underset{\sim}{\mu} \cdot \Delta \underset{\sim}{\chi})^2 \right).$$

For the terms S_{0N} in S_0 belonging to the $(N/2, N/2)$ representation of SU(2) × SU(2), we have

$$(\underset{\sim}{\mu} \cdot \Delta \underset{\sim}{\chi})^2 S_{00} = 0 \; ,$$
$$(\underset{\sim}{\mu} \cdot \Delta \underset{\sim}{\chi})^2 S_{01} = S_{01} \; ,$$
$$(\underset{\sim}{\mu} \cdot \Delta \underset{\sim}{\chi})^4 S_{02} = 4 (\underset{\sim}{\mu} \cdot \Delta \underset{\sim}{\chi})^2 S_{02}$$
$$(\underset{\sim}{\mu} \cdot \Delta \underset{\sim}{\chi})^4 S_{03} = \left[10 (\underset{\sim}{\mu} \cdot \Delta \underset{\sim}{\chi})^2 - 9 \right] S_{03}$$

and so on. This then gives

$$M S_{00} = S_{00} ,$$
$$M S_{01} = e^{-\lambda^2} (1 - 2\lambda^2) S_{01} ,$$
$$M S_{02} = \left[\tfrac{1}{3} + \tfrac{2}{3} e^{-4\lambda^2} (1 - 8\lambda^2) \right] S_{02} ,$$
$$M S_{03} = \left[\tfrac{1}{2} (1 - 2\lambda^2) e^{-\lambda^2} + \tfrac{1}{2} (1 - 18\lambda^2) e^{-9\lambda^2} \right] S_{03}$$

There still is a suppression of all terms in S_0 except S_{00} for the same reason as given previously. However, for $\lambda \to \infty$ the even-rank tensors $S_{02}, S_{04}, \cdots$ are now only suppressed by finite numerical factors. The consequences for the P and P' residues are more complicated than I had thought, but still well worth checking.

Local Operator Products and Renormalization in Quantum Field Theory

Wolfhart Zimmerman
New York University
New York, New York

TABLE OF CONTENTS

I. Introduction

The main topic of this course will be the problem of defining local operator products in quantum field theory. Let A_1, A_2 be two local field operators satisfying the usual postulates of quantum field theory. A_1, A_2 may be components of tensor or spinor fields. In general the operator product

$$A_1(x)\, A_2(y)$$

has a singularity near $x \sim y$ and for this reason it is not possible to define the product

$$A_1(x)\, A_2(x)$$

at the same point in a naive sense. It can easily be seen[1] that a singularity necessarily occurs for any scalar field $A(x)$, provided it has non-vanishing matrix elements between the vacuum Ω and one particle states Φ_k :

$$(\Omega, A(x)\, \Phi_k) \neq 0 \ , \quad P_\mu \Phi_k = k_\mu \Phi_k \ , \quad k^2 = m^2$$

By translation invariance

$$(\Omega, A(x)\, \Phi_k) = c\, e^{-ikx}$$

$$c = (\Omega, A(0)\, \Phi_k) \ .$$

Thus

$$(\Omega, A(x)^2 \Omega) \geq \int \frac{d_3 k}{2\omega} \left|(\Omega, A(x)\, \Phi_k)\right|^2 = \int \frac{d_3 k}{2\omega} |c|^2 ,$$

$$\omega = \sqrt{\vec{k}^2 + m^2} \ ,$$

which of course diverges.

The simplest example is a free scalar field $A_0(x)$. In this case the total singularity of $A_0(x) A_0(y)$ near $x \sim y$ is carried by its vacuum expectation value $\langle A_0(x) A_0(y) \rangle$. Subtracting the vacuum expectation value

$$: A_0(x)^2 : \; = \lim_{x \to y} \{ A_0(x) A_0(y) - \langle A_0(x) A_0(y) \rangle \} \qquad (I.1)$$

one obtains a well-defined local operator in the limit, which may be interpreted as the square of the free field.[2]

Unfortunately the situation is more involved for interacting fields. The problem is to analyze the singularities of the operator product $A_1(x) A_2(y)$ near $x \sim y$. Concerning these singularities Wilson[3, 4] has made a powerful hypothesis which states that any operator product allows an expansion

$$A_1(x+\xi) A_2(x-\xi) \sim E_1(\xi) B_1(x) + \cdots + E_n(\xi) B_n(x) \qquad (I.2)$$

up to terms vanishing for $\xi \to 0$ Here the B_j are local field operators (anti-)commuting with themselves and A_j, as well as any other local field operator θ of the model

$$\begin{aligned} &[B_i(x), B_j(y)]_\pm = 0 \\ &[B_i(x), \theta(y)]_\pm = 0 \end{aligned} \qquad \text{for } (x-y)^2 < 0 \qquad (I.3)$$

The E_j are functions of ξ. The B_j may represent tensor or spinor fields, accordingly the E_j are tensors or matrices in spin space.

If one is willing to make a number of assumptions one can indeed obtain such an expansion from general arguments. This will be discussed in section III, 1. The expansion will first be derived in the somewhat different form

$$A_1(x+\xi)A_2(x-\xi) = \sum_{j=1}^{m} f_j(\rho) . C_j(x,\eta) + R(x,\eta,\rho) \qquad (I.4)$$

where the operators C_j are allowed to depend on the direction η

$$\eta = \frac{\xi}{\rho} \qquad \rho = \sqrt{|\xi^2|} \qquad (I.5)$$

The coefficients f_j satisfy the following conditions

$$\lim_{\rho \to 0} \frac{f_j}{f_{j+1}} = \infty \qquad (I.6)$$

The last function f_m **has the limit**

$$\lim_{\rho \to c} f_m \neq 0 \text{ (it may be } \infty) \qquad (I.7)$$

while the remainder vanishes

$$\lim_{\rho \to 0} R = 0 \qquad \text{for } \eta \text{ constant.} \qquad (I.8)$$

Moreover the C_j are linearly independent operators

By (I. 6) the functions are ordered according to decreasing singularities. f_1 is the most singular coefficient, while f_m is the least singular one, which may even be finite It can be shown that the operators C_j are local in the sense

$$\left.\begin{aligned}&[C_i(x,\eta), C_j(y,\zeta)]_\pm = 0\\&[C_i(x,\eta), A_j(y)]_\pm = 0\end{aligned}\right\} \quad \text{for } (x-y)^2 < 0 \qquad (I.9)$$

Suppose the C_j are polynomials in η (which is of course a strong assumption) then (I. 4) becomes an expansion of the form (I. 2)

$$\begin{aligned}A_1(x+\xi)\, A_2(x-\xi) \sim \sum_j E_j(\xi)\, B^j(x) \; + \\ + \sum_{j,n} E_j^{\mu_1\cdots\mu_n}(\xi)\, B^j_{\mu_1\cdots\mu_n}(x)\end{aligned} \qquad (I.10)$$

with a finite number of terms. The coefficients are

$$E_j^{\mu_1\cdots\mu_n}(\xi) = \frac{\xi^{\mu_1}\cdots\xi^{\mu_n}}{|\xi^2|^{n/2}}\, f_j(\xi^2), \qquad (I.11)$$

$$B^j(x) = C^j(x,0)$$

With the expansion (I. 4) it is a straightforward matter to define local operator products associated with A_1A_2. Dividing (I. 4) by f_1 one finds

$$C_1 = \lim_{\xi\to 0} \frac{A_1(x+\xi)\, A_2(x-\xi)}{f_1} \qquad (I.12)$$

since

$$\lim_{\rho\to 0} \frac{f_j}{f_1} = 0, \; (j>1), \quad \lim_{\rho\to 0} \frac{R}{f_1} = 0$$

as consequence of (I. 5-7). Furthermore

$$C_2 = \lim_{\rho\to 0} \frac{A_1(x+\xi)\, A_2(x-\xi) - f_1 C_1}{f_2} \qquad (I.13)$$

and finally

$$C_m = \lim_{\rho \to 0} \frac{A_1(x+\xi)A_2(x-\xi) - \sum_{j=1}^{m-1} f_j C_j}{f_m} . \qquad (I.14)$$

The direction independent operators B_j, $B_j^{\mu_1 \cdots \mu_a}$ are obtained by expanding C_j with respect to the components of η. Which one of the operators B_j is called the product $A_1(x)A_2(x)$ is largely a matter of taste. Of course, one would exclude the identity which may appear as the first operator C_1.

As we have seen the operators C_j are uniquely determined for given coefficients f_j. The question arises how much freedom one has in choosing the C_j if the coefficients f_j are permitted to change. Let us assume that f_j, C_j and f'_j, C'_j are two sets of functions and operators satisfying the conditions I.4 - I.9. Then it can be shown (section III. 1) that the operators C_j are determined up to an arbitrary triangular transformation

$$\begin{aligned} C_1' &= a_{11} C_1 \\ C_2' &= a_{21} C_1 + a_{22} C_2 \qquad\qquad a_{\alpha\alpha} \neq 0 \\ &\cdots \\ C'_m &= a_{m1} C_1 + \cdots + a_{mm} C_m \qquad (I.15) \end{aligned}$$

There is a corresponding transformation of the coefficients f_j (see section III. 1).

Wilson[4] has extended his hypothesis to the statement that

$$A_1(x+\xi)\, A_2(x-\xi)$$

may even be equal to a series

$$A_1(x+\xi)\,A_2(x-\xi) = \sum_{j=1}^{\infty} E_j(\xi)\,B_j(x) \qquad (I.16)$$

with an infinite number of terms. Here only a finite number of coefficients become singular or different from zero at the origin while the remaining coefficients vanish for $\xi \to 0$. Of course, it can be arranged that the B_j are linearly independent. Moreover, Wilson makes a statement concerning the leading term of $E_j(\xi)$ near $\xi \sim 0$. Since this is related to scale invariance let us first assume a theory which is invariant under scale transformations

$$x \longrightarrow sx \quad .$$

In the quantum mechanical sense this means that there is a family of unitary transformations $U(s)$ with the property

$$U(s)^{-1}\,\theta(x)\,U(s) = s^{d(\theta)}\,\theta(sx) \qquad (I.17)$$

for any local field operator of the theory. The real number $d(\theta)$ is called the dimension of the operator θ. Using this transformation law Wilson determines the singularities of the coefficients in the following way. Applying the transformation $U(s)$ to the expansion (I.16) we get $s^{d(A_1)+d(A_2)}\,A_1(sx+s\xi)\,A_2(sx-s\xi) = \sum_j E_j(\xi)\,s^{d(B_j)}\,B_j(sx)$

Expanding A_1A_2 on the left hand side yields

$$s^{d(A_1)+d(A_2)} \sum_j E_j(s\xi)\, B_j(sx) =$$

$$= \sum_j E_j(\xi)\, s^{d(B_j)}\, B_j(sx)$$

Since the B_j are linearly independent we obtain

$$E_j(\xi) = s^{d(A_1)+d(A_2)-d(B_j)}\, E_j(s\xi)$$

as the scaling law of the coefficients. The exponent of s

$$d(E_j) = d(A_1) + d(A_2) - d(B_j) = d(A_1A_2) - d(B_j) \qquad (I.18)$$

is called the dimension of E_j and we have the result that in each term of the expansion the dimension of E_j and the dimension of B_j must add up to the total dimension of the left hand side

$$d(E_j) + d(B_j) = d(A_1A_2) \qquad (I.19)$$

The dimension of E_j indicates the behavior for $\xi \to 0$:

$$E_j(s\xi) = s^{-d(E_j)}\, E_j(\xi) \qquad (I.20)$$

or

$$E_j(\xi) = \rho^{-d(E_j)}\, E_j(\eta) \qquad (I.21)$$

with the variables (I. 5). As a consequence the E_j are singular or $\neq 0$ in the limit $\xi \to 0$ only if $d(B_j) \leq d(A_1A_2)$. (I.22)

Hence the singular and non-vanishing terms are provided by operators B_j of dimension less or equal to the dimension of the product $A_1 A_2$.

Unfortunately in all realistic theories scale invariance is broken; in fact any mass term breaks the scale invariance.[5] To make things worse, the indication is that scale invariance is not even recovered in the limit $m \to 0$. One might define the dimension d through the generator D of the scale transformations, but it may not even be possible to define D in a sensible way. So it is not clear how this part of Wilson's hypothesis should be formulated.

However, dimensional arguments work very well in perturbation theory.[3] In that case one may even use the conventional definition of dimension. As follows from the canonical commutation relations the dimensions of a scalar field A and a spin 1/2 field ψ in inverse units of length are

$$d(A) = 1 \;, \quad d(\psi) = 3/2 \qquad (I.23)$$

Each derivative increases the dimension by one ,

$$d(\partial_{\mu_1} \cdots \partial_{\mu_r} B) = r + d(B) \,, \qquad (I.24)$$

and the dimension of a product of operators is

$$d(A_1 A_2) = d(A_1) + d(A_2) \qquad (I.25)$$

For free fields of mass zero this dimension coincides with the dimension defined by scale transformations. This is not so in general; a counter example is the scale invariant Thirring model where both definitions disagree.[6, 7] In this case the dimension defined through scale invariance depends on the magnitude of the coupling constant.

In section III. 2 local operator products will be defined for renormalizable theories which are finite in every order of perturbation theory. These operator products will be called normal products and denoted by

$$N\{D_1(x) \cdots D_n(x)\} \qquad (I.26)$$

where $D_1, \ldots, D_n$ may be any derivatives of the basic field operators of the model. We will confirm that Wilson's expansion

$$A_1(x+\xi)\,A_2(x-\xi) = \sum_{j=1}^{\infty} E_j(\xi)\,B_j(x) \qquad (I.27)$$

is valid in perturbation theory (Section III 7). A_1 and A_2 are derivatives of the basic fields. For the B_j one takes all possible operator products (I. 26). Concerning the singularities of the coefficients we will find Wilson's rule

$$d(E_j) + d(B_j) \leq d(A_1 A_2)\,. \qquad (I.28)$$

Here the dimensions of the operators are the free field dimensions given by the rules (I. 22-24). The dimension of E_j is defined by the leading term near $\xi \sim 0$

$$E_j \sim |\xi^2|^{-\frac{1}{2}d(E_j)} \qquad (I.29)$$

up to logarithmic factors. Hence for renormalizable theories Wilson's hypothesis holds in perturbation theory with the conventional definition of dimension (with the possible exception of some mass zero theories). For the exact solution the situation may be quite different. This can be demonstrated in the Thirring model which is scale invariant. There Wilson's hypothesis holds in perturbation theory with the conventional dimension. But this is no longer true for the exact solution of the Thirring model for which the dimension defined by scale transformations becomes relevant.[7]

In Chapter IV local field equations in finite terms will be discussed as an application of the operator expansions. We briefly explain the method. For definiteness we consider the case of A^4-coupling. As interaction Lagrangian we take

$$\mathcal{L}_{INT} = -\lambda\left(\frac{1}{4}A^4 - \frac{1}{2}aA^2 - b\mathcal{L}_0\right) \quad , \qquad (I.30)$$

where $\mathcal{L}_0$ is the free Lagrangian ,

$$\mathcal{L}_0 = \frac{1}{2}\partial_\mu A\,\partial^\mu A - \frac{1}{2}m^2A^2$$

m^2, λ , a and b are finite constants.[8] The definition of the normal operator product $N\{A(x)^3\}$ leads to a surprisingly simple equation for the field operator

$$-(\Box+m^2)A(x)=\lambda N\{A(x)^3\}-aA(x)+b(\Box+m^2)A(x). \qquad (I.31)$$

(I. 31) resembles what one would formally obtain as equation of motion from the effective Lagrangian

$$\mathcal{L}_{EFF}=\mathcal{L}_0+\mathcal{L}_{INT} \qquad (I.32)$$

with the only difference that the normal product is taken for the interaction term. So far (I. 31) is a linear relation between $N\{A^3\}$ and the field A and its derivatives. In order to obtain a field equation for A alone the normal product should be expressed in terms of A . This can be done by using the expansion of

$$A(x+\xi)\,A(x)\,A(x-\xi)$$

near $\xi\sim 0$. The final result derived in section IV. 2 is

$$-(\Box+m^2)A(x)=$$

$$=\lambda\lim_{\xi\to 0}\frac{P(x,\xi)-\alpha(\xi)A(x)+\sigma_{\mu\nu}(\xi)\,\partial^\mu\partial^\nu A(x)}{f(\xi)}, \qquad (I.33)$$

$$\sigma_{\mu\nu}(\xi)=\frac{\xi_\mu\xi_\nu}{\xi^2}\,\sigma(\xi^2).$$

The operator product

$$P(x,\xi)=\,:A(x+\xi)\,A(x)\,A(x-\xi): \qquad (I.34)$$

is defined by taking trivial vacuum subtractions

$$:A(x_1)A(x_2)A(x_3): = A(x_1)A(x_2)A(x_3) - \langle A(x_1)A(x_2)\rangle A(x_3) - \langle A(x_2)A(x_3)\rangle A(x_1) - \langle A(x_3)A(x_1)\rangle A(x_2) \quad . \qquad (I.35)$$

In quantum electrodynamics a field equation of a similar form was first proposed by Valatin.[9, 10, 11] The limit form (I. 35) replaces the formal field equation of conventional renormalization theory which for this model reads

$$-(\Box + m^2)A = Z_4 Z_3^{-1} \lambda A^3 - \delta m^2 A \quad ; \qquad (I.36)$$

Z_3 is the renormalization constant of the field, Z_4 the renormalization constant of the vertex function. Since Z_3, Z_4 and δm^2 are divergent in perturbation theory the equation does not make sense. In the limit form (I. 33) the coordinates of the singular product A^3 are taken apart and the divergent constants are replaced by functions of ξ which become singular at the origin.

II. Renormalization[18]

1. Definition of the Finite Part of a Feynman Integral by Bogoliubov's Method.

In this section the finite part of a Feynman integral

$$F_\Gamma = \lim_{\epsilon \to 0^+} \int dk_1 \cdots dk_m \, I_\Gamma \qquad (\text{II}.1)$$

will be defined in the form

$$J_\Gamma = \lim_{\epsilon \to 0^+} \int dk_1 \cdots dk_m \, R_\Gamma \qquad (\text{II}.2)$$

where the modified integrand R_Γ is obtained from the original integrand I_Γ by subtracting a suitable number of counter terms. We first explain the notation which we use for the unrenormalized integrand I_Γ. Let Γ be a Feynman diagram belonging to an interaction described by a Lagrangian which is a local invariant polynomial of the fields and their derivatives. The vertices of the diagram are denoted by $V_1, \ldots, V_N$. The lines connecting the vertices V_a and V_b are denoted by $L_{ab\sigma}$. To each line $L_{ab\sigma}$ we assign an internal momentum

$$l_{ab\sigma} = - l_{ba\sigma}$$

To some of the vertices V_a external momenta $p_{a\sigma}$ are assigned. The total external momentum at V_a is

$$q_a = \sum_\sigma p_{a\sigma}$$

or

$$q_a = 0$$

if V_a carries no external momenta. The unrenormalized integrand is of the form

$$I_\Gamma = \prod_{ab\sigma} \Delta_F^{ab\sigma} \prod_a P_a \quad , \tag{II.3}$$

with the propagators

$$\Delta_F^{ab\sigma} = P_{ab\sigma}\left(l_{ab\sigma}^2 - \mu_{ab\sigma}^2 + i\epsilon(\vec{l}_{ab\sigma}^{\,2} + \mu_{ab\sigma}^2)\right)^{-1} ,$$
$$\varepsilon > 0 \ , \quad \mu_{ab\sigma}^2 > 0 \ , \tag{II.4}$$

and polynomials P_a of the momenta $l_{ab\sigma}$ joining at the vertex V_a. $P_{ab\sigma}$ is a polynomial in $l_{ab\sigma}$. The reason for the unconventional form of the ϵ-term in the propagator will be explained in Section II. 3.

The internal momenta $l_{ab\sigma}$ satisfy momentum conservation relations

$$\sum_{b\sigma} l_{ab\sigma} = q_a \tag{II.5}$$

at each vertex V_a. In (II. 4) the internal momenta are written as

$$l_{ab\sigma} = k_{ab\sigma} + q_{ab\sigma} \tag{II.6}$$

The $q_{ab\sigma}$ are linear combinations of the $p_{a\sigma}$ and form a particular solution of

$$\sum_{b\sigma} q_{ab\sigma} = q_a \tag{II.7}$$

The $q_{ab\sigma}$ are called basic internal momenta. The $k_{ab\sigma}$ are linear combinations of the integration variables $k_1, \ldots, k_m$ and represent the general solution of the homogeneous equations

$$\sum_{b\sigma} k_{ab\sigma} = 0 \qquad (II.8)$$

A diagram Γ is called proper if it is connected and cannot be separated in two parts by cutting a single line. Proper diagrams are also called one-particle-irreducible. A diagram Γ is called trivial if it consists of a single vertex and no line (Fig. 1). The dimension (or superficial divergence) of a proper diagram is defined by the dimension of the corresponding Feynman integral with respect to the integration variables. If Γ is proper we have

$$d(\Gamma) = \sum_{ab\sigma} d(\Delta_F^{ab\sigma}) + \sum_a d(P_a) + 4m\,, \qquad (II.9)$$

$d(\Delta_F^{ab\sigma})$ and $d(P_a)$ are the degrees of the propagator $\Delta_F^{ab\sigma}$ and the polynomial P_a. For a proper subdiagram γ we have the dimension (or superficial divergence)

$$d(\gamma) = \sum_{ab\sigma}^{\gamma} d(\Delta_F^{ab\sigma}) + \sum_a^{\gamma} d(P_a) + 4m(\gamma). \qquad (II\ 10)$$

The sums are restricted to lines and vertices of γ, $m(\gamma)$ is the number of independent integration vectors in the Feynman integral of γ. γ is called superficially divergent if $d(\gamma) \geq 0$.

We now turn to the definition of the renormalized integrand R_Γ. A simple example is Dyson's original prescription of constructing the finite part of a primitively divergent integral.[19] A diagram Γ is called primitively divergent if it is proper, superficially divergent (i.e. $d(\Gamma) \geq 0$) and becomes convergent if any internal line is broken up. In that case the finite part is simply

$$J_\Gamma = \int dk_1 \cdots dk_m \, (1 - t^\Gamma) \, I_\Gamma \, . \qquad (\text{II}.11)$$

Here $t^\Gamma I_\Gamma$ denotes the Taylor series with respect to the external variables $p_{a\sigma}$ around $p_{a\sigma} = 0$ up to order $d(\Gamma)$. We give the explicit form of the Taylor operator. Let $p_1, \ldots, p_M$ be the external momenta $p_{a\sigma}$ of Γ, consecutively numbered. Since Γ is connected, $M - 1$ of the external momenta are independent and we write

$$I_\Gamma = f(p_1, \cdots, p_{M-1}) \, .$$

Then the Taylor operator is given by

$$t_p^{d(\Gamma)} f(p_1, \cdots, p_{M-1}) = f(0, \cdots, 0) + \cdots +$$
$$+ \frac{1}{d!} \sum_{j_1 \nu_1, \cdots, j_d \nu_d} p_{j_1 \nu_1} \cdots p_{j_d \nu_d} \frac{\partial^d f}{\partial p_{j_1 \nu_1} \cdots \partial p_{j_d \nu_d}} \bigg|_{p_j = 0}$$

For more complicated diagrams overlapping divergencies occur which will be disentangled by using Bogoliubov's R-operation[22] for constructing the counter terms of the integrand. First we give some combinatorial definitions. Proper diagrams which are superficially divergent $\left(d(\gamma) \geq 0 \right)$ are called renormalization parts. Two subdiagrams γ_1, γ_2 are called disjoint,

$$\gamma_1 \cap \gamma_2 = \emptyset \, ,$$

if they have no line or vertex in common. Let $\{\gamma_1, \ldots, \gamma_c\}$ be a set of mutually disjoint connected subdiagrams of Γ. Then the reduced diagram

$$F = \Gamma / \{\gamma_1, \cdots \gamma_c\}$$

is defined by contracting each γ_j to a point.

We now give the definition of R_Γ. If Γ is not a renormalization part we set

$$R_\Gamma = \overline{R}_\Gamma \,. \qquad (\text{II}.12)$$

If Γ is a renormalization part we set

$$R_\Gamma = (1 - t^\Gamma)\overline{R}_\Gamma \,. \qquad (\text{II}.13)$$

The function $\overline{R}_\gamma$ is defined recursively by

$$\overline{R}_\gamma = I_\gamma + \sum_{\{\gamma_1 \cdots \gamma_c\}} I_{\gamma / \{\gamma_1 \cdots \gamma_c\}} \prod_{\tau=1}^{c} O_{\gamma_\tau} \,, \qquad (\text{II}.14)$$

$$O_\gamma = - t^\gamma \overline{R}_\gamma \,. \qquad (\text{II}.15)$$

The sum extends over all sets $\{\gamma_1, \cdots, \gamma_c\}$ of renormalization parts of γ which are mutually disjoint and different from γ;

$$\gamma_i \cap \gamma_j = \phi \text{ for } i \neq j \;; \quad \gamma_j \neq \gamma \,.$$

t^γ denotes a Taylor series in the external variables $p^\gamma_{a\sigma}$ of γ up to order $d(\gamma)$. We briefly explain how the variables $p^\gamma_{a\sigma}$, $k^\gamma_{ab\sigma}$ pertaining to a subdiagram γ are introduced. Let V_a be a vertex belonging to γ. The external momenta

$p^{\gamma}_{a\sigma}$ of V_a are the external momenta $p_{a\sigma}$ and in addition all vectors $-\ell_{ab\sigma}$ of lines joining at V_a but not belonging to γ. Then the total external momentum q_a at V_a becomes

$$q^{\gamma}_{a} = \sum_{a\sigma} p^{\gamma}_{a\sigma} = \sum_{a\sigma}{}' \ell_{ab\sigma} \qquad (\text{II}.16)$$

with the sum $\sum'$ extending over all lines $\ell_{ab\sigma}$ which join at V_a and belong to γ. The momenta $q^{\gamma}_{ab\sigma}$, $k^{\gamma}_{ab\sigma}$ are then introduced in a way analogous to the case of Γ. The internal momentum $\ell_{ab\sigma}$ is decomposed into

$$\ell_{ab\sigma} = k^{\gamma}_{ab\sigma} + q^{\gamma}_{ab\sigma} \qquad (\text{II}.17)$$

where the $q^{\gamma}_{ab\sigma}$ are particular solutions of

$$\sum_{b\sigma}{}^{\gamma} q^{\gamma}_{ab\sigma} = q^{\gamma}_{a} \qquad (\text{II}.18)$$

The solution $q^{\gamma}_{ab\sigma}$ should be chosen such that the $k^{\gamma}_{ab\sigma}$ become functions of $k_1, \ldots, k_m$ alone.[37]

Application of $-t^{\gamma}$ to Eq. (II. 14) yields a recursion formula for O_{γ},

$$O_{\gamma} = -t^{\gamma} I_{\gamma} - t^{\gamma} \sum_{\{\gamma_1 \cdots \gamma_c\}} I_{\gamma/\gamma_1 \cdots \gamma_c} \prod_{\tau=1}^{c} O_{\gamma_\tau}, \qquad (\text{II}.19)$$

which we will use in the following section.

Finally we check that for primitively divergent diagrams Dyson's prescription is obtained. In our terminology a proper diagram Γ is primitively divergent, if the dimension $d(\Gamma)$ is ≥ 0 and if Γ does not contain any renormalization part $\gamma \neq \Gamma$ Then

$$\overline{R}_\Gamma = I_\Gamma$$

and

$$R_\Gamma = (1 - t^\Gamma) I_\Gamma$$

in agreement with (II. 11).

Sometimes it is convenient to take more subtractions than would actually be necessary for convergence. To include this possibility we introduce a function $\delta(\gamma)$ which assigns to every proper diagram γ of Γ an integer larger or equal to the dimension of γ

$$\delta(\gamma) \geq d(\gamma) \quad . \qquad (\text{II}.20)$$

$\delta(\gamma)$ is called the degree of γ. In addition we require

$$\delta(\gamma) \geq d(\bar{\gamma}) + \sum_j \delta(\gamma_j) \qquad (\text{II}.21)$$

for any reduction of the diagram γ : i.e.

$$\bar{\gamma} = \gamma / \{\gamma_1, \cdots, \gamma_c\}$$

Proper diagrams of non-negative degree are called renormalization parts relative to $\delta(\gamma)$. Relative to $\delta(\gamma)$ the finite part and the functions R_γ, O_γ, $\overline{R}_\gamma$ are then defined by equations similar to (II. 13-15).

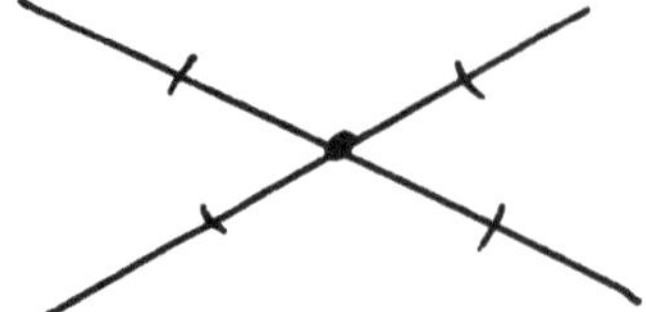

Figure 1.
A trivial diagram. A line marked with a bar does not belong to the diagram.

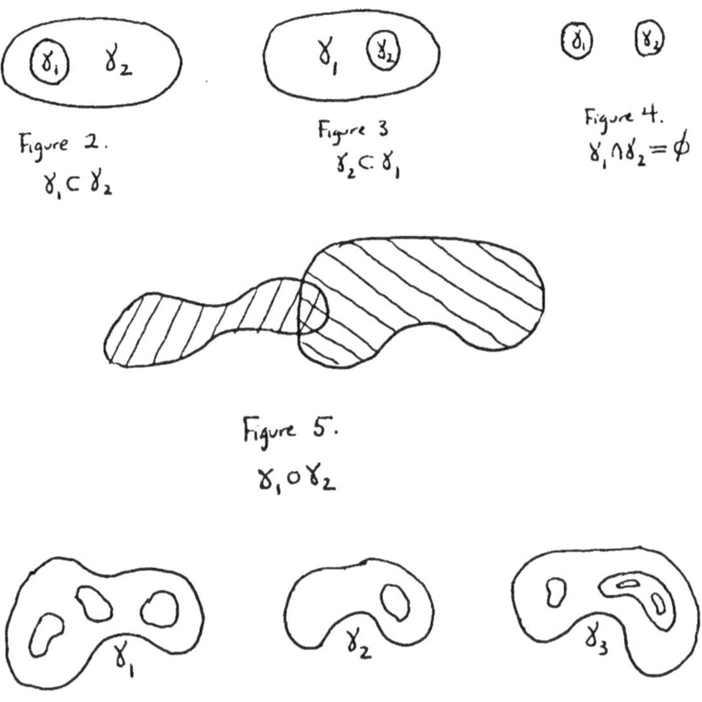

Figure 2.
$\gamma_1 \subset \gamma_2$

Figure 3
$\gamma_2 \subset \gamma_1$

Figure 4.
$\gamma_1 \cap \gamma_2 = \emptyset$

Figure 5.
$\gamma_1 \circ \gamma_2$

Figure 6.
A forest with three maximal elements: $\gamma_1, \gamma_2, \gamma_3$.

2. Explicit Form of the Renormalized Integrand

In this section we will derive an explicit formula for the function R_Γ. We begin with some combinatorial definitions concerning subdiagrams of a given diagram Γ. The diagrams γ_1, γ_2 are said to overlap

$$\gamma_1 \circ \gamma_2$$

if none of the following three relations holds

$$\gamma_1 \subseteq \gamma_2 , \quad \gamma_2 \subseteq \gamma_1 , \quad \gamma_1 \cap \gamma_2 = \emptyset$$

Otherwise γ_1, γ_2 are called non-overlapping

$$\gamma_1 \not\circ \gamma_2$$

Figs. 2-5 indicates the various possibilities.

Let Γ be any diagram. A Γ-forest U is a set of diagrams satisfying the following conditions

(i) the elements of U are renormalization parts of Γ,

(ii) any two elements γ', γ'' are non-overlapping,

(iii) U may also be the empty set.

A Γ-forest U containing Γ itself is called full, a Γ-forest U not containing Γ is called normal. If Γ is a renormalization part then there is a one-to-one correspondence between full Γ forests T and normal Γ-forests U given by

$$T = U \cup \{\Gamma\} . \qquad (\text{II}.22)$$

Note that the empty set $U = \emptyset$ corresponds to $T = \{\Gamma\}$.

We first prove the following explicit formula for O_γ:

$$O_\gamma = -t^\gamma \sum_U^{\text{Normal}} \prod_{\lambda \in U}{}' (-t^\lambda)\, I_\gamma \qquad (\text{II}.23)$$

or

$$O_\gamma = \sum_T^{\text{Full}} \prod_{\lambda \in T}{}' (-t^\lambda)\, I_\gamma$$

$\sum^{\text{Normal}}$ extends over all normal γ-forests, $\sum^{\text{Full}}$ over all full γ-forests. In the product $\prod' (-t^\lambda)$

the factors are ordered such that t^λ stands to the left of t^σ if $\lambda \supset \sigma$. For $\lambda \cap \sigma = \emptyset$ the order is irrelevant since then the differential operators t^λ and t^σ refer to independent variables.[66]

Both relations (II. 23) and (II. 24) are equivalent because of the one-to-one correspondence (II. 22) between full and normal forests.

For the proof of (II. 23-24) we denote the expression (II. 23-24) by the different symbol $\tilde{O}_\gamma$

$$\tilde{O}_\gamma = -t^\gamma \sum_U^{\text{Normal}} \prod_{\lambda \in U}{}' (-t^\lambda)\, I_\gamma \qquad (\text{II}.25)$$

or

$$\tilde{O}_\gamma = \sum_T^{\text{full}} \prod_{\lambda \in T} (-t^\lambda) I_\gamma \qquad (\text{II}.26)$$

Our aim is to prove $\tilde{O}_\gamma = O_\gamma$ by showing that $\tilde{O}_\gamma$ solves the recursive Eq. (II. 19).

If the normal γ -forest U is not empty it contains some maximal elements which we denote by $\gamma_1, \ldots, \gamma_c$. (See Fig. 6). Let T_j be the set of all $\gamma \in U$ with $\gamma \subseteq \gamma_j$ Then U is the union

$$U = T_1 \cup \cdots \cup T_c \quad . \qquad (\text{II}.27)$$

A tree is a forest which has only one maximal element. Apparently **each T_j is a tree with γ_j as the maximal element.** Hence (II. 27) **is the decomposition of the forest U into its trees $T_1, \ldots, T_c$. We can now rewrite (II. 25) as**

$$\tilde{O}_\gamma = -t^\gamma I_\gamma - t^\gamma \sum_{\{\gamma_1,\cdots,\gamma_c\}} \sum_U^{\{\gamma_1,\cdots,\gamma_c\}\text{Max}} \prod_{\lambda \in T_1}(-t^\lambda) \cdots \prod_{\lambda \in T_c}(-t^\lambda) I_\gamma$$

The first term on the right hand side is the contribution from the empty forest. The sum $\sum_U^{\{\gamma_1,\cdots,\gamma_c\}\text{Max}}$

extends over all forests U with the maximal elements $\gamma_1, \ldots, \gamma_c$. The sum $\sum_{\{\gamma_1,\cdots,\gamma_c\}}$ goes over all sets $\{\gamma_1, \ldots \gamma_c\}$ of renormalization parts of γ which are mutually disjoint and different from γ

T_j is a full γ_j' -forest. Replacing the sum over U by the sum over all T_j and using

$$I_\gamma = I_{\gamma/\{\gamma_1,\cdots,\gamma_c\}} I_{\gamma_1} \cdots I_{\gamma_c} \qquad (\text{II}.29)$$

we get

$$\tilde{O}_\gamma = -t^\gamma I_\gamma - t^\gamma \sum_{\{\gamma_1,\cdots,\gamma_c\}} I_{\gamma/\{\gamma_1,\cdots,\gamma_c\}} \times$$

$$\times \sum_{T_1}^{\text{Full }\gamma_1} \prod_{\lambda\in T_1} (-t^\lambda) I_{\gamma_1} \cdots \sum_{T_c}^{\text{Full }\gamma_c} \prod_{\lambda \in T_c} (-t^\lambda) I_{\gamma_c} . \qquad (\text{II}.30)$$

The sum $\sum\limits_{T_j}^{\text{full}}$ extends over all full γ_j -forests. With (II. 26) we have

$$\tilde{O}_\gamma = -t^\gamma I_\gamma - t^\gamma \sum_{\{\gamma_1,\cdots\gamma_c\}} I_{\gamma/\{\gamma_1,\cdots,\gamma_c\}} \tilde{O}_{\gamma_1} \cdots \tilde{O}_{\gamma_c} , \qquad (\text{II}.31)$$

which is the recursion formula (II. 19) for O_γ . Hence

$$\tilde{O}_\gamma = O_\gamma .$$

This completes the proof of (II. 23-24).

With this result it is easy to derive an explicit formula for R_Γ . Substituting (II. 24) for O_γ into (II. 14) we obtain

$$\bar{R}_\Gamma = I_\Gamma + \sum_{\{\gamma_1,\cdots,\gamma_c\}} I_{\Gamma/\{\gamma_1,\cdots,\gamma_c\}} \prod_{\tau=1}^{c} \sum_{T_\tau}^{\text{Full }\gamma_\tau} (-t^\lambda) I_{\gamma_\tau} =$$

$$= \sum_{U}^{\text{Normal}} \prod_{\lambda \in U} (-t^\lambda) I_\Gamma . \qquad (\text{II}.32)$$

This is an explicit formula for $\bar{R}_\Gamma$. If Γ is not a renormalization part no Γ-forest may contain Γ , thus any Γ-forest is normal. Hence

$$R_\Gamma = \bar{R}_\Gamma = \sum_U \prod_{\lambda \in U} (-t^\lambda) I_\Gamma$$

where now the sum extends over all possible Γ -forests. Next let Γ be a renormalization part. Then

$$R_\Gamma = \bar{R}_\Gamma - t^\Gamma \bar{R}_\Gamma = \sum_U^{\text{Normal}} \prod_{\lambda \in U} (-t^\lambda) I_\Gamma +$$

$$+ \sum_U^{\text{Full}} \prod_{\lambda \in U} (-t^\lambda) I_\Gamma =$$

$$= \sum_U \prod_{\lambda \in U} (-t^\lambda) I_\Gamma$$

Hence in all cases we get as final result the following explicit formula for the renormalized integrand

$$R_\Gamma = \sum_U \prod_{\lambda \in U} (-t^\lambda) I_\Gamma \qquad (\text{II}.33)$$

with the sum going over all Γ -forests.

Finally we consider the special case that Γ has no overlapping divergencies, i.e. any two renormalization parts of Γ are non-overlapping. In that case the terms of (II. 33) can be combined into the single product

$$R_\Gamma = \prod_\gamma (1 - t^\gamma) I_\Gamma \qquad (\text{II}.34)$$

which is taken over all renormalization parts of Γ. It is easily seen that (II. 34) follows from (II. 33) by multiplying out the factors $(1-t^{\gamma})$. (II. 34) represents Dyson's prescription for handling non-overlapping divergencies.[19]

3. Convergence of Renormalized Integrals.

With a common denominator the renormalized integrand (II. 33) may be written as

$$R_{\Gamma} = \frac{A}{B_1 B_2} \qquad (\text{II}.35)$$

$$B_1 = \prod_{ab\sigma} (l_{ab\sigma}^2 - \mu_{ab\sigma}^2 + i\epsilon(\vec{l}_{ab\sigma}^{\,2} + \mu_{ab\sigma}^2))$$

$$B_2 = \prod_{\gamma} \prod_{ab\sigma}^{\gamma} \{(k_{ab\sigma}^{\gamma 2} - \mu_{ab\sigma}^2 + i\epsilon(\vec{k}_{ab\sigma}^{\gamma 2} + \mu_{ab\sigma}^2))\}^{c_{ab\sigma}^{\gamma}},$$

where A is a polynomial in the integration variables and the external momenta. The denominators in B_2 come from differentiating with respect to the external momenta $p_{a\sigma}^{\gamma}$ of a renormalization part γ and setting $p_{a\sigma}^{\gamma} = 0$. (See Eq. (II. 17)). Hence the finite part has the general form

$$J_{\Gamma} = \int dk_1 \cdots dk_m \frac{P}{\prod_{j=1}^{n} (l_{jc}^2 - \vec{l}_j^{\,2} - \mu_j^2 + i\epsilon(\vec{l}_j^{\,2} + \mu_j^2))}; \qquad (\text{II}.36)$$

P is a polynomial in the integration variables and the external momenta. The four vectors l_j are linear combinations of the integration variables and the external momenta.

One would like to have absolute convergence for this integral. The natural way of checking this is power counting. But there is the well-known difficulty that Feynman integrals in Minkowski space are usually only conditionally convergent. We discuss this difficulty in the simple example

$$\int \frac{dk}{(k_0^2 - \vec{k}^2 - m^2 + i\epsilon)^3} \qquad (\text{II}.37)$$

The Euclidean counterpart of (II.37)

$$\int \frac{dk}{(k_0^2 + \vec{k}^2 + m^2)^3} \qquad (\text{II}.38)$$

is certainly absolutely convergent. But the Minkowski integral (II.37) is not, as can be seen in the following way. If it were absolutely convergent any subintegral

$$\int_R \frac{dk}{(k_0^2 - \vec{k}^2 - m^2 + i\epsilon)^3}$$

would have to be absolutely convergent too. Suppose we take for R the shell between two hyperboloids

$$a \le k^2 \le b$$

Then

$$|k^2 - m^2 + i\epsilon| \le A$$

is bounded in R . Hence

$$\int_R \frac{dk}{|(k_0^2-\vec{k}^2-m^2+i\epsilon)^3|} \geq \frac{1}{A^3}\int_R dk \quad \text{which diverges}$$

since the volume of the shell is infinite.

However, there is an easy way out of this difficulty. If one simply replaces $i\epsilon$ by $i\epsilon(\vec{k}^2+m^2)$ in (II.37)

$$\int \frac{dk}{\left(k_0^2-\vec{k}^2-m^2+i\epsilon\,(\vec{k}^2+m^2)\right)^3} , \qquad \text{(II.39)}$$

the new integral is majorized by the Euclidean integral (II.38) due to the inequality

$$\left|\frac{1}{k_0^2-\vec{k}^2-m^2+i\epsilon(\vec{k}^2+m^2)}\right| \leq \left(\frac{1}{\epsilon}+\sqrt{1+\frac{1}{\epsilon^2}}\right)\frac{1}{k_0^2+\vec{k}^2+m^2} . \qquad \text{(II.40)}$$

Hence also (II.36) is absolutely convergent provided the Euclidean version

$$\int dk_1\cdots dk_m \frac{P}{\prod_{j=1}^{n}(l_{j0}^2+\vec{l}_j^{\,2}+\mu_j^2)} \qquad \text{(II.41)}$$

is absolutely convergent. But to (II.41) we can apply Weinberg's form[38] of the power counting theorem, which states that (II.41) is absolutely convergent provided (II.41) and any subintegral

$$\int_H dV \frac{P}{\prod_{j=1}^{n}(l_{j\mu}^2+\vec{l}_j^{\,2}+\mu_j^2)}$$

have negative dimension. H is an arbitrary hyperplane in the 4m-dimensional space of $k_1, \ldots, k_m$. Thus the convergence proof is reduced to power counting. For a detailed proof we refer to [38,39].

The limit $\epsilon \to 0^+$ is difficult to perform for the momentum space integrals directly. A more convenient way is to rewrite (II.36) as a parametrized Feynman integral. Applying a method of Hepp[40] it can be shown that the parametrized integrals approach covariant distributions in the limit $\varepsilon \to 0^+$.[39]

4. Bogoliubov's Method of Renormalization in Coordinate Space.

Originally Bogoliubov[22] formulated the renormalization rules for the regularized integrands of Feynman integrals in coordinate space. Hepp has proved that these renormalized integrands approach well-defined distributions in the limit when the regularization is removed and $\epsilon \to 0^+$. [28]

In this section we discuss the equivalence between the two formulations in coordinate and momentum space.[41] We start out from the definition of the finite part in momentum space. For this purpose it is convenient to eliminate the polynomials P_a assigned to vertices by changing the polynomials of the adjacent lines accordingly. Moreover,

$$I_\Gamma = \prod_{ab\sigma} \Delta_{F\,ab\sigma} \qquad (\text{II}.42)$$

will be considered as function of the integration variables and the total external momenta q_a at each vertex. An external momentum will be introduced for all vertices including the internal ones. We regularize according to Pauli-Villars[42]

$$I_{\Gamma}^{REG} = \prod_{ab\sigma} \Delta_{F\,ab\sigma}^{REG} , \qquad (\text{II}.43)$$

where

$$\Delta_{F\,ab\sigma}^{REG} = i P_{ab\sigma}(\ell_{ab\sigma}) \Big\{ \frac{1}{\ell_{ab\sigma}^2 - \mu_{ab\sigma}^2 + i\epsilon(\vec{\ell}_{ab\sigma}^{\,2} + \mu_{ab\sigma}^2)} +$$

$$+ \sum_r \frac{c_r^{ab\sigma}(M)}{\ell_{ab\sigma}^2 - M^2 m_{ab\sigma}^{(r)2} + i\epsilon(\vec{\ell}_{ab\sigma}^{\,2} + M^2 m_{ab\sigma}^{(r)2})} \Big\} . \qquad (\text{II}.44)$$

The $m_{ab\sigma}^{(r)}$ are constants, the $c_r^{ab\sigma}(M)$ bounded functions of M which are arranged such that

$$\text{degree}\ \Delta_{F\,ab\sigma}^{REG} < -4 .$$

With this regularization the Fourier transform

$$\tilde{\Delta}_{F\,ab\sigma}^{REG}(x) = \int \frac{d^4p}{(2\pi)^4}\, e^{-ipx}\, \Delta_{F\,ab\sigma}^{REG}(p)$$

of the propagator is an ordinary function and already the unrenormalized Feynman integral J_{Γ} is absolutely convergent for $\epsilon > 0$. The Fourier transforms of J_{Γ}, the finite part F_{Γ}, and the corresponding integrals over R_{Γ}, O_{Γ} are

$$J_{\Gamma} = \int dk_1 \cdots dk_m \frac{dq_1 \cdots dq_n}{(2\pi)^4 \cdots (2\pi)^4}\, e^{-i\Sigma q_j x_j}\, \delta_{\Gamma}\, I_{\Gamma} , \qquad (\text{II}.45)$$

$$\mathcal{R}_\Gamma = \int dk_1 \cdots dk_m \frac{dq_1}{(2\pi)^4} \cdots \frac{dq_n}{(2\pi)^4} e^{-i\sum_j q_j x_j} \delta_\Gamma R_\Gamma , \qquad (\text{II}.46)$$

$$\bar{\mathcal{R}}_\Gamma = \int dk_1 \cdots dk_n \frac{dq_1}{(2\pi)^4} \cdots \frac{dq_n}{(2\pi)^4} e^{-i\sum_j q_j x_j} \delta_\Gamma \bar{R}_\Gamma , \qquad (\text{II}.47)$$

$$\mathcal{O}_\gamma = \int dk_1 \cdots \frac{dq_1}{(2\pi)^4} \cdots \frac{dq_{n(\gamma)}}{(2\pi)^4} e^{-i\sum_j q_j x_j} \delta(\Sigma q_j) O_\gamma . \qquad (\text{II}.48)$$

Here δ_Γ denotes the product of δ-functions expressing momentum conservation. Let $\Gamma_1, \ldots, \Gamma_c$ be the connected components of Γ_j then

$$\delta_\Gamma = \sum_{j=1}^{c} \delta(\sum_\alpha q_{j\alpha}) \qquad (\text{II}.49)$$

where the $q_{j\alpha}$ denote the external variables attached to Γ_j. (II. 45) is the integrand of the unrenormalized Feynman amplitude, (II. 46) the integrand of the renormalized amplitude of the diagram Γ in coordinate space, both in regularized form. $\mathcal{J}_\Gamma$ becomes just a product of Δ_F -functions

$$\mathcal{J}_\Gamma = \prod_{ab\sigma} \Delta^{REG}_{F\,ab\sigma}(x_a - x_b) . \qquad (\text{II}.50)$$

For $\overline{\mathcal{R}}_\Gamma$ one derives a recursion formula by taking the Fourier transform of (II. 14)

$$\overline{\mathcal{R}}_\Gamma = J_\Gamma + \sum_{\{\gamma_1,\dots,\gamma_c\}} J_{\Gamma\setminus\gamma_1\dots\gamma_c} \prod_{\tau=1}^{c} \mathcal{O}_{\gamma_\tau} \qquad (\text{II}.51)$$

The sum is taken over all sets $\{\gamma_1, \dots, \gamma_c\}$ of disjoint renormalization parts with $\gamma_j \neq \Gamma$. $\Gamma\setminus\gamma_1 \dots \gamma_c$ is the diagram obtained from Γ by omitting the lines belonging to $\gamma_1, \dots, \gamma_c$. $\mathcal{O}_{\gamma_\tau}$ is defined by (II. 48) with

$$\mathcal{O}_\gamma = - t_{q\gamma}^{d(\gamma)} \overline{\mathcal{R}}_\gamma \qquad (\text{II}.52)$$

or in Hepp's notation

$$\mathcal{O}_\gamma = - M_\gamma \overline{\mathcal{R}}_\gamma \quad . \qquad (\text{II}.53)$$

$\mathcal{R}_\Gamma$ is given by

$$\mathcal{R}_\Gamma = \overline{\mathcal{R}}_\Gamma \qquad (\text{II}.54)$$

if Γ is no renormalization part and

$$\mathcal{R}_\Gamma = (1 - M_\Gamma)\overline{\mathcal{R}}_\Gamma \qquad (\text{II}.55)$$

if Γ is a renormalization part.

Finally the regularization must be removed by taking the limit $M \to \infty$. Multiplying out all terms of $\Delta^{REG}_{Fab\sigma}$ in the definition of I^{reg} we obtain

$$I_\Gamma^{REG} = I_\Gamma + \sum_\tau I_\Gamma^\tau \qquad (II.56)$$

where

$$I_\Gamma^\tau = \prod_{ab\sigma} \Delta_{F\tau}^{ab\sigma} , \qquad (II.57)$$

$$\Delta_{F\tau}^{ab\sigma} = i P_{ab\sigma} \left\{ \frac{d^{ab\sigma\tau}}{l_{ab\sigma}^2 - m_{ab\sigma\tau}^2 + i\epsilon(\cdots)} \right\}$$

The masses $m_{ab\sigma\tau}$ are

either $= \mu_{ab\sigma\tau}$

or $= M m_{ab\sigma}^\tau$

where the second value involving the auxiliary mass M is attained for at least one line $l_{ab\sigma}$ of the diagram. Since Bogoliubov's R-operation is linear (see Eq. (II.33)) there is the similar relation

$$R_\Gamma^{REG} = R_\Gamma + \sum_\tau R_\Gamma^\tau \qquad (II.58)$$

with R_Γ^τ defined by applying the R-operation to I_Γ^τ. According to the results of the previous section

$$\int dk_1 \cdots dk_m \, R_\Gamma \qquad (II.59)$$

and

$$\int dk_1 \cdots dk_m \, R_\Gamma^\tau \qquad (II.60)$$

are separately absolutely convergent.

Since at least one factor $\Delta^{\tau}_{F ab \sigma}$ of I^{τ}_{Γ} contains M in the denominator it should be expected that the integrals (II. 60) vanish in the limit $M \longrightarrow \infty$. This can indeed be shown by appropriate estimates of the corresponding Euclidean integrals.[41] Thus

$$\lim_{M\to\infty} \int dk_1 \cdots dk_m R^{REG}_{\Gamma} = \int dk_1 \cdots dk_m R_{\Gamma}$$

$$\lim_{\epsilon\to 0^+} \lim_{M\to\infty} \int dk_1 \cdots dk_m R^{REG}_{\Gamma} = \lim_{\epsilon\to 0^+} \int dk_1 \cdots dk_m R_{\Gamma} \qquad \text{(II.61)}$$

This implies that $\mathcal{R}(x)$ approaches the Fourier transform of

$$\delta_{\Gamma} \lim_{\epsilon\to 0^+} \int dk_1 \cdots dk_m R_{\Gamma}$$

which is known to be a covariant distribution.

There are two minor differences between (II. 51-55) and the results of Bogoliubov, Parasiuk and Hepp.[22,28] In their work

(i) fewer subtractions are used

(ii) the ϵ-term is treated in a covariant manner even before the limits $M \to \infty$ and $\epsilon \to 0^+$ are taken.

It can be shown that the additional subtractions of (II. 51) are redundant (provided they are dropped before the limit $M \longrightarrow \infty$ is taken) and that the difference of the expressions is of order ϵ in the limit $\epsilon \to 0^+$. Thus there is complete equivalence between the two formulations of Bogoliubov's method.

5. Feynman Rules for A^4-Coupling

In this section the complete set of Feynman rules for the model of A^4-coupling will be given. As Lagrangian we take

$$\mathcal{L}_{EFF} = \mathcal{L}_0 + \mathcal{L}_{INT}$$
$$\mathcal{L}_0 = \frac{1}{2} \partial_\mu A \partial^\mu A - \frac{1}{2} m^2 A^2 \qquad (II.62)$$
$$\mathcal{L}_{INT} = -\lambda \left(\frac{1}{4} A^4 - \frac{1}{2} a A^2 - b \mathcal{L}_0(A) \right)$$

m, λ, a, b are finite constants.[8] (II. 62) is called the effective Lagrangian to distinguish it from the full Lagrangian involving divergent renormalization constants. The effective Lagrangian is used for defining the renormalized Green's functions as the finite part of the Gell-Mann Low series[43] :

$$\langle T A(x_1) \cdots A(x_n) \rangle =$$
$$= \mathrm{Fin} \langle T \{ e^{i \int \mathcal{L}^c_{INT} dz} A_0(x_1) \cdots A_0(x_n) \} \rangle . \qquad (II.63)$$

Here A_0 is a free field satisfying

$$(\Box + m^2) A_0(x) = 0 ,$$
$$[A_0(x), \dot{A}_0(y)] = i \delta_3(x-y) \quad \text{for } x^0 = y^0 \qquad (II.64)$$
$$[A_0(x), A_0(y)] = [\dot{A}_0(x), \dot{A}_0(y)] = 0 \quad \text{for } x^0 = y^0$$

$\mathcal{L}^0_{INT}$ is the interaction Lagrangian with A_c inserted

$$\mathcal{L}^0_{INT} = \mathcal{L}_{INT}(A_0) . \qquad (II.65)$$

The construction of the finite part is done according to the rules of the previous sections. We first expand the Fourier transform

$$\langle T\{\tilde{A}(p_1)\cdots\tilde{A}(p_n)\}\rangle = \\ = \int \frac{dx_1}{(2\pi)^2}\cdots\frac{dx_n}{(2\pi)^2}\, e^{+i\sum_j p_j x_j}\langle T\{A(x_1)\cdots A(x_n)\}\rangle \qquad (\text{II}.66)$$

of (II. 63) with respect to Fynman diagrams Γ

$$\langle T\{\tilde{A}(p_1)\cdots\tilde{A}(p_n)\}\rangle = \sum_{\Gamma}\langle T\{\tilde{A}(p_1)\cdots\tilde{A}(p_n)\}\rangle^{\Gamma}. \qquad (\text{II}.67)$$

The sum extends over all Feynman diagrams Γ with n external lines which can be built up using 4-vertices (Fig. 7) (corresponding to the A^4-term) and 2-vertices (Fig. 8) (corresponding to the quadratic terms of $\mathcal{L}_{INT}$). Connected components of Γ without external lines (i. e. disconnected closed loops) are not permitted. The contribution from the diagram Γ has the form

$$\langle T\{\tilde{A}(p_1)\cdots\tilde{A}(p_n)\}\rangle^{\Gamma} = \frac{1}{\gamma(\Gamma)}\,\delta(\Gamma)\int dk_1\cdots dk_m\, I_{\Gamma}\ ; \qquad (\text{II}.68)$$

δ_{Γ} is the product of δ-functions expressing energy-momentum conservation (one factor for each connected component). The symmetry number $\gamma(\Gamma)$ is [24]

$$\gamma(\Gamma) = g\,2^{\alpha}(3!)^{\beta}\,2^{\gamma}\ , \qquad (\text{II}.69)$$

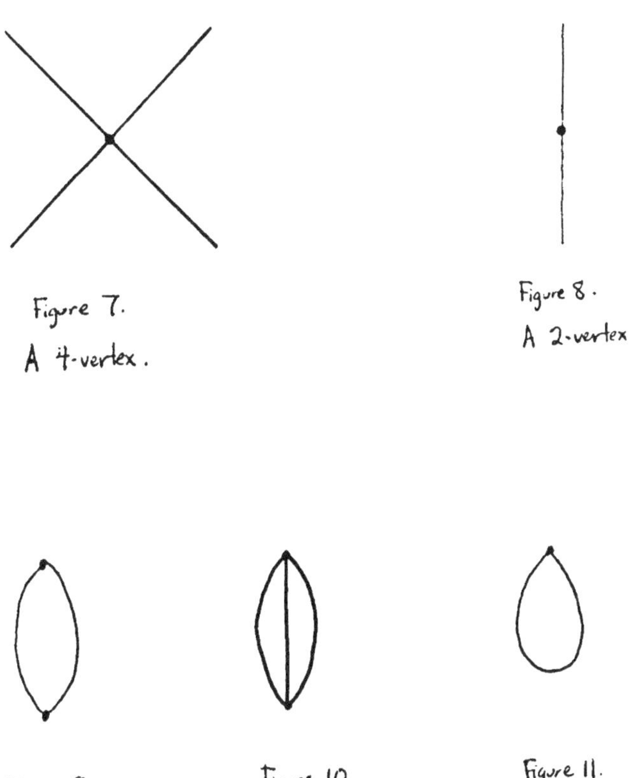

Figure 7. A 4-vertex.

Figure 8. A 2-vertex

Figure 9. Double line.

Figure 10. Triple Line.

Figure 11. A line connecting a vertex with itself.

where α is the number of double lines (i.e. pairs of vertices connected by two lines, Fig. 9), β is the number of triple lines (pairs of vertices connected by three lines, Fig. 10), γ is the number of lines connecting a vertex with itself (Fig. 11), and g is the number of automorphisms of the diagram, i.e. permutations of vertices which leave the diagram unchanged.

R_Γ is constructed from I_Γ according to the rules of section II.2-3. The insertion rules for the unrenormalized integral are:

For each line : $\dfrac{i}{\ell^2 - m^2 + i\epsilon(\vec{\ell}^{\,2} + m^2)}$

For each 4-vertex (Fig. 7): $-\dfrac{Gi\lambda}{(2\pi)^4}$ (II.70)

For each 2-vertex (Fig. 8): $i\lambda a + i\lambda b(\ell^2 - m^2)$

The finite constants a and b are used to adjust pole and residue of the propagator. For the propagator we use the notation

$$\Delta_F'(x-y) = \langle T\{A(x), A(y)\rangle ,$$
$$\hat{\Delta}_F'(p) = \int e^{+ipx} \Delta_F'(x)\, dx \qquad \text{(II.71)}$$

The function Π is introduced by

$$\hat{\Delta}_F' = \hat{\Delta}_F + \hat{\Delta}_F \Pi \hat{\Delta}_F' \qquad \text{(II.72)}$$

where $\hat{\Delta}_F$ is the free propagator

$$\hat{\Delta}_F = \frac{i}{p^2 - m^2 + i\epsilon} \quad .$$

$\hat{\Delta}'_F$ has a pole at $p^2 = m^2$ of residue i if Π and its derivative vanish at $p^2 = m^2$

$$\Pi(m^2) = 0 \, , \quad \left. \frac{\partial \Pi}{\partial p^2} \right|_{p^2 = m^2} = 0 \quad . \qquad (\text{II}.73)$$

According to (II. 72) Π is the sum over all contributions from proper self-energy diagrams ,

$$\Pi = \sum_S \frac{1}{\gamma(s)} F_s(p^2) \quad . \qquad (\text{II}.74)$$

Separating the contribution from the trivial diagram of Fig. 8 we get

$$\Pi(p^2) = i\lambda a + i\lambda b(p^2 - m^2) + \sum_s{}' \frac{1}{\gamma(s)} F_s(p^2) \, , \quad (\text{II}.75)$$

where now the sum is restricted to non-trivial proper self-energy parts. The conditions (II. 73) then lead to the relations

$$\begin{aligned} a &= -\frac{i}{\lambda} \sum_s{}' \frac{1}{\gamma(s)} F_s(m^2) \, , \\ b &= -\frac{i}{\lambda} \sum_s{}' \frac{1}{\gamma(s)} \left. \frac{\partial F_s}{\partial p^2} \right|_{p^2 = m^2} \, , \end{aligned} \qquad (\text{II}.76)$$

which recursively determine a and b as power series in λ with finite coefficients.

6. Unitarity Equations

For the model of A^4-coupling we defined the time ordered Green's functions of the field A by the finite part of the Gell-Mann Low series (see Eq. (II.63)). Similar definitions can be given for other renormalizable field theories.

The question arises whether the Green's functions thus defined fulfill the postulates of quantum field theory in every order of perturbation theory. This is not obvious since the Gell-Mann Low series were taken as a definition. Moreover, the Feynman integrals were mutilated by an ad hoc prescription for removing the divergencies.

Rigorous work has been done by Steinmann [27] and Epstein-Glaser [36] concerning the locality and unitarity properties of Green's functions in perturbation theory. Though not all points have been clarified in a completely rigorous fashion their work indicates that the renormalized Green's functions do indeed satisfy Wightman's postulates [44] as well as the unitarity of the S-matrix in all orders of perturbation theory.

Steinmann has constructed the Green's functions in every order starting from the unitarity equations of the retarded functions. [69,70] These equations already imply that the Green's functions describe a local field operator with a unitary S-matrix. However, it remains still to be shown that the expressions found by Steinmann are equivalent to the conventional Green's functions defined by Bogoliubov's rules.

In a completely different approach Epstein and Glaser have constructed generalized Green's functions (including the time ordered and retarded functions) which rigorously satisfy locality along with the other postulates of Wightman's formulation. But there are still difficulties with checking the unitarity of the S-matrix. Possibly a comparison of the methods of Steinmann and Epstein-Glaser may already provide the complete answer to these questions.

In this section a simple method of checking locality and unitarity is sketched which I used to convince myself that locality and unitarity are satisfied in renormalized perturbation theory. [47,48] The method is certainly not rigorous, in particular some limits are interchanged without justification.

We begin by checking the unitarity equations of the time ordered functions which on the mass shell represent the unitarity relations of the S-matrix elements. In general field theory these equations follow by taking the vacuum expectation value of the operator identity

$$\sum_{i_1 \cdots i} (-1)^\nu \,\overline{\overline{T}}(x_{i_1} \cdots x_{i_\nu})\, T(x_{i_{\nu+1}} \cdots x_{i_n}) = 0 \qquad (\mathrm{II}.77)$$

$$T(x_1 \cdots x_n) = T A(x_1) \cdots A(x_n)$$

The sum extends over all subsets $(x_{i_1} \ldots x_{i_\nu})$ of $(x_1 \ldots x_n)$. Summing over intermediate states and using the reduction technique one gets the unitarity relations

$$\sum_{i_1 \cdots i_\nu} (-1)^\nu \sum_{\ell=0}^{\infty} \frac{i}{\ell!} \int du_1 \cdots du_\ell \, dv_1 \cdots dv_\ell \, K_{u_1} \cdots K_{u_\ell} \, \tau^*(x_{i_1} \cdots x_{i_\nu} u_1 \cdots u_\ell)^* \times \prod_{\alpha=1}^{\ell} i\Delta^+(u_\alpha - v_\alpha) K_{v_1} \cdots K_{v_\ell} \, \tau(x_{i_{\nu+1}} \cdots x_{i_n} v_1 \cdots v_\ell) \; ; \qquad (\mathrm{II}.78)$$

$$\tau(x_1 \cdots x_n) = \langle T A(x_1) \cdots A(x_n) \rangle \, ,$$

for τ-functions. Our aim is to check that (II. 78) is satisfied by the Green's functions which we defined in the previous section. We consider the case of A^4-coupling. In coordinate space the unrenormalized integrand of the contribution from Γ to

$$i^n \langle T A(x_1) \cdots A(x_n) \rangle$$

is

$$J_\Gamma = \prod_a P_a \prod_{ab\sigma} \Delta_F^{REG}(z_a - z_b) \qquad (\mathrm{II}.79)$$

Here the propagators are regularized according to Pauli-Villars with the limit $\epsilon \longrightarrow 0^+$ already performed. The external coordinates $x_1 , \ldots , x_n$ and internal vertices $w_1, \ldots , w_m$ are labeled consecutively by $z_1, \ldots , z_n$. P_a is assigned to the vertex z_a and is

$P_a = i$ for an external vertex (i.e., endpoint x_j of an external line),

$P_a = -6i\lambda$ for a 4-vertex, (II.80)

and $P_a = ia\lambda + ib(\overleftarrow{\partial}_\mu \overrightarrow{\partial}^\mu - m^2)$ for a 2-vertex.

$\overleftarrow{\partial}_\mu$, $\overrightarrow{\partial}_\mu$ act on the coordinate z_a of the two propagators joining at V_a. We are going to derive an identity for $\mathcal{J}_r$ Modeled after the relation

$$\bar{T} A(x) B(y) - A(x) B(y) - B(y) A(x) + T A(x) B(y) = 0,$$

we set up an identity for two functions of two four vectors x_1, x_2. First define a "time ordered" function $g_{|12}$ by

$$g_{|12}(x_1 x_2) = \begin{cases} f_{12}(x_1 x_2) & \text{for } x_1^0 > x_2^0 \\ f_{21}(x_1 x_2) & \text{for } x_1^0 < x_2^0 \end{cases}$$

an antitime ordered function $g_{12|}$ by

$$g_{12|}(x_1 x_2) = \begin{cases} f_{21}(x_1 x_2) & \text{for } x_1^0 > x_2^0 \\ f_{12}(x_1 x_2) & \text{for } x_1^0 < x_2^0 \end{cases}$$

and "mixed" functions

$$g_{1|2}(x_1 x_2) = f_{12}(x_1 x_2)\ , \quad g_{2|1}(x_1 x_2) = f_{21}(x_1 x_2)$$

Then the identity

$$g_{|12} - g_{1|2} - g_{2|1} + g_{12|} = 0$$

holds for every x_1, x_2.

Before proceeding to the general case we give the corresponding identity for functions f_{ijk} of three four vectors x_1, x_2, x_3

$$-g_{123|} + g_{12|3} + g_{23|1} + g_{13|2} - g_{1|23} - g_{2|13} - g_{3|12} + g_{|123} = 0$$

with the functions g defined by

$$g_{|123}(x_1 x_2 x_3) = f_{ijk}(x_1 x_2 x_3) \quad \text{for } x_i^0 > x_j^0 > x_k^0$$

$$g_{123|}(x_1 x_2 x_3) = f_{ijk}(x_1 x_2 x_3) \quad \text{for } x_i^0 < x_j^0 < x_k^0$$

$$g_{1|23}(x_1 x_2 x_3) = f_{1jk}(x_1 x_2 x_3) \quad \text{for } x_j^0 > x_k^0, \ j, k = 2, 3$$

$$g_{12|3}(x_1 x_2 x_3) = f_{ij3}(x_1 x_2 x_3) \quad \text{for } x_i^0 < x_j^0, \ i, j = 1, 2$$

In the general case define

$$g_{|1\cdots n}(x_1 \cdots x_n) = f_{i_1 \cdots i_n}(x_1 \cdots x_n) \quad \text{for } x_{i_1}^0 > \cdots > x_{i_n}^0$$

$$g_{1\cdots n|}(x_1 \cdots x_n) = f_{i_1 \cdots i_n}(x_1 \cdots x_n) \quad \text{for } x_{i_1}^0 < \cdots < x_{i_n}^0$$

$$g_{j_1 \cdots j_\nu | j_{\nu+1} \cdots j_n}(x_1 \cdots x_n) = f_{i_1 \cdots i_n}(x_1 \cdots x_n) \qquad (\text{II}.81)$$

if $(i_1 \cdots i_\nu), (i_{\nu+1} \cdots i_n)$ are the permutations of $(j_1 \cdots j_\nu)$ and $(j_{\nu+1} \cdots j_n)$ with $x_{i_1}^0 < \cdots < x_{i_\nu}^0$, $x_{i_{\nu+1}}^0 > \cdots > x_{i_n}^0$. Then the identity

$$(-1)^n g_{1\cdots n|} + (-1)^{n-1} \sum_\alpha g_{\cdots|\alpha} + \cdots + \sum_{\alpha,\beta} g_{\alpha\beta|\cdots} - \sum_\alpha g_{\alpha|\cdots} + g_{|1\cdots n} = 0 \qquad (\text{II}.82)$$

follows with the abbreviation

$$g_{j_1 \cdots j_\nu | j_{\nu+1} \cdots j_n} = g_{j_1 \cdots j_\nu | \cdots} = g_{\cdots | j_{\nu+1} \cdots j_n}$$

We apply the identity (II. 82) to the integrand (II. 79) of the Feynman integral belonging to Γ. As functions f we take

$$f_{1\ldots n}(z_1 \cdots z_n) = \prod_a P_a \prod_{ab\sigma} i\, \Delta^{+}_{REG}(z_a - z_b)$$

(with the sign of the arguments in Δ^+ determined by a > b)

and

$$f_{i_1 \ldots i_n}(z_1 \cdots z_n) = \prod_a P_a \prod_{ab\sigma} i\, \Delta^{+}_{REG}(z_a - z_b), \quad \text{(II.83)}$$

with the sign of the arguments determined by $\alpha > \beta$ if $a = i_\alpha$, $b = i_\beta$. Applying the definitions (II.81) to the functions (II.83) we get

$$g_{1\ldots n} = \prod_a P_a \prod_{ab\sigma} \Delta^{F}_{REG}(z_a - z_b) = J_\Gamma, \quad \text{(II.84)}$$

$$g_{1\ldots n|} = (-1)^n \prod_a P_a^* \prod_{ab\sigma} \bar{\Delta}^{F}_{REG}(z_a - z_b) = (-1)^n J_\Gamma^*, \quad \text{(II.85)}$$

using the usual connections between Δ_F, $\bar{\Delta}_F$ and, Δ_+, which are retained in the Pauli-Villars regularization.[45] Next we determine the mixed functions g. Any subset $z_{j_1} \ldots z_{j_\nu}$ of $(z_1 \ldots z_n)$ defines a cut C of the diagram Γ which separates $z_{j_1} \ldots z_{j_\nu}$ from the remaining coordinates $z_{j_{\nu+1}} \ldots z_{j_n}$. Let Γ_+ be the subdiagram of Γ which contains the vertices $z_{j_1} \ldots z_{j_\nu}$ and any line connecting these vertices, Γ_- be the subdiagram with vertices $z_{j_{\nu+1}} \ldots z_{j_n}$ and connecting lines.

With this notation

$$g_{j_1 \cdots j_\nu | j_{\nu+1} \cdots j_n} = (-i)^\nu \mathcal{J}^*_{\Gamma_+} I_C \mathcal{J}_{\Gamma_-} \quad , \qquad (\text{II}.86)$$

where $\mathcal{J}_\pm$ denote the integrals associated with Γ_+ and Γ_-

$$\mathcal{J}_{\Gamma_\pm} = \prod_a{}^{\pm} P_a \prod_{ab\sigma}{}^{\pm} \Delta_F^{REG}(z_a - z_b) \ . \qquad (\text{II}.87)$$

The products are restricted to lines and vertices of $\Gamma_\pm$. I_c is the product over all Δ_+-functions associated with lines cut by C

$$I_C = \prod_{ab\sigma}{}^{C} i \Delta_+^{REG}(z_a - z_b) \ . \qquad (\text{II}.88)$$

With (II. 84-88) the equation (II. 82) already gives formal unitarity for the unrenormalized Feynman amplitude after integrating over the internal coordinates.

Before taking the limit $M \longrightarrow \infty$ we must make the necessary subtractions. Fortunately the counter terms have a structure similar to $\mathcal{J}_\Gamma$. They represent products of Δ_F-functions associated with reduced diagrams of Γ, times distributions which contribute for coinciding arguments only. Combining the identities (II. 82) for $\mathcal{J}_\Gamma$ and the subtraction terms one finds for the renormalized integrands

$$\sum_{(j_1 \cdots j_\nu)} R_{\Gamma_+} I_C R_{\Gamma_-} = 0 \qquad (\text{II}.89)$$

Integrating over the internal coordinates, taking the limit[46] $M \to \infty$ and finally summing over all Γ one obtains the unitarity relations, (II.78). The integration over the internal coordinates is best performed by considering the Fourier transform with respect to the z_j first. To avoid divergent or ambiguous expressions for momenta near the mass shell it is necessary to fulfill the conditions (II.73) by choosing a and b according to (II.76).[49]

This completes the check that the S-matrix is unitary in renormalized perturbation theory.

In order to see that the Green's functions define a local quantum field theory one has to show that a field operator $A(x)$ can be constructed which satisfies the usual postulates and has the time ordered functions (II.63). Necessary and sufficient for this are the unitarity equations of the retarded functions combined with the symmetry and retardedness requirements for the solutions. Since retarded functions will not be used in the work that follows we restrict ourselves to a brief description of the method.

The retarded operator product

$$R(x; x_1 \cdots x_n) = (-i)^n \sum \theta(x-x_1) \cdots \theta(x_{n-1}-x_n) [\cdots [A(x) A(x_1)] \cdots A(x_n)] \qquad (\text{II}.90)$$

can be expressed by (anti-) time ordered products

$$R(x;x_1\cdots x_n)=(-i)^n\sum_{i_1\cdots i_\alpha}(-1)^\alpha\,\bar{T}(x_{i_1}\cdots x_{i_\alpha})\,T(x_{i_{\alpha+1}}\cdots x_{i_n}\,x)\,. \qquad (\text{II}.91)$$

This relation suggests defining the contribution to

$$r(x;x_1\cdots x_n)=\langle R(x;x_1\cdots x_n)\rangle \qquad (\text{II}.92)$$

from the Feynman diagram Γ by the renormalized integrand

$$\mathcal{R}_{RET}(z_1\cdots z_N)=(-i)^N\sum_{i_1\cdots i_\alpha}(-1)^\alpha\,\mathcal{R}^*_{\Gamma_+}\,I_c\,\mathcal{R}_{\Gamma_-}\,, \qquad (\text{II}.93)$$

$z_1, \ldots , z_N$ denote the variables x, $x_1, \ldots , x_n$ and the internal coordinates $w_1, \ldots , w_m$. The sum goes over all subsets $(z_{i_1}, \ldots , z_{i_\alpha})$ not containing the variable x. Otherwise the notation is the same as in Eq. (II. 86-89). The retarded functions thus defined have the correct properties, such as symmetry in $x_1, \ldots x_n$, invariance, and retardedness.

The unitarity equations to be satisfied are

$$r(x;y\,x_1\cdots x_n)-r(y;x\,x_1\cdots x_n)=$$

$$=\sum_{i_1\cdots i_n}\sum_{k=0}^{n}\sum_{\ell=1}^{\infty}\int du_1\cdots du_\ell\,dv_1\cdots dv_\ell\,K_{u_1}\cdots K_{u_\ell}\,r(x;x_{i_1}\cdots x_{i_k}\,u_1\cdots u_\ell)\times$$

$$\times\,\Delta^+(u_1-v_1)\cdots\Delta^+(u_\ell-v_\ell)\,K_{v_1}\cdots K_{v_\ell}\,r(y;x_{i_{k+1}}\cdots x_{i_n}\,v_1\cdots v_\ell)+$$

$$-\,(x\longleftrightarrow y)\,. \qquad (\text{II}.94)$$

These relations can be verified by using the identity

$$\sum_1 (-1)^{\alpha} g_{i_1 \cdots i_\alpha | i_{\alpha+1} \cdots i_n} - \sum_2 (-1)^{\alpha} g_{i_1 \cdots i_\alpha | i_{\alpha+1} \cdots i_n} =$$

$$= \sum (-1)^{\alpha+\gamma} b_{j_1 \cdots j_\alpha | k_1 \cdots k_\beta | l_1 \cdots l_\gamma | m_1 \cdots m_\delta} - (z_1 \leftrightarrow z_2) \; ; \qquad \text{(II.95)}$$

$\sum_1$ goes over all $(z_{i_1} \ldots z_{i_\alpha})$ which do not contain z_1, $\sum_2$ over all sets $(z_{i_1} \ldots z_{i_\alpha})$ not containing z_2. The functions g are again defined by (II. 81, 83), the functions b by

$$b_{j_1 \cdots j_\alpha | k_1 \cdots k_\beta | l_1 \cdots l_\gamma | m_1 \cdots m_\delta}(z_1 \cdots z_n) = \qquad \text{(II.96)}$$

$$= f_{j_1' \cdots j_\alpha' | k_1' \cdots k_{\beta+1}' | l_1' \cdots l_\gamma' | m_1' \cdots m_{\delta+1}'}(z_1 \cdots z_n) \, .$$

The subscripts of b form a permutation of the numbers $(3, \ldots, n)$ with the following properties ($\nu < \mu$):

$j'_1 \ldots j'_\alpha$ is a permutation of $(j_1 \ldots j_\alpha)$ such that $(x_{j'_\nu})_0 < (x_{j'_\mu})_0$

$k'_1 \ldots k'_{\beta+1}$ is a permutation of $(k_1, \ldots k_\beta\ 1)$ such that $(x_{k'_\nu})_0 > (x_{k'_\mu})_0$

$l'_1 \ldots l'_\gamma$ is a permutation of $(l_1 \ldots l_\gamma)$ such that $(x_{l'_\nu})_0 < (x_{l'_\mu})_0$

$m'_1 \ldots m'_{\delta+1}$ is a permutation of $(m_1 \ldots m_\delta\ 2)$ such that $(x_{m'_\nu})_0 > (x_{m'_\mu})_0$

The sum on the right hand side of (II. 95) extends over all ordered partitions of $(3 \ldots n)$ into classes

$$(j_1 \cdots j_\alpha)(k_1 \cdots k_\beta)(l_1 \cdots l_\gamma)(m_1 \cdots m_\delta) \, .$$

III. Operator Product Expansions

1. General Derivation

We begin with a derivation of Wilson's operator product expansion in the general framework of quantum field theory. We will, however, have to make a number of restrictive assumptions. Moreover, some questions of mathematical rigor are put aside, in particular we will not concern ourselves with domain problems of operators in Hilbert space.

We consider a product of two operators

$$P(x\xi) = A_1(x+\xi)A_2(x-\xi) \qquad (\text{III}.1)$$

which may also be a suitably defined (anti-)time ordered product or any other type of operator product

$$P(x\xi) = T A_1(x+\xi)A_2(x-\xi) \text{ or } \bar{T} A_1(x+\xi)A_2(x-\xi). \qquad (\text{III}.2)$$

For suitable vectors Φ, Ψ the matrix elements

$$(\Phi, P(x\xi)\Psi)$$

are differentiable functions of x as is indicated by

$$\frac{\partial}{\partial x^\mu}(\Phi, P(x\xi)\Psi) = -i(\Phi, [P_\mu, P(x,\xi)]\Psi)$$

In general (III. 1-2) will be a distribution in ξ, but we assume that (III. 1-2) may be considered as an ordinary function off the light cone (i.e. $\xi^2 \neq 0$).

We are interested in two problems:

(i) The definition of local operator products associated with the formal product $A_1(x)\ A_2(x)$

(ii) The complete analysis of the singularities of $P(x, \xi)$ near $\xi = 0$.

Both problems are of course closely related to each other. We introduce the direction η and the length ρ of the vector ξ by

$$\eta = \frac{\xi}{\rho} \quad , \quad \rho = \sqrt{|\xi^2|} \ , \qquad (\text{III}.3)$$

and consider P henceforth as a function of x, η and ρ:

$$P(x, \xi) = P(x, \eta, \rho) \ .$$

If P diverges for $\rho \to 0$ we define an operator

$$C_1(x\eta) = \lim_{\rho \to 0} \frac{P(x\eta\rho)}{f_1(\rho)} \qquad \text{(weak limit)} \qquad (\text{III}.4)$$

dividing P by a suitable function f_1. The singularity of f_1 is restricted by the condition that the result be finite and different from zero. A suitable f_1 can be found if we make the assumption that there are some matrix elements of P

$$(\Phi', P(x'\eta'\rho)\Psi') \qquad (\text{III}.5)$$

which near $\rho = 0$ are as singular as or more singular than all other matrix elements of P.

In other words

$$\lim_{\rho \to 0} \frac{(\Phi, P(x\eta\rho)\Psi)}{(\Phi', P(x'\eta'\rho)\Psi')} \qquad (\text{III}.6)$$

exists for any Φ, Ψ, x, and η. Choosing any of these most singular matrix elements (III. 4) as the function f_1 the operator (III. 4) should have finite and non-vanishing matrix elements.

By (III. 3) the operator C_1 is local in the sense that

$$\begin{aligned} &[A_j(x), C_1(y,\eta)]_\pm = 0 \\ &[C_1(x,\eta), C_1(y,\eta')]_\pm = 0 \end{aligned} \qquad (\text{III}.7)$$

for $(x - y)^2 < 0$ with commutators or anticommutators taken appropriately. The (anti-)commutators vanish already before the limit provided ρ is small enough.

In many cases C_1 turns out to be a multiple of the identity. In perturbation theory it seems to be a general rule that the most singular part of a matrix element is given by

$$(\Phi, P(x\eta\rho)\Psi) \sim \langle P(x\eta\rho)\rangle (\Phi, \Psi), \qquad (\text{III}.8)$$

provided the vacuum expectation value does not vanish,

$$\langle P(x\eta\rho)\rangle \neq 0.$$

A rigorous theorem by de Mottoni and Genz concerning this point will be quoted later.

With (III.8) one obtains

$$\lim_{\rho \to 0} \frac{(\Phi, P(x\eta\rho)\Psi)}{\langle P(x\eta\rho)\rangle} = (\Phi, \Psi)$$

or

$$C_1 = 1$$

If the vacuum expectation value vanishes

$$\langle P(x\eta\rho)\rangle = 0 ,$$

(III. 4) may lead to a suitable definition, but we will see that in general there are more local operators which should be associated with P .

We now turn to the second problem of analyzing the singularities of the operator product. To this end we introduce the remainder

$$r_1(x\eta\rho) = \frac{P(x\eta\rho)}{f_1(\rho)} - C_1(x\eta) , \quad \text{(III.9)}$$

which vanishes in the limit

$$\lim_{\rho \to 0} r_1(x\eta\rho) = 0 , \quad \text{(III.10)}$$

because of (III. 4). In order to get some information about the singularities of P near $\xi = 0$ we multiply (III. 9) by $f_1(\rho)$ and solve for P

$$P(x\eta\rho) = f_1(\rho)\, C_1(x\eta) + P_2(x\eta\rho) , \quad \text{(III.11)}$$

with

$$P_2(x\eta\rho) = f_1(\rho)\, r_1(x\eta\rho)$$

Here $f_1 \to \infty$ and $r_1 \to 0$ for $\rho \to 0$ Hence nothing can be said in general about the limit of P_2. If

$$\lim_{\rho \to 0} P_2 = 0$$

(III. 12) already gives complete information on the singularities of P near $\xi = 0$:

$$P(x\xi) = f_1\left(\sqrt{|\xi^2|}\right) C_1\left(x, \frac{\xi}{\sqrt{|\xi^2|}}\right) + P_2(x\xi), \qquad \text{(III. 13)}$$

with

$$\lim_{\xi \to 0} P_2(x\xi) = 0 \quad \text{for} \quad \frac{\xi_\mu}{\sqrt{|\xi^2|}} \ \text{constant} \qquad \text{(III. 14)}$$

Hence the term $f_1 C_1$ carries the total singularity of P. If however P_2 diverges the product P has additional singularities near $\xi = 0$. If this is the case we repeat the procedure for

$$P_2(x\eta\rho) = P(x\eta\rho) - f_1(\rho)\, C_1(x\eta) \qquad \text{(III. 15)}$$

We choose one of the most singular matrix elements f_2 of P_2 and form

$$C_2(x\eta) = \lim_{\rho \to 0} \frac{P_2(x\eta\rho)}{f_2(\rho)} \qquad \text{(III. 16)}$$

With this we get more information on the singularities of P. Introducing the remainder

$$r_2(x\eta\rho) = \frac{P_2(x\eta\rho)}{f_2(\rho)} - C_2(x\eta), \qquad (III.17)$$

we have

$$\lim_{\rho\to 0} r_2(x\eta\rho) = 0. \qquad (III.18)$$

With

$$P_3 = r_2 f_2 \qquad (III.19)$$

we obtain

$$P_2 = f_2 C_2 + P_3. \qquad (III.20)$$

Inserting (III. 20) into (III. 11) we obtain

$$P(x\eta\rho) = f_1(\rho)\, C_1(x\eta) + f_2(\rho)\, C_2(x\eta) + P_3(x\eta\rho). \qquad (III.21)$$

Here f_2 is less singular than f_1,

$$\lim_{\rho\to 0} \frac{f_2(\rho)}{f_1(\rho)} = 0, \qquad (III.22)$$

which means that we have refined our analysis of the singularities of P. (III. 22) follows from

$$\lim_{\rho\to 0}\frac{f_2(\rho)}{f_1(\rho)} = \lim_{\rho\to 0}\frac{(\Phi, P_2(x\eta\rho)\Psi)/f_1(\rho)}{(\Phi, P_2(x\eta\rho)\Psi)/f_2(\rho)} = \lim_{\rho\to 0}\frac{(\Phi, r_1(x\eta\rho)\Psi)}{(\Phi, P_2(x\eta\rho)\Psi/f_2(\rho)} = 0,$$

because

$$\lim_{\rho\to 0} r_1 = 0 \quad (\text{Eq. III.10})$$

and

$$\lim_{\rho\to 0}\frac{P_2(x\eta\rho)}{f_2(\rho)} = C_2(x\eta) \neq 0, \quad (\text{III.16})$$

for at least one matrix element.

In this way we may proceed until we arrive at a term P_n which vanishes at $\xi = 0$. It is of course an assumption that the process terminates after a finite number of steps. As a result one has the expansion

$$P(x\eta\rho) = f_1(\rho)C_1(x\eta) + \cdots + f_n(\rho)C_n(x\eta) + R(x\eta\rho), \quad (\text{III.23})$$

with the following properties

$$\lim_{\rho\to 0}\frac{f_{i+1}(\rho)}{f_i(\rho)} = 0 \qquad i = 1, \cdots, n-1 \qquad (\text{III.24})$$

$$\lim_{\rho\to 0} f_n(\rho) = \infty \quad \text{OR} \quad \lim_{\rho\to 0} f_n(\rho) = c \neq 0. \quad (\text{III.25})$$

$$\lim_{\rho\to 0} R(x\eta\rho) = 0 \qquad (\text{III.26})$$

The operators C_j are given by

$$C_j = \lim_{\rho \to 0} \frac{P - \sum_{\alpha=1}^{j-1} f_\alpha C_\alpha}{f_j}, \tag{III.27}$$

and satisfy the locality conditions (I. 9).

Some arbitrariness in the expansion can be removed by requiring that the C_j are linearly independent. This can always be arranged without changing the general properties of the functions f_j. For if there is a linear relation

$$\sum_{\nu=1}^{n} a_\nu C_{j_\nu} = 0 \qquad (\text{no } a_\nu = 0)$$

we express the C_{j_ν} with the highest subscript $b = j_\nu$ by $C_1, \ldots, C_{b-1}$

$$C_b = -\sum_{k=1}^{b-1} \beta_k C_k$$

Then

$$P = \sum_{j=1}^{b-1} (f_j + \beta_j f_b) C_j + \sum_{j=b}^{n} f_j C_j .$$

The new coefficients again satisfy the conditions (III. 24-25). In this way all linear relations among the C_j can be removed until the finite set of operators is linearly independent. An expansion of P satisfying the conditions (III. 23-26) with linearly independent C_j is called a standard expansion.

If the operators C_j are polynomials in the components of η Eq. (III. 23) takes the form (I. 9-10) of Wilson's expansion. (I. 19) implies that the B_j, $B^j_{\mu_1 \cdots \mu_a}$ are local operators in the usual sense.

De Mattoni and Genz[71] proved a theorem concerning the most singular term $E_1 B_1$ in the expansion of

$$P(x\,\xi) = A(x+\xi)\,A(x-\xi)$$

where A is a (pseudo-)scalar field. They found that the vacuum expectation value

$$\langle B_1 \rangle \neq 0 \qquad \text{(III.28)}$$

is different from zero provided the Wightman functions have a power type behavior for coinciding points. In conjunction with the expansion (III. 23) the relation (III. 28) implies that the vacuum expectation value is the most singular matrix element of P.

Finally we quote without proof a uniqueness theorem which clarifies to what extent the operators C_j and coefficients f_j of a standard expansion can be changed.

Let

$$P(x\eta\rho) = \sum_{j=1}^{n} f_j(\rho)\, C_j(x\eta) + R(x\eta\rho)$$

and

$$P(x\eta\rho) = \sum_{j=1}^{n'} f'_j(\rho)\, C'_j(x\eta) + R'(x\eta\rho)$$

be two standard expansions of the same operator product. It then **follows that**

(i) $n' = n,\ C' = \Delta C$

with $C = (C_1 \ldots C_n)$, $C' = (C'_1 \ldots C'_n)$

and

$$\Delta = \begin{pmatrix} a_{11} & 0 & \cdots & 0 \\ a_{21} & a_{22} & \cdots & 0 \\ & & \cdot & \vdots \\ a_{n1} & a_{n2} & \cdots & a_{nn} \end{pmatrix}$$

(ii) $a_{\alpha\alpha} \neq 0$ for $\alpha = 1, \ldots, n$

(iii) The $f'_i(\rho)$ satisfy $\tilde{\Delta} f' = f - h$

where $\tilde{\Delta}$ is the transposed matrix of Δ and

$$\lim_{\rho \to 0} h_j(\rho) = 0$$

(iv) $R' = R + \sum_{j=1}^{n} h_j C_j$

On the other hand any set of operators C'_j and functions f'_j given by the transformations (i) and (iii) leads to a standard expansion of P. Δ may be any triangular matrix with non-vanishing diagonal elements, and the functions h_j are arbitrary provided

$$\lim_{\rho \to 0} h_j = 0$$

Unfortunately the general method of this section cannot be used in perturbation theory. One reason is that it is often impossible to decide whether a power series

$$P = a + C_1(\xi)g + C_2(\xi)g^2 + \cdots$$

is more singular than another one

$$Q = b + D_1(\xi)g + D_2(\xi)g^2 + \cdots$$

Let the coefficients C_K, D_K be logarithmically divergent at $\xi = 0$. Then, in general, both ratios P/Q, Q/P are power series with logarithmically divergent coefficients at $\xi = 0$. In this case it does not make sense to compare the strength of the singularities of P and Q in perturbation theory.

We will therefore proceed in a different way. First local operator products will be defined in renormalized perturbation theory. Expansions of time ordered operator products with respect to these local operators will then be verified to every order of perturbation theory.

2. Notation of Greens Functions

In this section we collect some definitions of Greens functions which will be used in the work that follows. The Fourier transform of a time ordered product will be denoted by

$$T\,\tilde{O}_1(p_1)\cdots\tilde{O}_n(p_n) =$$

$$= \int \frac{dx_1}{(2\pi)^2}\cdots\frac{dx_n}{(2\pi)^2}\, e^{i\sum p_j x_j}\, T O_1(x_1)\cdots O_n(x_n). \quad (\text{III}.29)$$

The O_j are components of local field operators. In particular O_j may be a component of a vector field A_μ, a spinor field ψ or a (pseudo-)scalar field A. Propagators are defined by

$$\Delta_F'(x-y) = \langle T A(x)\, A(y)\rangle, \quad \Delta'_{F\mu\nu}(x-y) = \langle T A_\mu(x) A_\nu(y)\rangle$$

$$S_F'(x-y) = \langle T \psi(x)\, \bar{\psi}(y)\rangle \quad (\text{III}.30)$$

$$\hat{\Delta}_F'(p) = \int dx\, e^{ipx}\Delta_F'(x), \text{ similarly for } \hat{\Delta}'_{F\mu\nu}, \hat{S}_F'.$$

The connected part of a time ordered function is defined recursively by

$$\langle T O_1(x_1)\cdots O_n(x_n)\rangle = \quad (\text{III}.31)$$

$$= \sum \langle T O_{i_{11}}(x_{i_{11}})\cdots\rangle^{\text{conn}} \cdots \langle T O_{i_{A1}}(x_{i_{A1}})\cdots\rangle^{\text{conn}}.$$

The sum extends over all partitions of $x_1, \ldots, x_n$ into classes

$$(x_{i_{11}}\cdots), \ldots, (x_{i_{A1}}\cdots)$$

In (III. 31) and some of the following formulae operators should always be ordered such that the over all permutation of the fermion operators is even. In perturbation theory the connected part is given by the sum over all connected Feynman diagrams. The Fourier transform of the connected part is of the form

$$\langle T\tilde{O}_1(p_1)\cdots\tilde{O}_n(p_n)\rangle^{CONN} = \delta(\Sigma p_j)\langle T\tilde{O}_1(p_1)\cdots\tilde{O}_n(p_n)\rangle^{TRUNC}. \tag{III.32}$$

A Fourier transformation with respect to p_1 yields

$$\langle T\tilde{O}_1(p_1)\cdots\tilde{O}(p_n)\rangle^{TRUNC} = (2\pi)^2\langle T O_1(0)\tilde{O}_2(p_2)\cdots\tilde{O}_n(p_n)\rangle^{CONN}. \tag{III.33}$$

The generalized Wick product

$$:O_1(x_1)\cdots O_n(x_n):$$

is defined recursively by

$$T O_1(x_1)\cdots O_n(x_n) = :O_1(x_1)\cdots O_n(x_n): + \tag{III.34}$$
$$+ \sum\langle T O_{i_{01}}(x_{i_{01}})\cdots\rangle :O_{i_{11}}(x_{i_{11}})\cdots:$$

The sum extends over all partitions of $(x_1, \ldots x_n)$ into classes

$$(x_{i_{01}}\cdots)\,(x_{i_{11}}\cdots)$$

where the second class may be empty.

The mixed product

$$T:O_1(x_1)\cdots O_n(x_n):O_1'(y_1)\cdots O_N'(y_N)$$

is defined by

$$T O_1(x_1)\cdots O_n(x_n)\,O_1'(y_1)\cdots O_N'(y_N) = \tag{III.35}$$
$$= T:O_1(x_1)\cdots O_n(x_n):O_1'(y_1)\cdots O_N'(y_N) +$$
$$+\sum\langle T O_{i_{01}}(x_{i_{01}})\cdots\rangle T:O_{i_{11}}(x_{i_{11}})\cdots:O_1'(y_1)\cdots O_N'(y_N).$$

In perturbation theory

$$\langle T: O_1(x_1)\cdots O_n(x_n): O_1'(y_1)\cdots O_N'(y_N)\rangle$$

is given by the sum of all terms where at least one y_j is attached to every connected component.

Next we introduce the factorization

$$\langle T\{O_1(x_1)\cdots O_n(x_n)\}O_1'(y_1)\cdots O_N'(y_N)\rangle^{FACT} =$$

$$= \sum \langle T O_{i_{01}}'(y_{i_{01}})\cdots\rangle\langle T O_1(x_1) O_{i_{11}}'(y_{i_{11}})\cdots\rangle^{CONN}\times\cdots\times$$

$$\times\langle T O_n(x_n) O_{i_{n1}}'(y_{i_{n1}})\cdots\rangle^{CONN} \qquad (III.36)$$

The sum extends over all ordered partitions of $y_1, \ldots, y_N$ into $n+1$ classes

$$(y_{i_{01}}\cdots)\ (y_{i_{11}}\cdots)\cdots(y_{i_{n1}}\cdots)$$

$(y_{i_{j1}}\cdots)$ is assigned to x_j, $j = 1, \ldots, n$. $(y_{i_{01}}\cdots)$ may be empty.

In perturbation theory the factorization is just the sum of all diagrams for which no pair of different variables (x_i, x_j) is connected by a path (Fig. 12). We give some simple properties of the factorization. For $N = n$

$$\langle T\{A(x_1)\cdots A(x_n)\}A(y_1)\cdots A(y_n)\rangle^{FACT} = \sum P\Delta_F'(x_{i_1}-y_1)\cdots\Delta_F'(x_{i_n}-y_n). \qquad (III.37)$$

with the sum taken over all permutations of the x_j.

$$\langle T\{O_1(x_1)\cdots O_n(x_n)\}O_1'(y_1)\cdots O_N'(y_N)\rangle^{FACT} = 0 \quad \text{for } n>N. \qquad (III.38)$$

$$\langle T\{O(x)\}O_1(y_1)\cdots O_N(y_N)\rangle^{FACT} = \langle T O(x) O_1(y_1)\cdots O_N(y_N)\rangle. \qquad (III.39)$$

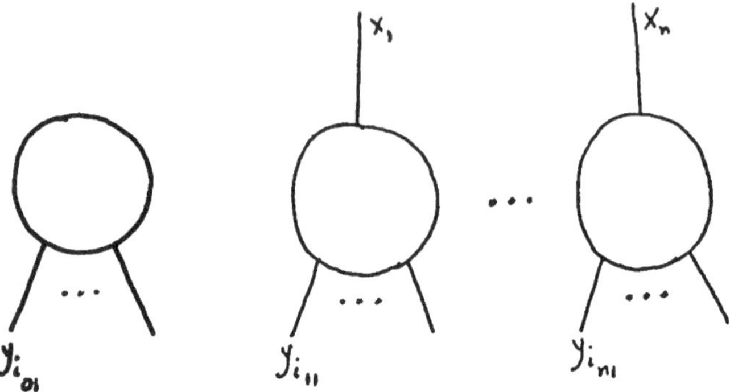

Figure 12.

Diagrams contributing to the factorization.

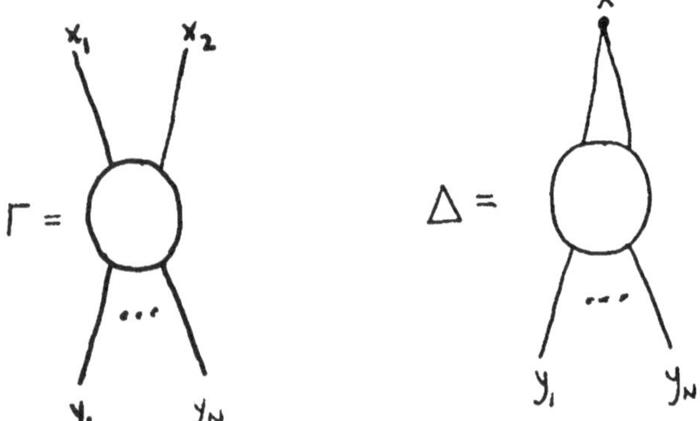

Figure 13.

Diagrams related by $\Delta = \tilde{\Gamma}$

In a model described by a single (pseudo-) scalar field A(x) we introduce the proper part

$$\langle T : A(x_1) \cdots A(x_n) : A(y_1) \cdots A(y_N) \rangle^{PROP} \qquad (III.40)$$

by the recursive formula

$$\langle T : A(x_1) \cdots A(x_n) : A(y_1) \cdots A(y_N) \rangle =$$

$$= \sum_{a=1}^{N} \frac{1}{a!} \int dz_1 \cdots dz_a \langle T : A(x_1) \cdots A(x_n) : A(z_1) \cdots A(z_a) \rangle^{PROP} \times$$

$$\times \langle T\{A(z_1) \cdots A(z_a)\} A(y_1) \cdots A(y_N) \rangle^{FACT}. \qquad (III.41)$$

For a unique definition we must require (III.40) to be symmetric in $y_1, \ldots, y_N$. In perturbation theory (III. 40) is the sum over all diagrams satisfying the following conditions

(i) At least one y_j is attached to each connected component

(ii) The diagram does not permit one particle cuts which would separate a group of coordinates $y_{i_1} \ldots y_{i_\alpha}$ from $x_1, \ldots, x_n$ and the remaining $y_{i_{\alpha+1}} \ldots y_{i_N}$. ($\alpha = N$ is permitted).

Condition (ii) may equivalently be stated as: The diagram should become proper after identifying $x_j = x$.

We give a few simple examples:

$$\langle T : A(x) : A(y) \rangle^{PROP} = \delta(x-y) \qquad (III.42)$$

$$\langle T : A(x) : A(y_1) A(y_2) \cdots A(y_N) \rangle^{PROP} = 0 \quad \text{for } N > 1. \qquad (III.43)$$

$$\langle T : \tilde{A}(p_1) \cdots \tilde{A}(p_n) : \tilde{A}(q) \rangle^{PROP} = \frac{\langle T : \tilde{A}(p_1) \cdots \tilde{A}(p_n) : \tilde{A}(q) \rangle}{\hat{\Delta}_F'(q)} \qquad (III.44)$$

We also introduce the proper part

$$\langle T\, B(x)\, A(y_1) \cdots A(y_N) \rangle^{PROP} \qquad (III.45)$$

for an arbitrary operator B(x). We define (III. 45) recursively by

$$\langle T\, B(x)\, A(y_1) \cdots A(y_N) \rangle =$$

$$= \sum_{a=1}^{N} \frac{1}{a!} \int dz_1 \cdots dz_a \langle T\, B(x) A(z_1) \cdots A(z_a) \rangle^{PROP} \times$$

$$\times \langle T\, A(z_1) \cdots A(z_a)\, A(y_1) \cdots A(y_N) \rangle^{FACT} . \qquad (III.46)$$

We briefly indicate how the proper part of a Greens function should be defined in other models. Consider a model which is described by a Lagrangian involving two elementary fields, a spin 1/2 field $\psi(x)$ and a (pseudo-)scalar field A(x) (for instance the pseudoscalar interaction). Let B be any field operator or a generalized Wick product of fields ψ and A

$$B = O_1(x)$$

or

$$B = :O_1(x_1) \cdots O_n(x_n): \qquad O_j = \begin{cases} \psi \\ A \end{cases}$$

Then the proper part

$$\langle T\, B\, \psi(x_1) \cdots \bar{\psi}(y_1) \cdots A(z_1) \cdots \rangle^{PROP} \qquad (III.47)$$

is defined recursively by

$$\langle T\, B\, C_1(u_1) \cdots C_N(u_N) \rangle =$$

$$= \sum_{\substack{a,b,c \\ a+b+c \le N}} \frac{1}{a!\, b!\, c!} \int dx_1 \cdots dx_a\, dy_1 \cdots dy_b\, dz_1 \cdots dz_c \langle T\, B\, \psi(x_1) \cdots \bar{\psi}(y_1) \cdots A(z_1) \cdots \rangle^{PROP} \times$$

$$\times \langle T \{ \bar{\psi}(x_1) \cdots \psi(y_1) \cdots A(z_1) \cdots \} C_1(u_1) \cdots C_N(u_N) \rangle^{FACT}$$

$$(III.48)$$

For a model involving a vector field A_μ and a spin 1/2 field Ψ as elementary fields (for instance, quantum electrodynamics) the proper part is defined by

$$\langle T\, B\, C_1(u_1)\cdots C_N(u_N)\rangle = \sum_{\substack{a,b,c\\ a+b+c\le N}} \frac{1}{a!\,b!\,c!}\int dx_1\cdots dx_a\, dy_1\cdots dy_b\, dz_1\cdots dz_c \times$$

$$\times \langle T\, B\, \Psi(x_1)\cdots\bar{\Psi}(y_1)\cdots A_{\mu_1}(z_1)\cdots\rangle^{PROP} \langle T\,\{\bar{\Psi}(x_1)\cdots\Psi(y_1)\cdots A^{\mu_1}(z_1)\cdots\}\, C_1(u_1)\cdots C_N(u_N)\rangle^{FACT}. \qquad (\text{III}.49)$$

Finally we need the elementary part

$$\langle T : O_1(x_1)\cdots O_n(x_n) : C_1(u_1)\cdots C_N(u_N)\rangle^{EL}$$

In perturbation theory the elementary part is defined by the sum of all connected diagrams contributing to the proper part. Accordingly

$$\langle T : O_1(x_1)\cdots O_n(x_n) : C_1(y_1)\cdots C_N(y_N)\rangle^{PROP} =$$

$$= \sum \langle T : O_{i_{11}}(x_{i_{11}})\cdots : C_{j_{11}}(y_{j_{11}})\cdots\rangle^{EL} \cdots \times$$

$$\times \langle T : O_{i_{A1}}(x_{i_{A1}})\cdots : C_{j_{A1}}(y_{j_{A1}})\cdots\rangle^{EL} \qquad (\text{III}.50)$$

The sum goes over all partitions of $x_1, \ldots, y_N$ into classes

$$(x_{i_{11}}\cdots y_{j_{11}}\cdots)\cdots(x_{i_{A1}}\cdots y_{j_{A1}}\cdots)$$

with at least one x_j and one y_j in every class We note the particularly simple relation between proper and elementary parts for $n = 2$ for a scalar field A

$$\langle T : A(x_1)A(x_2) : A(y_1) \cdot\cdot A(y_N)\rangle^{PROP} =$$

$$= \langle T : A(x_1)A(x_2) : A(y_1)\cdot\cdot A(y_N)\rangle^{EL} \quad \text{for } N > 2, \tag{III.51}$$

and

$$\langle T : A(x_1)\, A(x_2) : A(y_1)\, A(y_2)\rangle^{PROP} =$$

$$= \langle T : A(x_1)A(x_2) : A(y_1)A(y_2)\rangle^{EL} + \delta(x_1 - y_1)\,\delta(x_2 - y_2) + \delta(x_2 - y_1)\,\delta(x_1 - y_2). \tag{III.52}$$

For later reference we give the free field value of the proper parts:

$$\langle T : A_0(x_1)\cdot\cdot A_0(x_n) : A_0(y_1)\cdot\cdot A_0(y_N)\rangle^{PROP} =$$

$$= \delta_{NM} \sum_P P\, \delta(x_1 - y_{i_1})\cdot\cdot\,\delta(x_n - y_{i_n}) \tag{III.53}$$

with the sum extending over all permutations P of $y_1, \ldots, y_n$.

In the following section local field operators B(x) will be defined through their Green's functions

$$\langle T\ B(x)\, A(y_1)\cdot\cdot A(y_N)\rangle\ . \tag{III.54}$$

The matrix elements of B(x) between incoming and outgoing states can be expressed in terms of (III. 54) by using the reduction formulae

$$[\cdots [S\, B(x)\, \tilde{A}^*_{IN}(k_1)] \cdots \tilde{A}^*_{IN}(k_N)] =$$

$$= (-i)^N\, \epsilon(k_1)\cdot\cdot\,\epsilon(k_N)\,(2\pi)^N \prod_{j=1}^{N} (k_j^2 - m^2)\, \times \tag{III.55}$$

$$\times\ T\, B(x)\, \tilde{A}(k_1)\cdot\cdot\, \tilde{A}(k_N) \qquad \text{for } k_j^2 = m^2 .$$

$A_{IN}(x)$ denotes the incoming field. It is not obvious that the time ordered Green's functions of the operator B(x) given by (III. 55) are identical to the original functions (III. 54) (up to contributions from coinciding arguments). This requires the validity of some consistency conditions which are similar to the unitarity equations of the retarded functions. For the operators studied in the following sections these conditions are always satisfied as can be checked by the methods of section II. 6.

3. Normal Operator Products

We consider the model of a scalar field A(x) with A^4-coupling. Let $A_{(\mu)_1}, \ldots A_{(\mu)_n}$ denote derivatives of the field A

$$A_{(\mu)_j} = \partial_{(\mu)_j} A \,; \quad (\mu)_j = (\mu_{j_1}, \cdots, \mu_{j n(j)}) \,;$$

$$\partial_{(\mu)_j} = \partial_{\mu_{j_1}} \cdots \partial_{\mu_{j n(j)}} \qquad (\text{III}\ 56)$$

Then the normal **operator product**

$$B(x) = N\left\{ A_{(\mu)_1}(x) \cdots A_{(\mu)_n}(x) \right\} \qquad (\text{III}.57)$$

is defined by the Green's functions

$$\langle T\, B(x)\, A(y_1) \cdots A(y_N) \rangle =$$

$$= \mathrm{Fin} \left\langle T \left\{ \exp\left(i \int \mathcal{L}^{\circ}_{INT}\, dz\right) A^{\circ}_{(\mu)_1}(x) \cdots A^{\circ}_{(\mu)_n}(x) A^{\circ}(y_1) \cdots A^{\circ}(y_N) \right\} \right\rangle \qquad (\text{III}.58)$$

The symbol "Fin" means again that the finite part of each Feynman integral should be taken according to the rules of Section II. 2. The diagrams contributing to (III. 58) contain a new vertex V' with coordinate x which requires additional subtractions. The number of subtractions is determined by the degree $\delta(\gamma)$ which is assigned to proper parts containing V. For the operator product (III. 57) we choose $\delta(\gamma)$ as the dimension of the diagram γ:

$$\delta(\gamma) = d(\gamma).$$

We will also need operators

$$B_\delta(x) = N_\delta\{A_{(\mu)_1}(x)\cdots A_{(\mu)_n}(x)\} \qquad \text{(III. 59)}$$

which are constructed with additional subtractions. The degree δ of the operator may be any integer which is greater or equal to the dimension d of B

$$d = n + \sum \#(\mu)_j$$

$$\#(\mu)_j = n(j) = \text{degree of } \partial_{(\mu)_j} \qquad \text{(III. 60)}$$

$$\delta \geq d\,. \quad \delta = d + c, \quad c = 0, 1, 2, \cdots$$

The Green's functions

$$\langle T B_\delta(x) A(y_1) \cdots A(y_N) \rangle \qquad \text{(III. 61)}$$

are defined by (III. 58) but with a larger number of subtractions. To any proper part γ containing V we assign the degree

$$\delta(\gamma) = d(\gamma) + c\,, \quad c = \delta - d$$

where $d(\gamma)$ is the dimension (or superficial divergence) of γ. For $\delta = d$ we have

$$B = B_d$$

The Feynman rules for constructing the unrenormalized integrand of

$$< \widetilde{B}_\delta(p)\, \widetilde{A}(q_1) \cdots \widetilde{A}(q_N) >^\Delta \qquad (III.63)$$

are conveniently derived in the following way. Let Γ be a diagram with external lines $x_1, \ldots x_n, y_1, \ldots y_N$ contributing to the Green's function

$$< T\, A(x_1) \cdots A(x_n)\, A(y_1) \cdots A(y_N) > .$$

Identifying the coordinates

$$x_i = x$$

we obtain a diagram Δ contributing to (III. 61). The diagrams Γ and Δ are related by a one-to-one correspondence which we denote by (Fig 13)

$$\Delta = \widetilde{\Gamma} \quad , \; \Gamma = \hat{\Delta} \quad . \qquad (III.64)$$

Let Γ have the following connected components :

$$\Gamma_1 , \cdots , \Gamma_c , \Gamma_1' , \cdots , \Gamma_{c'}' , \Gamma_1'' , \cdots \Gamma_{c''}''$$

To Γ_j at least one x and one y are attached. To Γ_j' only y are attached and to Γ_j'' only x . With this the unrenormalized contribution from Γ becomes

$$\begin{aligned} &< T\, \widetilde{A}(p_1) \cdots \widetilde{A}(p_n)\, \widetilde{A}(q_1) \cdots \widetilde{A}(q_N) >_u^\Gamma = \\ &= \delta_\Gamma \lim_{\epsilon \to 0^+} \int dk_1 \cdots dk_m\, I_\Gamma(p_1 \cdots p_n\, q_1 \cdots q_N\, k_1 \cdots k_m) . \end{aligned} \qquad (III.65)$$

$$\delta_\Gamma = \prod_j \delta\Big(\sum_\alpha p_{j\alpha} + \sum_\alpha q_{j\alpha}\Big) \prod_j \delta\Big(\sum_\alpha q'_{j\alpha}\Big) \prod_j \delta\Big(\sum_\alpha p''_{j\alpha}\Big) .$$

$p_{j\alpha}, q_{j\alpha}, q'_{j\alpha}, p''_{j\alpha}$ denote the external momenta belonging to $\Gamma_j, \Gamma'_j,$ or Γ''_j respectively. A Fourier transformation with respect to the x_j yields [67]

$$\langle T\, A_{(\mu)_1}(x_1) \cdots A_{(\mu)_n}(x_n)\, \tilde{A}(q_1) \cdots \tilde{A}(q_N) \rangle_u^{\Gamma} = \qquad (\text{III}.66)$$

$$= \frac{1}{(2\pi)^{2n}} \int dp_1 \cdots dp_n\, e^{-i\sum p_j x_j}\, P_{\{\mu\}}\, \delta_\Gamma \lim_{\epsilon \to 0^+} \int dk_1 \cdots dk_m\, I_\Gamma$$

with

$$P_{\{\mu\}} = (-i)^{\sum \#(\mu)_j}\, p_{1(\mu)_1} \cdots p_{n(\mu)_n}$$
$$\{\mu\} = (\mu)_1, \cdots (\mu)_n . \qquad (\text{III}.67)$$

Introducing new variables of integration

$$p_{j\alpha} = \frac{p^{(j)}}{n_j} + K_{j\alpha}, \quad p''_{j\alpha} = K''_{j\alpha}$$

$$p^{(j)} = -\sum_\alpha q_{j\alpha}$$

we obtain

$$\langle T\, A_{(\mu)_1}(x_1) \cdots A_{(\mu)_n}(x_n)\, \tilde{A}(q_1) \cdots \tilde{A}(q_N) \rangle_u^{\Gamma} = \qquad (\text{III}.68)$$

$$= \frac{1}{(2\pi)^{2n}} \int \prod dK_{j\alpha} \prod dK''_{j\alpha}\, e^{-i\sum p_j x_j}\, \delta_\Gamma P_{\{\mu\}} \lim_{\epsilon \to 0^+} \int \prod dk_j\, I_\Gamma ,$$

where in the new variables

$$\delta_\Gamma = \delta(\Sigma K_{j\alpha})\, \delta(\Sigma K''_{j\alpha})\, \delta(\Sigma q'_{j\alpha}) . \qquad (\text{III}.69)$$

Setting $x_j = x$ and taking a Fourier transform with respect to x we get

$$\langle T\, \tilde{B}_\delta(p)\, \tilde{A}(q_1) \cdots \tilde{A}(q_N) \rangle_u^{\Delta} =$$
$$= \delta_\Delta \lim_{\epsilon \to 0^+} \int \prod dK_{j\alpha}\, dK''_{j\alpha}\, dk_j\, \delta'_\Delta\, I_\Delta , \qquad (\text{III}.70)$$

where
$$I_\Delta = \frac{1}{(2\pi)^{2n-2}} P_{\{\mu\}} I_\Gamma$$
$$\delta_\Delta = \delta(p + \Sigma q_{j\alpha}) \prod_j{}' \delta(\Sigma q'_{j\alpha}) \qquad (\text{III}.71)$$
$$\delta'_\Delta = \prod_j{}' \delta(\Sigma K_{j\alpha}) \prod_j{}' \delta(K''_{j\alpha}) \; .$$

This formula represents the explicit construction of the unrenormalized integral. The finite part of the integral is given by

$$\langle T \tilde{B}_\delta(p) \tilde{A}(q_1) \cdots \tilde{A}(q_N) \rangle^\Delta =$$
$$= \delta_\Delta \lim_{\varepsilon \to 0^+} \int \prod dK_{j\alpha} \, dK''_{j\alpha} \, dk_j \, \delta'_\Delta \, R_\Delta^{(\delta)} \qquad (\text{III}.72)$$

Here R^δ_Δ is obtained by applying the R-operation with the degree function (III. 62) to (III. 71). Finally we sum over all diagrams

$$\langle T \tilde{B}_\delta(p) \tilde{A}(q_1) \cdots \tilde{A}(q_N) \rangle =$$
$$= \sum_\Delta \frac{1}{\gamma(\Delta)} \langle T \tilde{B}_\delta(p) \tilde{A}(q_1) \cdots \tilde{A}(q_N) \rangle^\Delta \qquad (\text{III}.73)$$

with

$$\gamma(\Delta) = \gamma(\Gamma)$$

It is often convenient to have a normal product

$$N_\delta\{A_{(\mu)_1}(x_1) \cdots A_{(\mu)_n}(x_n)\} \qquad (\text{III}.74)$$

available which approaches the local product (III. 59) in the limit $x_j \to x$. For the Green's functions this means that the contribution $\langle T N_\delta\{A_{(\mu)_1}(x_1) \cdots A_{(\mu)_n}(x_n)\} \tilde{A}(q_1) \cdots \tilde{A}(q_N) \rangle^\Gamma \qquad (\text{III}.75)$

should approach the Fourier transform of (III. 70)

$$\langle T B_\delta(x)\, \tilde{A}(q_1) \cdots \tilde{A}(q_N) \rangle^\Delta =$$

$$= \frac{1}{(2\pi)^2} \int \prod dK'_{j\alpha}\, dK''_{j\alpha}\, e^{-ipx}\, \delta_\Gamma \lim_{\epsilon \to 0^+} \int \prod dk_j\, R^\delta_\Delta$$

$$p = -\sum q_j \qquad \text{(III.76)}$$

in the limit $x_j \to x$. (III. 76) suggests defining

$$\langle T N_\delta \{ A_{(\mu)_1}(x_1) \cdots A_{(\mu)_n}(x_n) \} \tilde{A}(q_1) \cdots \tilde{A}(q_N) \rangle^\Delta =$$

$$= \frac{1}{(2\pi)^2} \int dK'_{j\alpha}\, dK''_{j\alpha}\, e^{-i\sum p_j x_j}\, \delta_\Gamma \lim_{\epsilon \to 0^+} \int dk_1 \cdots dk_m\, R^\delta_\Delta$$

$$\text{(III.77)}$$

which indeed approaches (III. 76) in a suitable limit. However, due to the presence of the ϵ -limit it is not easy to study the limit $x_j \to x$ rigorously. A careful analysis shows that the limit must be taken such that all quantities

$$\frac{x_i - x_j}{\sqrt{|(x_i - x_j)^2|}}$$

stay bounded. Otherwise the limit may diverge since (III. 77) in general has singularities on the light cone. In the following sections we will study in detail the relation between (III. 77) and the ordinary Green's functions [67]

$$\langle T A_{(\mu)_1}(x_1) \cdots A_{(\mu)_n}(x_n)\, \tilde{A}(q_1) \cdots \tilde{A}(q_N) \rangle^\Gamma = \qquad \text{(III.78)}$$

$$= \frac{1}{(2\pi)^{2n}} \int \prod dK'_{j\alpha} \prod dK''_{j\alpha}\, e^{-i\sum p_j x_j} \lim_{\epsilon \to 0^+} \int \prod dk_j\, P_{\{\mu\}}\, R_\Gamma \,.$$

Eventually these investigations will lead to a proof of Wilson's hypothesis in perturbation theory.

We finally note that at different points x_j the normal products can be defined for any integer δ. However, for $\delta < d$ the limit $x_j \rightarrow x$ does not exist. In particular we have

$$N_{-1}\{A_{(\mu)_1}(x_1)\ldots A_{(\mu)_n}(x_n)\} = T A_{(\mu)_1}(x_1)\cdots A_{(\mu)_n}(x_n)$$

since for $\delta = -1$ no subtractions are performed for proper parts containing V.

In the remainder of this section we will work out some normalization conditions which are satisfied by normal operator products. Let Δ be a diagram which contributes to (III. 73) and is a renormalization part. An external line of Δ which is directly attached to V is called trivial. We assume that Δ has no trivial lines. The relation

$$R_\Delta^\delta = (1 - t_q^{\delta(\Delta)})\,\overline{R}_\Delta^\delta$$

implies

$$\partial_{(\nu)_1}^{q_1}\cdots\partial_{(\nu)_N}^{q_N} R_\Delta^\delta = 0 \qquad , \quad q + \Sigma q_i = 0$$

provided

$$\Sigma\, \#(\nu)_j \leq \delta(\Delta) = \delta - N\,.$$

The condition also holds for renormalization parts containing trivial external lines, but the proof is more involved in that case. Differentiating (III. 77) with respect to q we obtain [68]

$$\langle T N_\delta \{A_{(\mu)_1}(x_1) \cdot\cdot A_{(\mu)_n}(x_n)\} \tilde{A}_{(\nu)_1}(0) \cdot\cdot \tilde{A}_{(\nu)_N}(0)\rangle^\Delta = 0$$

Summing over all proper diagrams we obtain the final form of the normalization condition

$$\langle T N_\delta \{A_{(\mu)_1}(x_1) \cdot\cdot A_{(\mu)_n}(x_n)\} \tilde{A}_{(\nu)_1}(0) \cdot\cdot \tilde{A}_{(\nu)_N}(0)\rangle^{PROP} =$$

$$= \langle T : A^0_{(\mu)_1}(x_1) \cdot\cdot A^0_{(\mu)_n}(x_n) : \tilde{A}^0_{(\nu)_1}(0) \cdot\cdot \tilde{A}^0_{(\nu)_N}(0)\rangle^{PROP}, \quad (III.80)$$

provided

$$N + \sum \#(\nu)_j \leq \delta \qquad (III.81)$$

The right hand side of (III. 80) is the contribution from the trivial vertex V. A_0 denotes a free field; the value of (III. 80) is determined from (III. 53).

For $x_j = 0$ (III. 80) becomes a normalization condition of the local operator product

$$\langle T B_\delta(0) \tilde{A}_{(\nu)_1}(0) \cdot\cdot \tilde{A}_{(\nu)_N}(0)\rangle^{PROP} =$$

$$= \langle T : A^0_{(\mu)_1}(0) \cdot\cdot A^0_{(\mu)_n}(0) : \tilde{A}^0_{(\nu)_1}(0) \cdot\cdot \tilde{A}^0_{(\nu)_N}(0)\rangle^{PROP}$$

$$\text{if } N + \sum \#(\nu)_j \leq \delta . \qquad (III.82)$$

4. Algebraic Identities.[50]

In this section we will investigate the relation between R_Γ and the renormalized integrand R^δ_Γ which we used for defining normal operator products, in the case of A^4-coupling. More generally let Γ be a diagram belonging to a Lagrangian which is a local, invariant polynomial in the fields and their derivatives. $V_1, \dots, V_n$ are some vertices of Γ, each of them connected to the remainder of the diagram by a single line L_j with momentum ℓ_j. Δ is the diagram obtained from Γ by identifying the vertices $V_1, \dots, V_n$ to a new vertex V to which a polynomial P in the ℓ_j is assigned. R_Γ is then constructed according to Bogoliubov's rules with the dimension $d(\gamma)$ as degree function for renormalization parts. R_Δ^δ is constructed with $d(\gamma)$ for renormalization parts not containing V and

$$\delta(\gamma) = d(\gamma) + c \qquad (\text{III } 83)$$

for renormalization parts containing V. The relation between δ and c is

$$\delta = d + c$$

$$d = n + \text{degree } P$$

Using the explicit formulae derived in section III, 5 R_Γ and R_Δ^δ may be written as

$$R_\Gamma = \sum_{U \in \mathcal{U}(\Gamma)} \prod_{\gamma \in U} (-t^\gamma) I_\Gamma \qquad (\text{III}.84)$$

$$R_\Delta^\delta = \sum_{U \in \mathcal{U}(\Delta)} \prod_{\gamma \in U} (-t^\delta) I_\Delta \qquad (\text{III}.85)$$

with $I_\Delta = P I_\Gamma$

$\mathcal{U}(\Gamma)$ is the set of all Γ -forests, $\mathcal{U}(\Delta)$ the set of all Δ-forests. Apparently

$$\mathcal{U}(\Delta) = \mathcal{U}(\Gamma) + \Phi$$

where Φ is the set of all Δ -forests with at least one renormaliza tion part γ containing V From (III. 85)

$$R^S_\Delta = P R_\Gamma + X \qquad \text{(III. 86)}$$

where

$$X = \sum_{U \in \Phi} \prod_{\gamma \in U} (-t^\gamma) I_\Delta \qquad \text{(III 87)}$$

The problem will be to write X **in a convenient form. First we** note that any two elements γ , γ' of U containing V must satisfy

$$\gamma \subset \gamma' \text{ or } \gamma' \subset \gamma$$

For $\gamma \cap \gamma' \neq \phi$ since γ, γ' have V in common. Hence **among** all elements $\gamma \in U$ containing V there is a smallest one which we call τ. Then

$$U = U_1 \cup U_2 \cup \{\tau\} \qquad \text{(III.88)}$$

U_1 is the set of all $\gamma \in U$ $(\gamma \neq \tau)$ with $\tau \subset \gamma$ or $\tau \cap \gamma = \emptyset$.
U_2 is the set of all $\gamma \in U$ $(\gamma \neq \tau)$ with $\gamma \subset \tau$. (III. 88) implies

$$\prod_{\gamma \in U} (-t^\gamma) = \prod_{\gamma \in U_1} (-t^\gamma)(-t^\tau) \prod_{\gamma \in U_2} (-t^\gamma) \qquad \text{(III.89)}$$

Let σ be the subdiagram of Γ which corresponds to the subdiagram τ of Δ . In other words, $\tau = \tilde{\sigma}$

is obtained from σ by identifying the vertices $V_1, \ldots, V_N$. Apparently U_2 is a forest of σ.

With this information (III. 87) becomes

$$X = \sum_{\tau \in T} \sum_{U_1 \in \mathcal{M}_\tau} \sum_{U_2 \in \mathcal{U}(\sigma)} \prod_{\gamma \in U_1} (-t^\gamma)(-t^\tau) \prod_{\gamma \in U_2} (-t^\gamma) I_\Delta \qquad \text{(III.90)}$$

Here T is the set of all renormalization parts τ of Δ which contain V. $\mathcal{M}_\tau$ is the set of all Δ -forests U_1 having the property that each $\gamma \in U_1$ satisfies

$$\tau \subset \gamma \quad \text{or} \quad \tau \cap \gamma = \phi .$$

$\mathcal{U}(\sigma)$ is the set of all σ-forests. (III. 87) and (III. 90) represent the final form of the algebraic identities which will be used in the following sections for checking Wilson's hypothesis in perturbation theory.

We shortly discuss a generalization of (III. 87), (III. 90) which will be needed in section III. 7 for the relation between normal products of different degree. Suppose R^φ_Δ and R^δ_Δ are formed with different degrees

$$\varphi < \delta .$$

We have

$$R^\delta_\Delta = \sum_{U \in \mathcal{U}(\Delta)} \prod_{\gamma \in U} (-t^\gamma_\delta) I_\Delta \qquad \text{(III.91)}$$

$$R_\Delta^\varphi = \sum_{U \in \mathcal{U}(\Delta)} \prod_{\gamma \in U} (-t_\varphi^\gamma) I_\Delta \qquad (\text{III}.92)$$

with t_ς^γ t_φ^γ denoting the appropriate Taylor operators. For any γ containing V we split

$$t_\varsigma^\gamma = t_\varphi^\gamma + (t_\varsigma^\gamma - t_\varphi^\gamma) \qquad (\text{III}.93)$$

while

$$t_\varsigma^\gamma = t_\varphi^\gamma = t_{q\gamma}^{d(\gamma)}$$

for γ which do not contain V. Substituting (III. 93) into (III. 91) we obtain

$$R_\Delta^\delta = \sum_{U \in \mathcal{U}} \sum_{F \in \mathcal{F}(U)} \prod_{\gamma \in U} F_\gamma I_\Delta$$

Here $\mathcal{F}(U)$ is the family of functions with the property

F_γ either $= -t_\varphi^\gamma$ OR $-(t_\varsigma^\gamma - t_\varphi^\gamma)$
if γ contains V

or $F_\gamma = -t_\varsigma^\gamma = -t_\varphi^\gamma$
if γ does not contain V.

For any U there is a function F_0 in $\mathcal{F}(U)$ which assigns

$$F_\gamma \equiv -t_\varphi^\gamma \qquad \text{to any } \gamma \in U$$

Taking out all terms with $F = F_0$ we find

$$R_\Delta^\delta = R_\Delta^\varphi + X$$

$$X = \sum_{U \in \mathcal{U}'} \sum_{F \in F(U), F \neq F_0} \prod_{\gamma \in U} F_\gamma I_\Delta$$

$\mathcal{U}'$ is the set of all forests having at least one element γ which contains V We have to rearrange X For $F \neq F_0$ there is a smallest element τ in U which contains V and satisfies

$$F_\tau = -(t_\delta^\tau - t_\varphi^\tau)$$

For given U and F we decompose

$$U = U_1 + U_2 + \{\tau\}$$

where U_1 and U_2 are defined as before. As generalization of (III.87), (III.90) we finally obtain

$$R_\Delta^\delta = R_\Delta^\varphi + X$$

$$X = \sum_{\tau \in T} \sum_{U_1 \in \mathcal{M}_\tau} \sum_{U_2 \in \mathcal{U}(\sigma)} \prod_{\gamma \in U_1} (-t_\delta^\gamma) \times \left(-(t_\delta^\tau - t_\varphi^\tau)\right) \prod_{\gamma \in U_2} (-t_\varphi^\gamma) I_\Delta . \qquad \text{(III.94)}$$

T, $\mathcal{M}_T$, σ are defined as in the previous case.

5. Singularities of $TA(x_1)\,A(x_2)$ at Short Distances

In this and the following sections we will use the identities of the renormalized Feynman integrands for the derivation of Wilson's expansion formulae We begin with the case of $TA(x_1)A(x_2)$ and try to determine the singularities near $x_1 \sim x_2$.

First we recall the definition of the Green's functions

$$\begin{aligned} &\langle T : A(x_1) A(x_2) : A(y_1) \cdot\cdot A(y_N) \rangle \\ &\langle T N\{ A(x)^2 \} A(y_1) \cdot\cdot A(y_N) \rangle \qquad N \geq 2 \\ &\langle T N\{ A(x_1) A(x_2) \} A(y_1) \cdot\cdot A(y_N) \rangle \end{aligned} \tag{III.95}$$

in perturbation theory.

Let Γ be a connected diagram with $N + 2$ external vertices labeled by $x_1\, x_2\, y_1 \cdot\cdot\, y_N$. Let

$$\Delta = \tilde{\Gamma}$$

be the diagram which is obtained from Γ by identifying (Fig. 13)

$$x_1 = x_2 = x .$$

The new vertex at x is called V. The contributions from Γ , Δ to (III. 95) are

$$\langle T : A(x_1) A(x_2) : \tilde{A}(q_1) \cdot\cdot \tilde{A}(q_N) \rangle^{\Gamma} =$$

$$= \frac{1}{(2\pi)^4} \int dK_1 dK_2\, \delta(K_1 + K_2)\, e^{-i(p_1 x_1 + p_2 x_2)} \times$$

$$\times \lim_{\epsilon \to 0^+} \int dk_1 \cdot\cdot dk_m\, R_\Gamma (p_1 p_2 q_1 \cdot\cdot q_N k_1 \cdot\cdot k_m) ; \tag{III.96}$$

$$\langle T N\{ A(x)^2 \} \tilde{A}(q_1) \cdot\cdot \tilde{A}(q_N) \rangle^{\Delta} = \frac{1}{(2\pi)^4} e^{-ipx} \lim_{\epsilon \to 0^+} \int dK_1 dK_2 \times$$

$$\times\, \delta(K_1 + K_2) \int dk_1 \cdot\cdot dk_m\, R_\Delta^{(2)} (p_1 p_2 q_1 \cdot\cdot q_N k_1 \cdot\cdot k_m) , \tag{III.97}$$

$$\langle T\, N\{A(x_1)A(x_2)\}\tilde{A}(q_1)\cdots\tilde{A}(q_N)\rangle^{\Gamma} = \frac{1}{(2\pi)^2}\int dk_1 dk_2\, \delta(k_1+k_2)\times$$

$$\times e^{-i(p_1x_1+p_2x_2)} \lim_{\epsilon\to 0^+} \int dk_1\cdots dk_m\, R_{\Delta}^{(2)}(p_1 p_2 q_1\cdots q_N k_1\cdots k_m). \quad (\text{III}.98)$$

In the preceding section we proved that the integrands R_{Γ} and $R_{\Delta}^{(2)}$ appearing in (III. 96 - 98) are related by

$$R_{\Delta}^{(2)} = \frac{1}{(2\pi)^2} R_{\Gamma} + X \qquad (\text{III}.99)$$

with

$$X = \sum_{\tau\in T}\ \sum_{U_1\in \mathcal{M}_\tau}\ \sum_{U_2\in \mathcal{F}(\sigma)}\ \prod_{\gamma\in U_1}(-t^{\gamma})(-t^{\tau})\prod_{\gamma\in U_2}(-t^{\gamma})\, I_{\Delta} \qquad (\text{III}.100)$$

In the work that follows we will study the expression X in some detail. First we should determine all possible diagrams τ to be summed over in (III. 100). The set T was introduced as the set of all renormalization parts containing V . These are the proper and non-trivial subdiagrams of Δ which contain the vertex V and have dimension

$$d(\tau) \geq 0$$

Now

$$d(\tau) = 2 - a$$

where a is the number of lines connecting τ with the remainder of Δ (see Fig. 14). Hence for $N > 0$ only diagrams with a = 2 are eligible (Fig. 15). Their dimension is $d(\tau) = 0$

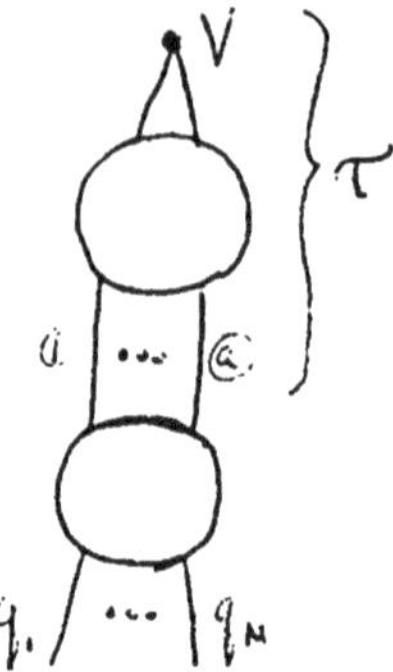

Figure 14
Subdiagram τ containing V, connected by a lines to the rest of the diagram.

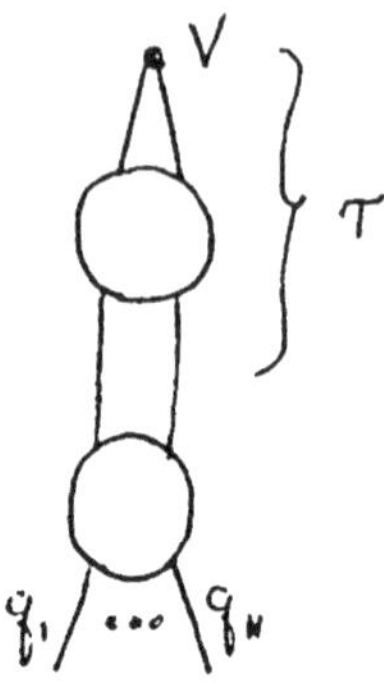

Figure 15.
Renormalization part τ containing V.

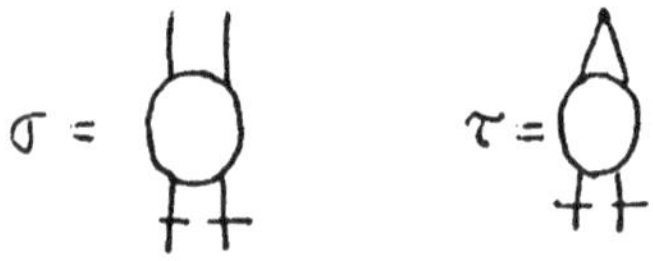

$\sigma =$ $\tau =$

Figure 16. Renormalization part τ and corresponding diagram σ related by (III.101).

Figure 17.
Trivial proper diagram containing V.

Therefore T is the set of all proper, non-trivial subdiagrams τ of Δ which contain the vertex V and are connected to the remainder of Δ by two lines. The Taylor operator becomes

$$t^{\tau} = t^{d(\tau)}_{q_\tau} = t^{0}_{q_\tau}$$

and prescribes merely that the external momenta of τ be set equal to zero.

The diagram σ was defined by

$$\tau = \tilde{\sigma} \qquad \text{(III.101)}$$

i.e. τ is obtained from σ by identifying $x_1 = x_2$. (Fig. 16) The set of all such σ is called S. (III. 101) is a one-to-one correspondence between S and T. For any subdiagram σ of Γ the unrenormalized Feynman integrands factorize

$$I_\Gamma = I_{\Gamma/\sigma} I_\sigma \qquad \text{(III.102)}$$

or

$$(2\pi)^2 I_\Delta = (2\pi)^2 I_{\Delta/\tau} I_\sigma \qquad \text{(III.103)}$$

due to the convention

$$(2\pi)^2 I_\Delta = I_\Gamma , \quad (2\pi)^2 I_{\Delta/\tau} = I_{\Gamma/\sigma}$$

We shall see that also X factorizes. To this end we recall that the sum

$$\sum_{U_i \in \mathcal{M}_\tau} \qquad \text{(III.104)}$$

extends over all forests of renormalization parts γ of Γ which satisfy $\gamma \supset \tau$ or $\gamma \cap \tau = \phi$.

Accordingly (III. 104) can be replaced by the sum over all forests U' of Δ/τ with elements $\gamma' = \gamma/\tau$. The sum

$$\sum_{U_2 \in \mathcal{F}(\sigma)}$$

extends over all renormalization parts of σ Combining this information with (III. 100) and (III. 103) we obtain

$$\begin{aligned} X &= -\sum_{\tau \in T} R^{(2)}_{\Delta/\tau} \mathring{R}_{\sigma} \\ \mathring{R}_{\sigma} &= R_{\sigma} \quad \text{for } q_{\tau} = 0 \end{aligned} \qquad (\text{III. } 105)$$

where

$$R^{(2)}_{\Delta/\tau} = \sum_{U \in \mathcal{M}_{\tau}} \prod_{\gamma \in U}' (-t^{\gamma}) I_{\Delta/\tau} = \sum_{U' \in \mathcal{F}(\Delta/\tau)} \prod_{\gamma' \in U'}' (-t^{\gamma'}) I_{\Delta/\tau}$$

$$R_{\sigma} = \sum_{U \in \mathcal{F}(\sigma)} \prod_{\gamma \in U}' (-t^{\gamma}) I_{\sigma}$$

Hence X is a sum of terms which factorize into the renormalized integrand of the reduced diagram Δ/τ and the renormalized integrand of σ. Thus

$$\frac{1}{(2\pi)^2} R_{\Gamma} = R^{(2)}_{\Delta} + \sum_{\tau \in T} R^{(2)}_{\Delta/\tau} \mathring{R}_{\sigma} \qquad (\text{III. } 106)$$

Multiplying by

$$\delta(K_1 + K_2) e^{-i(p_1 x_1 + p_2 x_2)} = \delta(K_1 + K_2) e^{-ipx} e^{-i(K_1 - K_2)\xi}$$

with $\quad x_1 = x + \xi \, , \; x_2 = x - \xi \qquad$ (III. 107) and integrating over K_i and k_j we obtain

$$\langle T : A(x_1) A(x_2) : \tilde{A}(q_1) \cdots \rangle^{\Gamma} =$$

$$= \langle T N\{A(x_1) A(x_2)\} \tilde{A}(q_1) \cdots \rangle^{\Gamma} + \qquad (\text{III } 108)$$

$$+ (2\pi)^4 \sum_{\tau \in T} \langle T N\{A(x)^2\} \tilde{A}(q_1) \cdots \rangle^{\Delta/\tau} \langle T A(\xi) A(-\xi) \tilde{A}(0)^2 \rangle^{\sigma}$$

The same relation holds for disconnected Γ. Summing over all possible Γ one finds

$$\langle T : A(x_1) A(x_2) : \tilde{A}(q_1) \cdot\cdot \tilde{A}(q_N) \rangle =$$

$$= \langle T N\{A(x_1) A(x_2)\} \tilde{A}(q_1) \cdot\cdot \tilde{A}(q_N) \rangle + \qquad (\text{III}.109)$$

$$+ \frac{(2\pi)^4}{2} G(\xi) \langle T N\{A(x)^2\} \tilde{A}(q_1) \cdot\cdot \tilde{A}(q_N) \rangle$$

with

$$G(\xi) = \sum_{\tau \in T} \frac{1}{\gamma(\sigma)} \langle T A(\xi) A(-\xi) \tilde{A}(0)^2 \rangle^{\sigma} =$$

$$= \langle T : A(\xi) A(-\xi) : \tilde{A}(0)^2 \rangle^{PROP} - \frac{2}{(2\pi)^4} . \qquad (\text{III}.110)$$

Multiplying (III. 109) by factors $q^2_j - m^2$ and setting $q^2_j = m^2$ one obtains the operator equation

$$: A(x_1) A(x_2) : = N\{A(x_1) A(x_2)\} +$$

$$+ \frac{(2\pi)^4}{2} G(\xi) N\{A(x)^2\} \qquad (\text{III}.111)$$

as a consequence of the reduction formulae. Introducing the remainder $R(x_1, x_2) = N\{A(x_1) A(x_2)\} - N\{A(x)^2\}$ (III.112)

we can write (III. 111) in the equivalent form

$$T A(x+\xi) A(x-\xi) = E_1(\xi) + E_2(\xi) N\{A(x)^2\} + R(x,\xi) \tag{III.113}$$

with

$$E_1(\xi) = \langle T A(\xi) A(-\xi) \rangle, \quad E_2(\xi) = \frac{(2\pi)^4}{2} \langle T : A(\xi) A(-\xi) : \tilde{A}(0)^2 \rangle^{PROP} \tag{III.114}$$

The remainder vanishes in the limit

$$R(x,\xi) \to 0 \quad \text{for } \xi \to 0 \text{ and } \frac{\xi_\mu}{\sqrt{|\xi^2|}} \text{ bounded}$$

Applying the reduction technique to some of the q_j only one obtains

$$T : A(x+\xi) A(x-\xi) : A(y_1) \cdots A(y_N) =$$

$$= E_1(\xi) T A(y_1) \cdots A(y_N) + E_2(\xi) T N\{A(x)^2\} A(y_1) \cdots A(y_N) +$$

$$+ R(x\,\xi\, y_1 \cdots y_N). \tag{III.115}$$

$$R \to 0 \quad \text{for } \xi \to 0 \text{ and } \frac{\xi_\mu}{\sqrt{|\xi^2|}} \text{ bounded.}$$

Finally we check Wilson's statement concerning the leading singularities of the coefficients E_j. Since the diagrams contributing to E_1 have superficial divergence 2 the function E_1 is at most quadratically divergent. The diagrams contributing to E_2 have superficial divergence O, hence E_2 is at most logarithmically divergent. It can be checked in lowest order that the indicated

singularities are actually present. Hence Wilson's rule is satisfied with

$$d(E_1) = 2$$

$$d(E_2) + d(A^2) = 2$$

In perturbation theory we found that the coefficients E_j equal certain Green's functions. These relations can be obtained more easily from the expansion formulae (III. 115) with $N = 0$, 2 by using the normalization conditions of $N\{A^2\}$, which we found in the preceding section. Details will not be given here since Chapter IV will contain numerous examples for this method of determining the coefficients E_j.

6. Expansion of $TA(x_1)A(x_2)$ with Respect to Normal Products of Constant Degree.

In the last section we determined the singularities of $TA(x_1)A(x_2)$ near $x_1 \sim x_2$ in perturbation theory. We are interested in the following more general problems:

(i) Higher order terms in Wilson's expansion (I. 27)

(ii) Singularities of derivatives of $TA(x_1)A(x_2)$

It will be seen that both problems are closely related to each other. As a first step towards the solution of these problems we will derive a formula which expresses $TA(x_1)A(x_2)$ by local operators of a given degree δ . This will be a generalization of Eq. (III. 113) which expressed $TA(x_1)A(x_2)$ in terms of operators of degree $\delta = 2$.

We consider the Green's function

$$\langle T N_\delta \{ A(x_1) A(x_2) \} A(y_1) \cdots A(y_N) \rangle \qquad (\text{III. 116})$$

of the normal product

$$N_\delta \{ A(x_1) A(x_2) \}$$

of degree $\delta \geq 2$. The contribution from a connected diagram to (III. 116) is given by

$$\langle T N_\delta \{ A(x_1) A(x_2) \} \tilde{A}(q_1) \cdots \tilde{A}(q_N) \rangle^\Gamma =$$

$$= \frac{1}{(2\pi)^2} \int dk_1 dk_2 \, \delta(k_1 + k_2) e^{-i(p_1 x_1 + p_2 x_2)} \times$$

$$\times \lim_{\epsilon \to 0^+} \int dk_1 \cdots dk_m \, R_\Delta^{(\delta)} (p_1 p_2 q_1 \cdots q_N k_1 \cdots k_m) \, . \qquad (\text{III. 117})$$

Again we use the identity (III. 86), (III. 90). This time T is the set of all proper, non-trivial subdiagrams of Δ which contain V and which have degree

$$\delta(\tau) = d(\tau) + c = \delta - a \geq 0$$

Here

$$c = \delta - d$$

denotes the number of additional subtractions used in the definition of N_δ . Hence the number of lines connecting τ with the remainder of Δ is restricted by

$$a \leq \delta, \quad a \text{ even.}$$

The number of lines joining at the reduced vertex of Δ/τ is a.
The factorization (III.102) leads to

$$X = -\sum_{\tau \in T} (2\pi)^{2a-4} \sum_{U_1 \in \mathcal{M}_\tau} \prod_{\gamma \in U_1} (-t^\gamma) I_{\Delta/\tau} \times$$

$$\times \, t^\tau \left\{ \sum_{U_2 \in \mathcal{F}(\sigma)} \prod_{\gamma \in U_2} (-t^\gamma) I_\sigma \right\} \qquad \text{(III.118)}$$

The π-factor arises from

$$(2\pi)^2 I_\Delta = I_\Gamma , \quad (2\pi)^{2n-2} I_{\Gamma/\sigma} = I_{\Delta/\tau}$$

The explicit form of the Taylor operator is

$$t^\tau = \sum_{a + \Sigma \#(\rho)_j \leq d} \frac{1}{(\Sigma \#(\rho)_j)!} \, q^\tau_{1\,(\rho)_1} \cdots q^\tau_{a\,(\rho)_a} \, \partial^{(\rho)_1}_{q^\tau_1} \cdots \partial^{(\rho)_a}_{q^\tau_a}$$

with $q^\tau_j = 0$ in the differentiated expression. By definition the external momenta q^τ_j of τ equal

$$q^\tau_1 = -\ell_1 \quad \cdots , \quad q^\tau_a = -\ell_a$$

where the $\ell_1, \ \ , \ell_a$ denote the total momenta of the lines connecting τ with the remainder of the diagram Δ. Introducing

$$I'_{\Delta/\tau} = I_{\Delta/\tau} (-1)^{\#(\rho)_1} \ell_{1(\rho)_1} \quad (-1)^{\#(\rho)_a} \ell_{a(\rho)_a} \qquad \text{(III.119)}$$

we obtain the following factorization of X :

$$X = -\sum_{\tau \in T} (2\pi)^{2a-4} \sum_{(\rho)_1 \cdots (\rho)_a} \frac{(-i)^{\sum \#(\rho)_j}}{(\sum \#(\rho)_j)!} \times$$

$$\times R^{(\delta)}_{(\rho)_1 \cdots (\rho)_a} \left\{ \partial^{(\rho)_1}_{q_1^\tau} \cdots \partial^{(\rho)_a}_{q_a^\tau} R_\sigma \right\}_{q^\tau = 0} \qquad \text{(III. 12c)}$$

$$R^{(\delta)}_{(\rho)_1 \cdots (\rho)_a} = \sum_{U_1 \in \mathcal{M}_\tau} \prod_{\gamma \in U_1} (-t^\gamma) I'_{\Delta/\tau}$$

$$R_\sigma = \sum_{U \in F(\sigma)} \prod_{\gamma \in U} (-t^\gamma) I_\sigma$$

$R^{(\delta)}_{(\rho)_1 \cdots (\rho)_a}$ **is a contribution to the Green's function**

$$\langle T N_\delta \{ A_{(\rho)_1}(x) \cdots A_{(\rho)_a}(x) \} \tilde{A}(q_1) \cdots \tilde{A}(q_N) \rangle$$

Thus we obtain from (III. 86)

$$\frac{1}{(2\pi)^2} R_\Gamma = R^{(\delta)}_\Delta - X \qquad \text{(III. 119')}$$

with X given by (III. 118). Multiplying (III. 119') by (III. 107), integrating over the internal momenta and summing over Γ we obtain

$$\langle T N_\delta \{ A(x+\xi) A(x-\xi) \} \tilde{A}(q_1) \cdots \tilde{A}(q_N) \rangle =$$

$$= \langle T : A(x+\xi) A(x-\xi) : \tilde{A}(q_1) \cdots \tilde{A}(q_N) \rangle +$$

$$- \sum_{a=1}^{\delta} \sum_{(\rho)_1 \cdots (\rho)_a} G_{(\rho)_1 \cdots (\rho)_a}(\xi) \langle T N_\delta \{ A_{(\rho)_1}(x) \cdots A_{(\rho)_a}(x) \} \tilde{A}(q_1) \cdots \tilde{A}(q_N) \rangle \qquad \text{(III. 120')}$$

The sum is restricted to values of a, $(\rho)_1, \ldots (\rho)_a$ for which

$$a + \sum \#(\rho)_j \leq d \qquad \text{and a even.} \qquad (\text{III}.121)$$

The function G is given by

$$G_{(\rho)_1 \cdots (\rho)_a} = \frac{(2\pi)^{2a}}{a!} \frac{(-i)^{\sum \#(\rho)_j}}{(\sum \#(\rho)_j)!} \langle T : A(\xi)A(-\xi) : \tilde{A}^{(\rho)_1}(0) \cdots \tilde{A}^{(\rho)_a}(0)\rangle^{EL} \qquad (\text{III}.122)$$

The elementary part $\langle \; \rangle^{EL}$ was defined (Eq. (III. 50)) by the sum over all connected diagrams σ which contribute to the proper part $\langle \; \rangle^{PROP}$. This corresponds to the sum over all non-trivial proper parts $\tau = \tilde{\sigma}$.

Using the reduction formulae we obtain the following operator form of (III. 120)

$$: A(x+\xi)A(x-\xi) : = N_\delta \{A(x+\xi)A(x-\xi)\} + \sum_{a=1}^{\delta} \sum_{\substack{(\rho)_1 \cdots (\rho)_a \\ a+\sum \#(\rho)_j \leq d}} G^{(\rho)_1 \cdots (\rho)_a}(\xi) N_\delta \{A_{(\rho)_1}(x) \cdots A_{(\rho)_a}(x)\} \qquad (\text{III}.123)$$

with the sum restricted by (III. 121).

Unfortunately (III. 123) does not meet the requirements of Wilson's expansion. Indeed the non-vanishing coefficients G all diverge for $\xi \to 0$ while in (I. 27) it is required that the coefficient of

$$N\{A_{(\rho)_1}(x) \cdots A_{(\rho)_a}(x)\}$$

decrease like

$$s^{2-a-\Sigma \#(\rho)_j} \qquad s = \sqrt{|\xi^2|}$$

The reason for this is that (III. 123) uses operator products of constant degree δ. But δ is different from the dimension d of the operator product for the higher terms. In the following section we will derive a modified expansion formula in which the degree equals the dimension of the local operator products occurring in the expansion.

It should be remarked, however, that expansions of constant degree are sufficient for the discussion of field equations in renormalizable theories.

7. Wilson's Expansion of $TA(x_1)A(x_2)$ and Derivatives.[51] Light Cone Singularities

We first establish some relations between normal products of different degree. We consider the Green's functions

$$\begin{aligned} &\langle T N_\delta \{A(x_1) A(x_2)\} A(y_1) \cdots A(y_n)\rangle \\ &\langle T N_4 \{A(x_1) A(x_2)\} A(y_1) \cdots A(y_N)\rangle \end{aligned} \qquad \text{(III. 124)}$$

and assume $q < \delta$. The contributions from a connected diagram to (III. 124) are given by (III. 117). The two renormalized integrands R^{δ}_{Δ} and R^{q}_{Δ} are related by identity (III. 94). Applying the methods developed in the preceding section we obtain

$$N_q\{A(x+\xi)A(x-\xi)\} = N'_\delta\{A(x+\xi)A(x-\xi)\} + \\ + \sum_{\{\rho\}}{}' H^{\{\rho\}}_{q\delta}(\xi)\, B^{\delta}_{\{\rho\}}(x) \qquad (\text{III}.125)$$

with the abbreviation

$$\{\rho\} = (\rho)_1 \cdots (\rho)_a$$

The sum $\sum'$ goes over all values of $\{\rho\}$ which satisfy

$$q < a + \Sigma \#(\rho)_j \leq \delta \qquad (\text{III}.126)$$

The functions H are given by

$$H^{\{\rho\}}_{q\delta} = \frac{(2\pi)^{2a}}{a!}\frac{(-i)^{\Sigma\#(\rho)_j}}{(\Sigma\#(\rho)_j)!}\langle TN_q\{A(\xi)A(-\xi)\}\tilde{A}^{(\rho)_1}(0)\cdots\tilde{A}^{(\rho)_a}(0)\rangle^{EL}. \qquad (\text{III}.127)$$

Here the elementary part $\langle\ \rangle^{EL}$ denotes the sum over all contributions to $\langle\ \rangle^{PROP}$ excluding the contribution from the trivial diagram Fig. 17. Accordingly the relation between $\langle\ \rangle^{EL}$ and $\langle\ \rangle^{PROP}$ is

$$\langle TN_q\{A(x_1)A(x_2)\}\tilde{A}(q_1)\cdots\tilde{A}(q_N)\rangle^{PROP} =$$

$$= \langle TN_q\{A(x_1)A(x_2)\}\tilde{A}(q_1)\cdots\tilde{A}(q_N)\rangle^{EL} + \\ + \langle T:A_0(x_1)A_0(x_2):\tilde{A}_0(q_1)\cdots\tilde{A}_0(q_N)\rangle^{PROP} \qquad (\text{III}.128)$$

The second term on the right hand side is the proper part of a free field A_0 which is given by (III. 53).

Setting $\rho = \delta - 1$ in (III. 125) we get

$$N_{\delta-1}\{A(x+\xi)A(x-\xi)\} := N_\delta\{A(x+\xi)A(x-\xi)\} + \\ + \sum_{\{\rho\}}{}'' H_{\delta-1}^{\{\rho\}}(\xi)\, B_{\{\rho\}}^{\delta}(x) \qquad \text{(III.129)}$$

Here the sum $\sum''$ is restricted by

$$a + \sum \#(\rho)_j = \delta \qquad \text{(III. 130)}$$

In this relation all local operators have equal degree and dimension. For $\delta \geq 3$ the coefficients are finite in the limit $\xi \to 0$, but still do not vanish. However, the appropriate behavior of the coefficients can be achieved by introducing

$$M_\delta\, A(x+\xi)A(x-\xi) = (1 - t_\xi^c)\, N_\delta\, A(x+\xi)A(x-\xi) \\ c = \delta - 2 \qquad \text{(III.131)}$$

for $c < 0$, i.e. $\delta < 2$ we set

$$M_\delta\, A(x_1)A(x_2) = N_\delta\, A(x_1)A(x_2)$$

We first check that the new products behave as

$$M_\delta A(x+\xi)A(x-\xi) \sim s^{c+1} \ln^\alpha s \qquad s = \sqrt{|\xi^2|} \qquad \text{(III.132)}$$

for $\xi \longrightarrow 0$ and $\frac{\xi_\mu}{\sqrt{|\xi^2|}}$ bounded.

Since N_δ exists with all derivatives of order $\leq$ c at $\xi = 0$ we may use the Taylor formula

$$(1-t_\xi^{c-1}) N_\delta = \sum_{\#(\mu)=c} \frac{\xi^{(\mu)}}{\#(\mu)!} \partial_{(\mu)} N_\delta(\theta\xi) \qquad \text{(III.133)}$$

with

$$0 < \theta < 1$$

Hence

$$(1-t_\xi^{c}) N_\delta = \sum_{\#(\mu)=c} \frac{\xi^{(\mu)}}{\#(\mu)!} \left\{ \partial_{(\mu)} N_\delta(\theta\xi) - \partial_{(\mu)} N_\delta(0) \right\}$$

Since the expression in the bracket behaves like $s \lg^\alpha s$ for $s \to 0$ we find

$$(1-t_\xi^{c}) N_\delta \sim s^{c+1} \lg^\alpha s \qquad \text{(III 134)}$$

We will now replace N_δ by M_δ in (III. 127). Application of $1-t_\xi^{\delta-3}$ to (III. 129) yields

$$M_{\delta-1}\{A(x_1)A(x_2)\} = (1-t_\xi^{\delta-3}) N_\delta\{A(x_1)A(x_2)\} + \\ + \sum_{\{\rho\}} H_{\{\rho\}}^{\delta-1} B^{\{\rho\}}(x) \qquad \text{(III.135)}$$

Here

$$(1-t_\xi^{\delta-3}) N_\delta\{A(x_1)A(x_2)\} = M_\delta\{A(x_1)A(x_2)\} + \\ + \sum_{\#(\mu)=\delta-2} \frac{\xi^{(\mu)}}{\#(\mu)!} \partial_{(\mu)}^{\xi} N_\delta\{A(x_1)A(x_2)\}\Big|_{\xi=0} \qquad \text{(III.136)}$$

It will be seen that the second term of (III. 136) cancels if we simultaneously replace the elementary part by the proper part in (III. 129). Indeed

$$\partial^{\xi}_{(\mu)} N_{\delta}\, A(x_1) A(x_2) = N_{\delta}\, \partial^{\xi}_{(\mu)} A(x_1) A(x_2) +$$

$$+ \sum_{\#(\mu)=\delta-2} \frac{\xi^{(\mu)}}{\#(\mu)!} N_{\delta}\, \partial^{\xi}_{(\mu)} T A(x_1) A(x_2) =$$

$$= \sum_{(\mu)_1 (\mu)_2} \frac{\xi^{(\mu)_1} (-\xi)^{(\mu)_2}}{\#(\mu)_1 !\; \#(\mu)_2 !} N_{\delta}\, A_{(\mu)_1}(x_1) A_{(\mu)_2}(x_2)$$

(III. 137)

with

$$\#(\mu)_1 + \#(\mu)_2 = \delta - 2$$

On the other hand we use (III. 128). For $a > 2$ there is no change since

$$\langle T\{A_0(x_1) A_c(x_2)\} \tilde{A}_c(q_1) \cdot\cdot \tilde{A}_c(q_N)\rangle^{prop} = 0$$

For $a = 2$ we have

$$\langle T\{A_0(\xi) A_c(-\xi)\} \tilde{A}_0^{(\rho)_1}(0)\, \tilde{A}_c^{(\rho)_2}(0)\rangle^{prop} =$$

$$= \frac{i^{[\#(\rho)_1 + \#(\rho)_2]}}{(2\pi)^4} \left\{ \xi_{(\rho)_1} (-\xi)_{(\rho)_2} + \xi_{(\rho)_2} (-\xi)_{(\rho)_1} \right\}$$

Since this expression is homogenous of degree $\#(\rho)_1 + \#(\rho)_2$ the operator $1 - t_{\xi}^{\delta-2}$ has no effect. For

$$\delta - 3 = \Sigma\, \#(\rho)_j - 1$$

Therefore, replacing the elementary part by the proper part in (III. 135) we get a contribution which exactly cancels (III. 137). The final result is

$$M_{\delta-1}\{A(x_1)A(x_2)\} = M_\delta\{A(x_1)A(x_2)\} + \\ + \sum_{\{\rho\}} E_{\{\rho\}}(\xi)\, B^{\{\rho\}}(x) ,$$

$$E_{\{\rho\}}(\xi) = \frac{(2\pi)^{2a}}{a!} \cdot \frac{(-1)^{\sum \#(\rho)_j}}{(\sum \#(\rho)_j)!} \langle T M_{\delta-1}\{A(\xi)A(-\xi)\} \times \\ \times \tilde{A}^{(\rho)_1}(0) \ldots \tilde{A}^{(\rho)_a}(0) \rangle^{PROP} \qquad (III.138)$$

The sum is restricted by (III. 130). Eq. (III. 138) is a recursion formula for M_δ. For $\delta = 0$ it takes the form

$$T A(x_1)A(x_2) = :A(x_1)A(x_2): + \langle T A(\xi)A(-\xi)\rangle$$

By induction we arrive at the following form of Wilson's expansion formula

$$T A(x+\xi)A(x-\xi) = E_0(\xi) + \sum_{a=1}^{e} \sum_{\{\rho\}} E_{\{\rho\}}(\xi)\, B_d^{\{\rho\}}(x) + \\ + M_e\{A(x+\xi)A(x-\xi)\} \qquad (III.139)$$

The sum extends over all $\{\rho\}$ with

$$a + \sum \#(\rho)_j \le e$$

d is the dimension of the local operator product

$$d = a + \sum \#(\rho)_j$$

For $\xi \to 0$ the coefficients behave at most like

$$M_{d-1}\{A(\xi)A(-\xi)\} \sim s^{c+1} \lg^{\alpha} s \qquad c = d-3 \qquad (\text{III. }140)$$

Hence the sum of the dimensions of the coefficients E and the corresponding operator product does not exceed two

$$d + (-(d-2)) = 2$$

as required by Wilson's hypothesis. The remainder behaves at most as

$$M_e\{A(x+\xi)A(x-\xi)\} \sim s^{e-2} \lg^{\alpha} s \qquad (\text{III. }141)$$

This implies that (III. 139) gives the short distance behavior of $TA(x_1)A(x_2)$ to any degree of accuracy provided e is chosen large enough.

Formula (III. 139) also gives the singularities of $TA(x+\xi)A(x-\xi)$ near the light cone $\xi^2 = 0$, in agreement with results previously derived by Brandt[51]. It should be noted that an infinite number of terms of the form

$$\xi^{\mu_1} \cdot\cdot\, \xi^{\mu_n} \lg^{\alpha} \xi^2 \; B_d^{\{e\}}(x)$$

contribute to the light cone singularity. The structure of singularities near the light cone thus appears far more complex than near the origin.

With the results of this section it is easy to determine the singularities of the derivative

$$\partial^{x_1}_{(\mu)_1} \partial^{x_2}_{(\mu)_2} T A(x_1) A(x_2) \qquad (III.142)$$

of a time ordered product. The notation used here is

$$(\mu)_j = \mu_{j_1} \cdots \mu_{j_{n(j)}}$$

$$\partial^{x_j}_{(\mu)_j} = \partial^{x_j}_{\mu_{j_1}} \cdots \partial^{x_j}_{\mu_{j_{n(j)}}}$$

Since differentiation with respect to $x = x_1 + x_2$ does not produce new singularities it is convenient to express (III. 142) in terms of derivatives with respect to x and ξ. The singularities of

$$\partial^{\xi}_{(\mu)} T A(x+\xi) A(x-\xi) \qquad (III.143)$$

are then simply obtained by differentiating the expansion (III. 139) with respect to ξ :

$$\partial^{\xi}_{(\mu)} T A(x+\xi) A(x-\xi) = \partial^{\xi}_{(\mu)} E_0(\xi) +$$

$$+ \sum_{a=1}^{e} \sum_{\{\rho\}} \partial^{\xi}_{(\mu)} E_{\{\rho\}}(\xi) B_a^{\{\rho\}}(x) +$$

$$+ \partial^{\xi}_{(\mu)} H_e \{ A(x+\xi) A(x-\xi) \} \qquad (III.144)$$

The coefficients in the expansion of different derivatives (III. 143) appear thus to be related by simple differentiation rules

8. Singularities of $TA(x_1)A(x_2)A(x_3)$

The derivation of Wilson's expansion for products of more than two operators is a rather complicated problem. In this section we restrict ourselves to determining the singularities of $TA(x_1)A(x_2)A(x_3)$, the only case involving more than two operators which will be needed in the following chapter.

Again we use the identity (III. 86), (III. 90) for contributions ,

$$\langle T : A(x_1) A(x_2) A(x_3) : \tilde{A}(q_1) \cdots \tilde{A}(q_N) \rangle^{\Gamma}$$

In momentum space the external momenta of Γ are labeled as indicated in Fig. 18. We decompose the set T of all renormalization parts containing V into

$$T = \sum_{j=1}^{5} T_j$$

where $T_1, \ldots T_5$ are defined as indicated in Figs 18'(1-5). Correspondingly X is decomposed into

$$X = \sum_{j=1}^{5} X_j$$

where for $j = 1, 2, 3$

$$X_j = \sum_{\tau \in T_j} (2\pi)^4 R^{(3)}_{\Delta/\tau} \mathring{R}_{\sigma}$$

and

$$X_4 = \sum_{\tau \in T_4} \frac{(2\pi)^6}{3!} R^{(2)}_{\Delta/\tau} \mathring{R}_{\sigma}$$

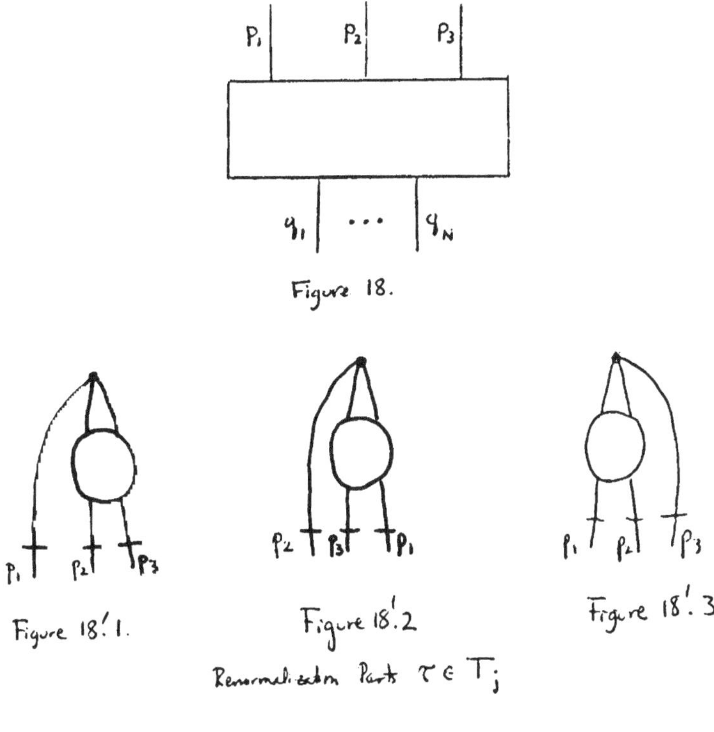

Figure 18.

Figure 18'.1. Figure 18'.2 Figure 18'.3

Renormalization Parts $\tau \in T_j$

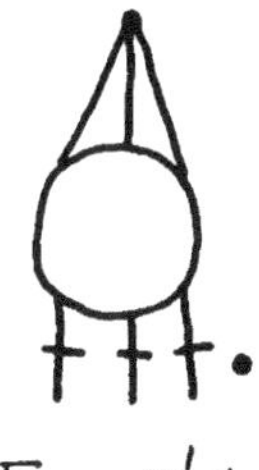

Figure 18'.4.

Figure 18'.5

X_5 is more complicated due to the quadratic superficial divergence of the renormalization parts $\tau \in T_5$.

$$X_5 = \sum_{\tau \in T_5} (2\pi)^2 R_{\Delta/\tau} \mathring{R}_\sigma +$$

$$+ \sum_{\tau \in T_5} (2\pi)^2 (-i)\, p_\rho\, R_{\Delta/\tau}\, \partial^\rho R_\sigma \Big|_{q^\tau = 0} +$$

$$+ \sum_{\tau \in T_5} (2\pi)^2 \frac{(-i)^2}{2}\, p_{\rho_1} p_{\rho_2}\, R_{\Delta/\tau}\, \partial^{\rho_1} \partial^{\rho_2} R_\sigma \Big|_{q^\tau = 0}$$

Note that in Δ/τ the reduced vertex W is connected to the remainder of the diagram by a single line. Accordingly no additional subtractions are taken in $R_{\Delta/\tau}$

After application of the integral operator

$$\int dp_1 dp_2 dp_3\, e^{-i(p_1 x_1 + p_2 x_2 + p_3 x_3)}\, \delta(p + \sum q_j) \lim_{\epsilon \to 0^+} \int dk_1 \cdots dk_m$$

to the identity (III. 86) and summing over all Γ one obtains

$$\langle T : A(x_1) A(x_2) A(x_3) : \tilde{A}(q_1) \cdots \rangle =$$

$$= \langle T N \{ A(x_1) A(x_2) A(x_3) \} \tilde{A}(q_1) \cdots \rangle +$$

$$+ G_1 \langle T N \{ A(x_1) A^2(x_{23}) \} \tilde{A}(q_1) \cdots \rangle + \text{cycle permutations} +$$

$$+ G_4 \langle T N \{ A(x)^3 \} \tilde{A}(q_1) \cdots \rangle +$$

$$+ G_5 \langle T\, A(x)\, \tilde{A}(q_1) \cdots \rangle +$$

$$+ G_6^{\rho}\, \partial_\rho^x \langle T\, A(x)\, \tilde{A}(q_1) \cdots \rangle +$$

$$+ G_7^{\rho_1 \rho_2}\, \partial_{\rho_1}^x \partial_{\rho_2}^x \langle T\, A(x)\, \tilde{A}(q_1) \cdots \rangle \tag{III.145}$$

We are using the variables

$$x_j = x + \xi_j \qquad j = 1, 2, 3$$

$$x = \frac{1}{3} \sum_{j=1}^{3} x_j$$

$$x_j = x_{jk} + \xi_{jk}$$

$$j \neq k$$

$$x_k = x_{jk} - \xi_{jk}$$

The coefficients in (III. 145) are

$$G_1 = \frac{(2\pi)^4}{2} \langle T : A(\xi^{23})\, A(-\xi^{23}) : \tilde{A}(0)^2 \rangle^{EL}$$

$$G_4 = \frac{(2\pi)^6}{3!} \langle T : A(\xi_1) \cdots A(\xi_3) : \tilde{A}(0)^3 \rangle^{EL}$$

$$G_5 = (2\pi)^2 \langle T : A(\xi_1) \cdots A(\xi_3) : \tilde{A}(0) \rangle^{EL}$$

$$G_6^{\rho} = (2\pi)^2(-i)\langle T : A(\xi_1)\cdots A(\xi_3) : \tilde{A}^{\rho}(0)\rangle^{EL}$$

$$G_7^{\rho_1\rho_2} = (2\pi)^2 \frac{(-i)^2}{2} \langle T : A(\xi_1)\cdots A(\xi_3) : \tilde{A}^{\rho_1\rho_2}(0)\rangle^{EL}$$

$$\tilde{A}^{(\rho)}(p) = \partial_p^{(\rho)} \tilde{A}(p) \qquad (\text{III}.146)$$

Setting

$$x_1 = x + \xi, \quad x_2 = x, \quad x_3 = x - \xi$$

and applying the reduction technique to $q_1, \cdots, q_N$ we obtain

$$:A(x+\xi)\,A(x)\,A(x-\xi): =$$

$$= E_1(\xi)\,A(x) + E_2^{\rho}(\xi)\,\partial_\rho A(x) + E_3^{\rho\sigma}(\xi)\,\partial_\rho\partial_\sigma A(x) +$$

$$+ E_4(\xi)\,N\{A(x)^3\} + R(x\,\xi), \qquad (\text{III}.147)$$

with

$$R(x\,\xi) \to 0 \quad \text{for} \quad \xi \to 0 \quad \text{and} \quad \frac{\xi_\mu}{\sqrt{|\xi^2|}} \quad \text{bounded.}$$

The coefficients are

$$E_1 = (2\pi)^2 \langle T : A(\xi)\,A(0)\,A(-\xi) : \tilde{A}(0)\rangle^{PROP}$$

$$E_2^{\rho} = -i\,(2\pi)^2 \langle T : A(\xi)\,A(0)\,A(-\xi) : \tilde{A}^{\rho}(0)\rangle^{PROP}$$

$$E_3^{\rho\sigma} = -\tfrac{1}{2}(2\pi)^2 \langle T : A(\xi)\,A(0)\,A(-\xi) : \tilde{A}^{\rho\sigma}(0)\rangle^{PROP}$$

$$E_4 = \tfrac{1}{6}(2\pi)^6 \langle T : A(\xi)\,A(0)\,A(-\xi) : \tilde{A}(0)^3\rangle^{PROP}$$

$$(\text{III}.148)$$

Here

$$E_2^\rho = 0$$

since it is even in ξ and of the form

$$E_2^\rho = \xi^\rho f(\xi^2)$$

Applying the reduction technique to only some of the $q_1, \cdots, q_N$ we obtain similar expansions for the time-ordered products

$$T : A(x+\xi)\, A(x)\, A(x-\xi) : A(y_1) \cdots A(y_N)$$

The dimensions of the coefficients E_j are

$$d(E_1) = 2$$

$$d(E_3^{\rho\sigma}) = d(E_4) = 0$$

in agreement with Wilson's rule.

IV. Local Field Equations[53]

1. General Remarks.

In the preceding chapter we obtained a general prescription for defining local operator products in perturbation theory. The method will now be applied to set up finite forms of local field equations for renormalizable theories.

We first outline the procedure for the general case. We consider a renormalizable theory with a Lagrangian density

$$\mathcal{L} = \mathcal{L}\{\varphi_r, \partial_\mu \varphi_r\} \qquad (\text{IV}.1)$$

which is a local invariant polynomial in the components of the renormalized fields and their derivatives

$$\varphi_r = (\varphi_{r1}, \cdots \varphi_{rn}) \qquad \partial_\mu \varphi_r = \frac{\partial \varphi_r}{\partial x^\mu}$$

For simplicity **we assume that the coupling terms of** $\mathcal{L}$ contain no derivatives and have dimensionless coupling constants. The interaction part $\mathcal{L}_{INT}$ is defined by

$$\mathcal{L} = \mathcal{L}_0 + \mathcal{L}_{INT} \qquad (\text{IV}\ 2)$$

where $\mathcal{L}_0$ is the free Lagrangian of the renormalized fields. In perturbation theory the time ordered Green's functions of the renormalized fields are given by Gell-Mann Low expansion [43]

$$\langle T \varphi_{j_1}(x_1) \cdots \varphi_{j_n}(x_n) \rangle = \langle T \exp(i \int \mathcal{L}^0_{INT} dz)\, \varphi^c_{j_1}(x_1) \cdots \varphi^c_{j_n}(x_n) \rangle . \qquad (\text{IV}\ 3)$$

where the φ_j^0 denote free fields belonging to $\mathcal{L}_0$ and

$$\mathcal{L}^0_{INT} = \mathcal{L}_{INT}\{\varphi^0_i\}$$

The main theorem of renormalization theory states that in perturbation theory either

(i) the divergencies of the Feynman integrals occurring in (IV.3) can be absorbed into the parameters of the Lagrangian (IV.1) or

(ii) $\mathcal{L}$ can be modified by adding a finite number of counter terms such that all infinities may be absorbed by the parameters of the new Lagrangian.

With the appropriate infinite values of the Lagrangian the Green's functions (IV.3) become finite in perturbation theory.

This renormalization theorem was first proved by Dyson in the case of electrodynamics.[19] The most elegant proof has been given by Bogoliubov.[22] His method even applies to non-renormalizable theories in which case however an infinite number of counter terms must be admitted.

Since a Lagrangian involving divergent constants is not meaningful the theorem is of heuristic value only. We will therefore avoid using the formal Lagrangian of conventional renormalization theory. Instead we will introduce an effective Lagrangian

$$\mathcal{L}_{EFF} = \mathcal{L}_{EFF}\{\varphi_r , \partial_\mu \varphi_r\} \tag{IV.4}$$

which is used for constructing the finite part of the Green's functions. In contrast to $\mathcal{L}$ the effective Lagrangian depends on finite parameters but otherwise has the same form as $\mathcal{L}$. We decompose $\mathcal{L}_{EFF}$ into

$$\mathcal{L}_{EFF} = \mathcal{L}_0 + \mathcal{L}_{INT} \quad , \quad \mathcal{L}_{INT} = \mathcal{L}_1 + \mathcal{L}_2 \tag{IV.5}$$

where $\mathcal{L}_0$ is the free Lagrangian of the renormalized fields, $\mathcal{L}_1$ is the interaction part consisting of non-derivative terms with dimensionless constants, and $\mathcal{L}_2$ is a sum of bilinear counter-terms with finite coefficients. The Greens functions of the theory are then defined by[43]

$$\langle T \varphi_{j_1}(x_1) \cdots \varphi_{j_n}(x_n) \rangle = \mathrm{Fin} \langle T e^{i\int \mathcal{L}^0_{INT} dz} \varphi^0_{j_1}(x_1) \cdots \varphi^0_{j_n}(x_n) \rangle . \tag{IV.6}$$

The finite part of the Gell-Mann Low series is constructed according to the rules given in the preceding chapter. Though all subtractions are made at zero momentum the poles and residues of the propagator can be adjusted by properly choosing the parameters of $\mathcal{L}_2$.

Definition (IV.6) implies the following relations

$$N\left\{\frac{\partial \mathcal{L}}{\partial \varphi_{r\alpha}}\right\} - N\left\{\frac{\partial}{\partial x^\mu} \frac{\partial \mathcal{L}}{\partial(\partial_\mu \varphi_{r\alpha})}\right\} = 0 \tag{IV.7}$$

or

$$\frac{\partial(\mathcal{L}_0 + \mathcal{L}_2)}{\partial \varphi_{r\alpha}} - \frac{\partial}{\partial x^\mu} \frac{\partial(\mathcal{L}_0 + \mathcal{L}_2)}{\partial(\partial_\mu \varphi_{r\alpha})} = -N\left\{\frac{\partial \mathcal{L}_1}{\partial \varphi_{r\alpha}}\right\} \tag{IV.8}$$

Hère N indicates that the occurring operator products should be taken as normal products. We do not give a general proof of relation (IV.8). The validity of the relation will be obvious for the examples discussed in the following sections.

(IV.8) is a linear relation between the kinetic terms of the field equation and the normal products of the interaction terms. Equation (IV.8) may thus be interpreted as a finite form of the field equation. In order to obtain an equation for the fields alone the normal products must be expressed in terms of the field operator. To this end one takes the operator products occurring in $\frac{\partial \mathcal{L}_1}{\partial \varphi_{r\alpha}}$ at different points and applies Wilson's expansion to the corresponding time ordered products. In this way one can write $N\left\{\frac{\partial \mathcal{L}_1}{\partial \varphi_{r\alpha}}\right\}$ as a limit of operator products involving renormalization functions which may become singular at the origin.

In the following section the method will be illustrated in the models of A^4-coupling, pseudoscalar meson-nucleon interaction and neutral vector meson theories, including quantum electrodynamics.

2. A^4-Coupling

2a. Field Equations in General Form

The effective Lagrangian and the Feynman rules for the model of A^4-coupling were given in section II.5. It is easy to derive the following system of field equations for the time ordered Green's functions

$$
\begin{aligned}
&-(\Box_x + m^2)\langle T\, A(x)\, A(y_1)\cdots A(y_N)\rangle = \\
&= \lambda \langle T\, N\{A(x)^3\}\, A(y_1)\cdots A(y_N)\rangle - \lambda a \langle T\, A(x)\, A(y_1)\cdots A(y_N)\rangle + \\
&\quad + \lambda b\, (\Box_x + m^2)\langle T\, A(x)\, A(y_1)\cdots A(y_N)\rangle + \\
&\quad + i \sum_{j=1}^{N} \delta(x - y_j)\langle T\, A(x)\, A(y_1)\cdots \not{A}(y_j) \cdots A(y_N)\rangle .
\end{aligned}
\qquad (IV.9)
$$

We recall that the Green's functions of $N\{A^3(x)\}$ were defined by the finite part of the Gell-Mann Low series (equ. (III.58))

$$
\begin{aligned}
&\langle T\, N\{A(x)^3\}\, A(y_1)\cdots A(y_N)\rangle = \\
&= \mathrm{Fin} \langle T\, e^{i\int \mathcal{L}^0_{INT} dz}\, A_0(x)^3 A_0(y_1)\cdots A_0(y_N)\rangle_c
\end{aligned}
\qquad (IV.10)
$$

The first term on the left hand side of equ. (IV.9) comes from diagrams where the vertex x is connected by a line to a 4-vertex

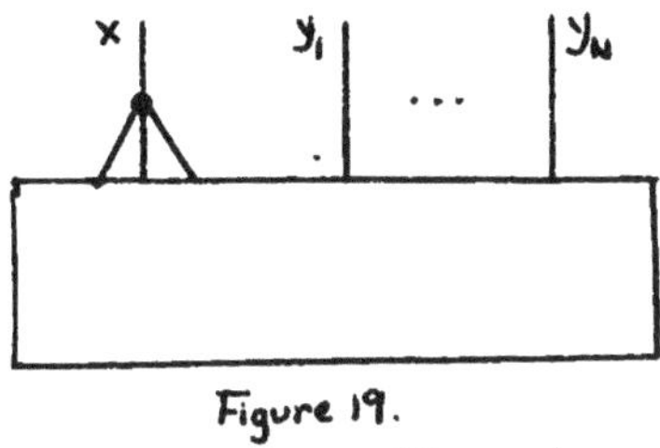

Figure 19.
X Connected To Far-vertex.

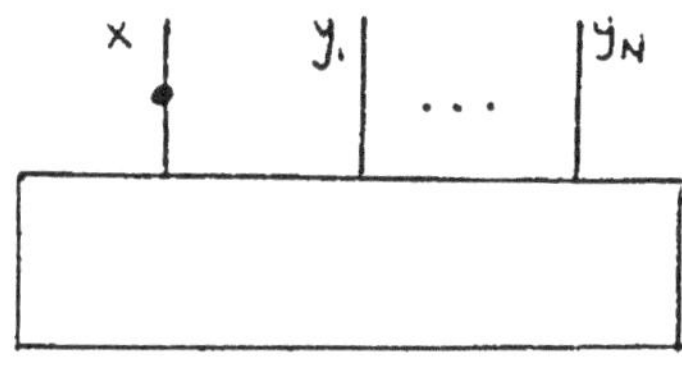

Figure 20.
X Connected To Two-vertex

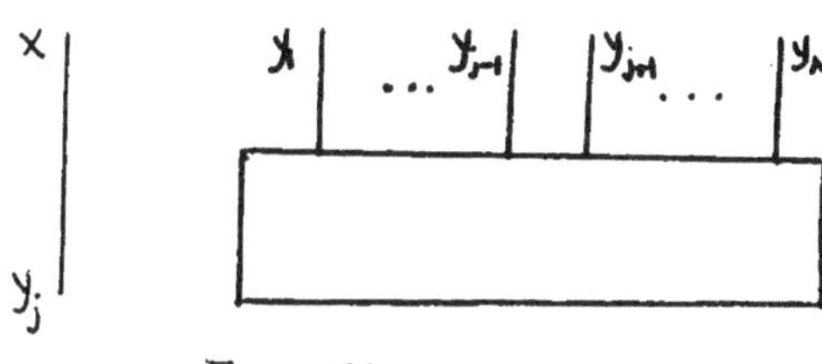

Figure 21.
X Directly Connected To Coordinate y_j

(Fig. 19). The second and third term represent the contributions from diagrams of type Fig. 20. The last term is due to diagrams where the vertex x is connected by a line to a vertex y_j (Fig. 21).

Applying the reduction formulae to some of the y_j we obtain (IV.9) in operator form

$$-(\Box_x + m^2)\, T\, A(x) A(y_1) \cdots A(y_N) =$$

$$= \lambda\, T\, N\{A(x)^3\} A(y_1) \cdots A(y_N) - \lambda a\, T\, A(x) A(y_1) \cdots A(y_N) +$$

$$+ \lambda b\, (\Box_x + m^2)\, T\, A(x) A(y_1) \cdots A(y_N) +$$

$$+ i \sum_{j=1}^{N} \delta(x - y_j)\, T\, A(x) A(y_1) \cdots \cancel{A(y_j)} \cdots A(y_N) \qquad \text{(IV.11)}$$

For N = 1 we have the field equation

$$-(\Box + m^2) A(x) = \lambda \{ N A(x)^3 - a A(x) + b (\Box + m^2) A(x) \} \qquad \text{(IV.12)}$$

which is equ. (IV.8) for the Lagrangian (II.62). So far (IV.12) is a linear relation between NA^3 and the field A . NA^3 can be expressed in terms of the field A by using the Wilson expansion (equ. (III.147)) of

$$P(x\, \xi) = :A(x+\xi) A(x) A(x-\xi):$$

$$P(x\, \xi) \sim E_1(\xi) A(x) + E_2^{\mu}(\xi) \partial_\mu A(x) +$$

$$+ E_3^{\mu\nu}(\xi) \partial_\mu \partial_\nu A(x) + E_4(\xi) N\{A(x)^3\} \qquad \text{(IV.13)}$$

up to terms which vanish for $\xi \to 0$ provided $\frac{\xi^\mu}{\sqrt{|\xi^2|}}$ is bounded. We further have the following expansion of time ordered products $T\, P(x\xi)\, A(y_1)\cdots A(y_N) \sim$

$$\sim E_1(\xi)\, T\, A(x) A(y_1)\cdots A(y_N) +$$
$$+ E_2^{\mu}(\xi)\, \partial_\mu^x\, T\, A(x)\, A(y_1)\cdots A(y_N) +$$
$$+ E_3^{\mu\nu}(\xi)\, \partial_\mu^x \partial_\nu^x\, T\, A(x)\, A(y_1)\cdots A(y_N) +$$
$$+ E_4(\xi)\, T\, N\{A(x)^3\}\, A(y_1)\cdots A(y_N) \qquad (IV.14)$$

In a given order of perturbation theory the functions E_j behave no stronger than

$$E_1 \sim \frac{\lg^\alpha s}{s^2} \qquad E_2 \sim \frac{\lg^\alpha s}{s} \qquad s = \sqrt{|\xi^2|}$$
$$E_3 \sim \lg^\alpha s \qquad E_4 \sim \lg^\alpha s \qquad (IV.15)$$

for $\xi \to 0$ with ξ^μ/s bounded.

Dividing (IV.13) by E_4 and taking the limit $\xi \to 0$ we can express $N\{A^3\}$ in terms of the field[54]

$$N\{A^3\} = \lim_{\xi\to 0} \frac{P - E_1 A - E_2^{\mu}\partial_\mu A - E_3^{\mu\nu}\partial_\mu\partial_\nu A}{E_4} \qquad (IV.16)$$

The relations (IV.12) and (IV.13) represent the dynamics of the system. Combined with the general principles of quantum field theory (IV.12) and (IV.13) give a description of the model of A^4-coupling in finite terms.[65]

The remaining part of section 2 will deal with the technical problem of finding a more convenient form of the current operator (IV.16). First the renormalization conditions of the theory will be worked out in section 2.b. This will lead to a more natural definition of the normal product $\{A^3\}$. In section 2.c the ξ-dependence of the coefficients E_j will be investigated. Finally a limit form of the field equation is obtained which will be compared with the formal field equation of conventional renormalization theory (Section 2d).

2b. Renormalization Conditions.

Since the subtractions for vertex parts were made at zero momentum we expect the renormalized coupling constant λ to be related to the zero momentum value of the vertex function. In order to find the precise relation we introduce the vertex function

$$\Lambda(p_1 \cdots p_4) = (2\pi)^4 \frac{\langle T\, \tilde{A}(p_1) \cdots \tilde{A}(p_4) \rangle^{\text{TRUNC}}}{\hat{\Delta}'_F(p_1) \cdots \hat{\Delta}'_F(p_4)} \qquad (\text{IV}.17)$$

The perturbative expansion is

$$\Lambda(0000) = -6i\lambda + \sum_{\Gamma} \frac{1}{\gamma(\Gamma)} \Lambda_{\Gamma}(0000)$$

The sum extends over all non-trivial proper vertex parts Γ with four (amputated) external lines. Each Γ contributes

$$\Lambda_{\Gamma}(p_1 \cdots p_4) = \int dk_1 \cdots dk_m\, R_{\Gamma}(p_1 \cdots p_4\, k_1 \cdots k_m)$$

where

$$R_{\Gamma} = \bar{R}_{\Gamma} - \bar{R}_{\Gamma}\Big|_{p_j = 0}$$

This implies $\quad \Lambda_{\Gamma}(0000) = 0.$

The renormalized coupling constant is thus given by

$$\Lambda(0000) = -6i\lambda \qquad \text{(IV.18)}$$

Recursive formulae for the constants a and b were already obtained in Section II. (Equ. II.76). We will now derive a more elegant form of these conditions which does not directly refer to perturbation theory. To this end we use the field equation of the propagator (equ. (IV.11) for N = 1)

$$-(\Box_x + m^2)\,\Delta_F'(x-y) = \lambda\langle T\,N\{A(x)^3\}A(y)\rangle +$$

$$-\lambda\,(a - b(\Box_x - m^2))\,\Delta_F'(x-y) + i\,\delta(x-y)$$

Taking the Fourier transform with respect to y we get

$$(p^2 - m^2)\,\hat{\Delta}_F'(p) = \lambda(2\pi)^2\langle T\,N\{A(0)^3\}\tilde{A}(p)\rangle +$$

$$-\lambda\,(a + b(p^2 - m^2))\,\hat{\Delta}_F'(p) + i \qquad \text{(IV.19)}$$

We recall that the conditions

$$\Pi(m^2) = 0 \qquad \left.\frac{\partial\Pi}{\partial p^2}\right|_{p^2=m^2} = 0 \qquad \text{(IV.20)}$$

are equivalent to the requirement

$$(p^2 - m^2)\,\hat{\Delta}_F'(p) = 0 \quad \text{at} \quad p^2 = m^2 \qquad \text{(IV.21)}$$

if Π is defined by

$$\hat{\Delta}_F' = \hat{\Delta}_F + \hat{\Delta}_F\,\Pi\,\hat{\Delta}_F'$$

For Π we obtain from (IV.19)

$$i\Pi\hat{\Delta}_F' = (p^2 - m^2)\hat{\Delta}_F' - i = \qquad \text{(IV.22)}$$

$$= \lambda(2\pi)^2\langle T\,N\{A(0)\}^3\tilde{A}(p)\rangle - \lambda\,(a + b(p^2 - m^2))\,\hat{\Delta}_F'(p)$$

The proper part $\langle T\,N\{A(0)\}^3\tilde{A}(p)\rangle^{PROP}$ is defined by

$$\langle T\,N\{A(0)\}^3\tilde{A}(p)\rangle = \langle T\,N\{A(0)^3\}\tilde{A}(p)\rangle\;\hat{\Delta}_F'(p)$$

according to (III. 46). Hence dividing (IV. 21) by $\hat{\Delta}_F'$ we obtain

$$i\Pi = \lambda \left\{ (2\pi)^2 \langle TN\{A(0)^3\} \tilde{A}(p) \rangle^{PROP} - a - b(p^2 - m^2) \right\}$$

This shows that (IV. 20) can be satisfied by putting

$$a = (2\pi)^2 \langle TN\{A(0)^3\} \tilde{A}(p) \rangle^{PROP}_{p^2=m^2} \; ; \; b = \frac{\partial}{\partial p^2} \langle TN\{A(0)^3\} \tilde{A}(p) \rangle^{PROP}_{p^2=m^2} \qquad \text{(IV. 23)}$$

As follows from (III. 82) the normal product $N\{A^3\}$ satisfies the normalization conditions

$$\langle TN\{A(0)^3\} \tilde{A}(0) \rangle^{PROP} - \partial_p^{\nu} \partial_{\nu}^{p} \langle TN\{A(0)^3\} \tilde{A}(p) \rangle^{PROP}_{p=0} = 0$$

$$\langle TN\{A(0)^3\} \tilde{A}(0)^3 \rangle^{PROP} = \frac{6}{(2\pi)^6} \qquad \text{(IV. 24)}$$

It is convenient to introduce a new normal product which has normalization conditions on the mass shell. To this end we define the operator product

$$\{A(x)^3\} = N\{A(x)^3\} - a\,A(x) + b(\Box + m^2)\,A(x) \qquad \text{(IV. 25)}$$

and time ordered products

$$T\{A(x)^3\}\, A(y_1) \cdots A(y_N) = TN\{A(x)^3\}\, A(y_1) \cdots A(y_N) +$$

$$- \left(a - b(\Box_x + m^2)\right) T\, A(x)\, A(y_1) \cdots A(y_N) \qquad \text{(IV. 26)}$$

The new operator product satisfies the following normalization conditions

$$\langle T\{A(0)^3\} \tilde{A}(p) \rangle^{PROP}_{p^2=m^2} = 0$$

$$\frac{\partial}{\partial p^2} \langle T\{A(0)^3\} \tilde{A}(p) \rangle^{PROP}_{p^2=m^2} = 0 \qquad \text{(IV. 27)}$$

$$\langle T\{A(0)^3\} \tilde{A}(0)^3 \rangle^{PROP} = \frac{6}{(2\pi)^6} \qquad \text{(IV. 28)}$$

as follows from (IV.23-25). In (IV.28) we used that

$$\langle T\, A(x)\, \tilde{A}(0)^3 \rangle^{PROP} = 0$$

because of (III.43).

With the modified operator product $\{A^3\}$ the field equations take the particularly simple form

$$-(\Box + m^2)\, A(x) = \lambda \{A(x)^3\} \qquad (IV.29)$$

$$-(\Box_x + m^2)\, T\, A(x)\, A(y_1) \cdots A(y_N) = \lambda T \{A(x)^3\} A(y_1) \cdots A(y_N) +$$

$$+ i \sum_{j=1}^{N} \delta(x - y_j)\, T\, A(x)\, A(y_1) \cdots \widehat{A(y_j)} \cdots A(y_N) \qquad (IV.30)$$

with m and λ being the renormalized mass and coupling constant.

2c. The Current Operator

In the expansion (IV.13) we replace $N\{A^3\}$ by the new operator product $\{A^3\}$ (equ. (IV.25))

$$P(x\xi) \sim E_1{}' A + E_2{}'^{\mu} \partial_\mu A + E_3{}'^{\mu\nu} \partial_\mu \partial_\nu A +$$

$$+ E_4{}' \{A^3\} \qquad (IV.31)$$

We want to express the coefficients by Green's functions of A . It is not necessary to use the formulae for E_j which we derived in perturbation theory. The coefficients $E_j{}'$ can be found directly from (IV.31) using the normalization conditions of $\{A^3\}$.

We first determine $E_4{}'$ from

$$\langle T\, P(x\xi)\, \tilde{A}(0)^3 \rangle^{PROP} \sim E_1{}' \langle T\, A(x)\, \tilde{A}(0)^3 \rangle^{PROP} +$$

$$+ E_2{}'^{\mu} \partial_\mu^x \langle T\, A(x)\, \tilde{A}(0)^3 \rangle^{PROP} + E_3{}'^{\mu\nu} \partial_\mu^x \partial_\nu^x \langle T A(x)\, \tilde{A}(0)^3 \rangle^{PROP} +$$

$$+ E_4{}' \langle T \{A(x)^3\}\, \tilde{A}(0)^3 \rangle^{PROP}$$

which follows by taking the proper part of the expansion of $\langle T\, P(x\xi)\tilde{A}(0)^3\rangle$.

With (III. 43), (IV. 28) we get

$$E_4'(\xi) \sim \gamma(\xi^2) = \frac{(2\pi)^6}{6} \langle T P(0\xi)\tilde{A}(0)^3\rangle^{PROP} \qquad (IV.32)$$

In order to determine E_1', E_2' and E_3' we consider the expansion of

$$\langle T\, P(x\xi)\tilde{A}(p)\rangle^{PROP} \sim E_1'\langle T A(x)\tilde{A}(p)\rangle^{PROP} +$$

$$+ E_2'^{\mu}\partial_\mu^x \langle T A(x)\tilde{A}(p)\rangle^{PROP} + E_3'^{\mu\nu}\partial_\mu^x\partial_\nu^x\langle T A(x)\tilde{A}(p)\rangle^{PROP} +$$

$$+ \gamma \langle T\{A(x)^3\}\tilde{A}(p)\rangle^{PROP}$$

with (III. 42) we get

$$(2\pi)^2\langle T\, P(0\xi)\tilde{A}(p)\rangle^{PROP} \sim E_1' + i p_\mu E_2'^{\mu} +$$

$$- p_\mu p_\nu E_3'^{\mu\nu} + (2\pi)^2\gamma\langle T\{A(0)^3\}\tilde{A}(p)\rangle^{PROP} \qquad (IV.33)$$

Setting $p = 0$ we find

$$E_1'(\xi) \sim E_1''(\xi^2)$$

$$E_1'' = (2\pi)^2\langle T P(0\xi)\tilde{A}(0)\rangle^{PROP} - (2\pi)^2\gamma\langle T\{A(0)^3\}\tilde{A}(0)\rangle^{PROP} \qquad (IV.34)$$

Differentiating with respect to p^μ we find

$$i E_2'^{\mu} \sim (2\pi)^2 \partial_\mu^p \langle T P(0\xi)\tilde{A}(p)\rangle^{PROP}_{p=0} +$$

$$- (2\pi)^2\gamma\,\partial_\mu^p\langle T\{A(0)^3\}\tilde{A}(0)\rangle^{PROP}_{p=0}$$

For invariance reasons the right hand side is of the form

$$\xi_\mu f(\xi^2)$$

Since on the other hand $P(0,\xi)$ is even in ξ it follows

$$i E_2'^{\mu} \sim 0 \qquad (IV.35)$$

Hence $E_2'^{\mu}$ vanishes in the limit.

Finally

$$E_3'^{\mu\nu} \sim -\frac{1}{2}(2\pi)^2 \partial_p^{\mu} \partial_p^{\nu} \langle T\, P(0\xi)\tilde{A}(p)\rangle^{PROP}_{p=0} +$$

$$+ \frac{1}{2}\delta(\xi^2)\, \partial_p^{\mu} \partial_p^{\nu} \langle T\{A(0)^3\}\tilde{A}(p)\rangle^{PROP}_{p=0}$$

For invariance reasons the right hand side is of the form

$$E_3'^{\mu\nu}(\xi) \sim g^{\mu\nu} E_3''(\xi^2) - \frac{\xi^{\mu}\xi^{\nu}}{\xi^2}\sigma(\xi^2) \qquad (IV\ 36)$$

Inserting the results (IV.34-36) back into expansion (IV.33) we find

$$(2\pi)^2 \langle T\, P(0\xi)\tilde{A}(p)\rangle^{PROP} \sim E_1''(\xi^2) - p^2 E_3''(\xi^2) +$$

$$+ \frac{(\xi p)^2}{\xi^2}\sigma(\xi^2) + \delta(\xi^2) f(p^2),$$

$$f(p^2) = \langle T\{A(0)^3\}\tilde{A}(p)\rangle^{PROP}$$

Hence

$$(2\pi)^2 \langle T : A(\xi)\, A(0)\, A(-\xi) : \tilde{A}(p)\rangle^{PROP}$$

consists of a directional dependent part

$$\sim \frac{(\xi p)^2}{\xi^2}\sigma(\xi^2)$$

and an invariant part

$$(2\pi)^2 \, \mathrm{Inv} < T\, P(o\xi)\tilde{A}(p)>^{PROP} =$$

$$= E_1''(\xi^2) - p^2 E_3''(\xi^2) + \gamma(\xi^2) f(p^2) \qquad (\mathrm{IV}.37)$$

which depends on ξ^2 and p^2 only. We may therefore define

$$\alpha(\xi^2) = (2\pi)^2 \, \mathrm{Inv} < T\, P(o\xi)\tilde{A}(p)>^{PROP}_{p^2=m^2} \qquad (\mathrm{IV}.38)$$

$$\beta(\xi^2) = (2\pi)^2 \frac{\partial}{\partial p^2} \mathrm{Inv} < T\, P(o\xi)\tilde{A}(p)>^{PROP}_{p^2=m^2} \qquad (\mathrm{IV}.39)$$

Then

$$\alpha = E_1'' - m^2 E_3'' \ , \quad \beta = - E_3'' \qquad (\mathrm{IV}.40)$$

because of the normalization conditions (IV.27). Combining (IV.32), (IV.34-36) with (IV.40) we have as final expressions for the coefficients E_j'

$$E_1' \sim \alpha - m^2\beta \qquad E_4' \sim \gamma$$

$$E_2'^{\mu} \sim 0$$

$$E_3'^{\mu\nu} \sim -g^{\mu\nu}\beta - \frac{\xi^\mu\xi^\nu}{\xi^2}\sigma$$

Hence the expansion (IV.31) becomes

$$P \sim \alpha A - \beta(\Box + m^2) A +$$

$$- \frac{\xi^\mu \xi^\nu}{\xi^2} \sigma \, \partial_\mu \partial_\nu A + \gamma \{A^3\} \qquad (\mathrm{IV}.41)$$

Dividing by γ **we get the following limit relation for the current** operator

$$\{A^3\} = \lim_{\xi \to 0} \frac{P - \alpha A + \beta(\Box + m^2)A - \frac{\xi^\mu\xi^\nu}{\xi^2}\sigma\,\partial_\mu\partial_\nu A}{\gamma} \qquad (\mathrm{IV}.42)$$

(IV.29) and (IV.42) provide a field equation for A alone.

However, it is not quite satisfactory that second order time derivatives of A appear in the current operator. In the directional dependent term this can be avoided by setting $\xi_0 = 0$. In order to eliminate the wave operator in (IV.42) we go back to (IV.41) and insert (IV.29) for $(\Box + m^2)A$

$$P \sim \alpha A - \frac{\xi^\mu \xi^\nu}{\xi^2}\sigma + (\lambda\beta + \gamma)\{A^3\} \qquad (\text{IV}.43)$$

Dividing by $\lambda\beta + \gamma$ we obtain a more suitable expression for the current

$$\{A^3\} = \lim_{\xi \to 0} \frac{P - \alpha A + \frac{\xi^\mu \xi^\nu}{\xi^2}\sigma\, \partial_\mu \partial_\nu A}{\lambda\beta + \gamma} \qquad (\text{IV}.44)$$

Here it was used that $(\lambda\beta + \gamma)^{-1}$ diverges logarithmically while the remainder of (IV.43) vanishes linearly in the limit up to logarithmic factors. Similar formulae follow for the time ordered products

$$T\{A(x)^3\}A(y_1)\cdots A(y_N) =$$

$$= \lim_{\xi \to 0} \frac{P_T - \alpha A_T + \sigma^{\mu\nu}\partial_\mu^x \partial_\nu^x A_T}{\lambda\beta + \gamma}$$

$$P_T = T\, P(x\,\xi)\,A(y_1)\cdots A(y_N) \qquad (\text{IV}.45)$$

$$A_T = T\, A(x)\,A(y_1)\cdots A(y_N)$$

2d. Discussion of the Field Equation

With the expression (IV. 44) for the current operator the field equation (IV. 29) takes the form

$$-(\Box + m^2)\, A(x) =$$

$$= \lambda \lim_{\xi \to 0} \frac{P(x\,\xi) - \alpha(\xi^2)\, A(x) + \sigma_{\mu\nu}(\xi)\, \partial^\mu \partial^\nu A(x)}{f(\xi^2)}$$

$$P(x\,\xi) = :A(x+\xi)\, A(x)\, A(x-\xi): \qquad \text{(IV 46)}$$

$$\sigma_{\mu\nu} = \frac{\xi_\mu \xi_\nu}{\xi^2}\, \sigma(\xi^2) \qquad\qquad f = \lambda\beta + \gamma$$

The functions α , β , γ and σ are given by (IV. 38), (IV. 39), (IV. 32) and (IV. 36).

It does not seem quite satisfactory that the field equation (IV. 46) involves a time ordered product. There are several ways of eliminating the time ordered product. One possibility is to restrict the limit to spacelike ξ . In that case P may be replaced by

$$Q(x\,\xi) = A(x_1)\, A(x_2)\, A(x_3) - \langle A(x_1) A(x_2) \rangle\, A(x_3) - \text{cyclic permutations}$$

$$x_1 = x + \xi\ , \quad x_2 = x\ , \quad x_3 = x - \xi$$

On the other hand there is a field equation similar to (IV. 46) which uses the anti-time ordered product. Using both types of field equations, depending on the sign of ξ_0 , one obtains a new form of the field equation which involves ordinary operator products only

$$-(\Box + m^2)\, A(x) =$$
$$= \lambda \lim_{\xi \to 0} \frac{Q(x\xi) - \alpha'(\xi) A(x) + \sigma'_{\mu\nu}(\xi)\, \partial^\mu \partial^\nu A(x)}{f'(\xi)}$$

It should be remarked that there is an ambiguity in defining β depending on how the directional dependent singularities are separated in equ. (IV.36). Another possibility would be to set

$$E_3'^{\mu\nu} \sim g^{\mu\nu} \rho'(\xi^2) + \sigma'^{\mu\nu}(\xi^2)$$

with

$$\sigma'^{\mu\nu} = \left(\frac{\xi^\mu \xi^\nu}{\xi^2} - \frac{g^{\mu\nu}}{4} \right) \sigma(\xi^2)$$

This definition of the directional dependent term is distinguished by the property that $\sigma'^{\mu\nu}$ is a traceless tensor.

Our choice (IV.36) has the property that for special vectors

$$\xi = (0, \xi_1, \xi_2, \xi_3)$$

no time derivatives of the field occur in the formula for the current. This may be more convenient for a discussion of the field equation.

Combining (IV.26) and (IV.45) field equations for the time ordered products follow

$$-(\Box_x + m^2)\, T A(x) A(y_1) \ldots A(y_N) =$$
$$= \lambda \lim_{\xi \to 0} \frac{N(x\xi y_1 \ldots y_N)}{\lambda \beta(\xi^2) + \gamma(\xi^2)} \qquad (IV.47)$$

$$N = T\,P(x\xi)\,A(y_1)\cdots A(y_N) +$$

$$+\left(\sigma_{\mu\nu}(\xi)\,\partial_x^{\mu}\partial_x^{\nu} - \alpha(\xi^2)\right) T\,A(x)\,A(y_1)\cdots A(y_N) +$$

$$+\, i \sum_j \gamma(\xi^2)\,\delta(x-y_j)\,T\,A(x)\,A(y_1)\cdots \widehat{A(y_j)}\cdots A(y_N)\,.$$

It is interesting to compare this with the formal field equation of conventional renormalization theory. The formal Lagrangian of A^4-coupling is

$$\mathcal{L} = \tfrac{1}{2}\,\partial_\mu A_u\,\partial^\mu A_u - \tfrac{1}{2} m_0^2 A_u^2 - \tfrac{1}{4}\lambda_0 A_u^4$$

where A_u denotes the unrenormalized field satisfying the following field and commutation relations

$$-(\Box + m_0^2)\,A_u = \lambda_0 A_u^3$$

$$\left.\begin{aligned} &[A_u(x), \dot A_u(y)] = i\,\delta_3(x-y) \\ &[A_u(x), A_u(y)] = [\dot A_u(x), \dot A_u(y)] = 0 \end{aligned}\right\}\ \text{at } x_0 = y_0$$

Renormalized quantities are introduced by

$$m^2 = m_0^2 + \delta m^2 \qquad A_u = Z_3^{1/2} A \qquad \lambda_0 = Z_4 Z_3^{-2}\lambda$$

In terms of the renormalized quantities field equation and commutation relations read

$$\begin{aligned} &-(\Box + m^2)\,A = Z_4 Z_3^{-1}\lambda A^3 - \delta m^2 A \\ &[A(x), \dot A(y)] = i\,Z_3^{-1}\delta_3(x-y) \quad \text{for } x_0 = y_0 \end{aligned} \qquad (\text{IV}.48)$$

The field equations of the time ordered products are

$$-(\Box_x + m^2)\, T A(x) A(y_1) \cdots A(y_N) =$$

$$= Z_4 Z_3^{-1} \lambda\ T\ A(x)^3 A(y_1) \cdots A(y_N) +$$

$$- \delta m^2\ T A(x) A(y_1) \cdots A(y_N) +$$

$$+ i Z_3^{-1} \sum_{j=1}^{N} \delta(x - y_j)\, T A(y_1) \cdots \not{A}(y_j) \cdots A(y_N)$$

$$\text{(IV.49)}$$

Comparing these equations with (IV. 46-47) we can express the renormalization constants by the values of the functions α , β and γ at the origin

$$Z_3 = 1 + \lambda \frac{\beta}{\gamma}\ , \quad Z_4 = \gamma^{-1}\ , \quad \delta m^2 = \lambda \frac{\alpha}{\lambda\beta + \gamma} \qquad \text{(IV.50)}$$

In perturbation theory these values are infinite, but nothing can be said at present how the renormalization functions behave for the exact solutions. Introducing ξ -dependent renormalization functions Z_3 , Z_4 and δm^2 by (IV.50) we may rewrite (IV.46) in the form

$$-(\Box + m^2) A(x) = \lim_{\xi \to 0} \{ Z_4(\xi^2) Z_3^{-1}(\xi^2) \lambda P(x\xi) +$$

$$+ (\tau_{\mu\nu}(\xi) \partial^\mu \partial^\nu - \delta m^2(\xi^2)) A(x) \} \qquad \text{(IV.51)}$$

$$\tau_{\mu\nu} = \frac{\xi_\mu \xi_\nu}{\xi^2} \tau(\xi^2) \qquad \tau = \frac{\sigma}{\gamma}$$

Similar relations hold for time ordered products.

It has not been studied rigorously how the wave function renormalization function $Z_3(\xi)$ is related to the constant appearing in the sum rule

$$Z_3^{-1} = 1 + \int_{4m^2}^{\infty} d\kappa^2 \rho(\kappa^2) \qquad 0 \leq Z_3 \leq 1$$

derived from the Källen-Lehmann representation of the propagator with weight function ρ One should expect the following relation to hold

$$\lim_{\xi \to 0} Z_3(\xi^2) = \lim_{M \to \infty} \frac{1}{1 + \int_{4m^2}^{M^2} d\kappa^2 \rho(\kappa^2)}$$

However, the ambiguity in defining β should be noted in this context. The condition $Z_3 \leq 1$ implies that the ratio

$$\lim_{\xi \to 0} \frac{\beta(\xi)}{\gamma(\xi)}$$

must be finite for the exact solution. In terms of β and γ the condition for divergent wave function renormalization is

$$\lim_{\xi \to 0} \frac{\beta(\xi)}{\gamma(\xi)} = - \frac{1}{\lambda}$$

We finally comment on the form (IV.42) of the current which we derived first. In terms of the renormalization functions Z_3, Z_4 and δm^2 the field equation formed with this current is

$$-(\Box + m^2)A(x) = \lim_{\xi \to 0} \{ \lambda Z_4 P - \delta m^2 A + (Z_3 - 1)(\Box + m^2) A + \\ + \tau_{\mu\nu} \partial^\mu \partial^\nu A \} \qquad (IV.52)$$

If Z_3^{-1} is divergent we have

$$\lim_{\xi \to 0} Z_3(\xi) = 0$$

and (IV.51) is equivalent to

$$\lim_{\xi \to 0} \{ \lambda Z_4(\xi) P(x,\xi) - Z_3(\xi) \delta m^2(\xi) A(x) + \tau_{\mu\nu}(\xi) \partial^\mu \partial^\nu A(x) \} = 0 \qquad (IV.53)$$

Hence for divergent wave function renormalization the equation (IV.52) does not contain the kinetic term in a non-trivial way. (IV.53) is certainly a valid equation which also follows from (IV.51). But it is not directly possible to recover (IV.51) from (IV.52-53) since this amounts to dividing by a function Z_3 which vanishes in the limit.

3. Pseudoscalar Meson-Nucleon Interaction.

In this section we discuss the derivation of field equations for the model of pseudo-scalar meson-nucleon interaction. A new feature will occur for the meson field equation where two independent finite currents A^3 and $\bar{\psi} \gamma_5 \psi$ can be defined. We begin with the notation of conventional renormalization theory.

3a. Formal Lagrangian.

The formal Lagrangian for pseudo-scalar meson-nucleon

interaction is

$$\mathcal{L} = \tfrac{1}{2}\partial_\mu A_u \partial^\mu A_u \quad \tfrac{1}{2} m_0^2 A_u^2 \; +$$
$$+ \tfrac{1}{2} i \left(\bar{\psi}_u \gamma^\mu \partial_\mu \psi_u - \partial_\mu \bar{\psi}_u \gamma^\mu \psi_u \right) - M_0 \bar{\psi}_u \psi_u +$$
$$- g_0 \bar{\psi}_u \gamma_5 \psi_u A_u - \tfrac{1}{4} \lambda_0 A_u^4$$

which leads to the field equations and commutation relations

$$(i\gamma_\mu \partial^\mu - M_0)\psi_u = g_0 A_u \gamma_5 \psi_u$$
$$-(\Box + m_0^2) A_u = \lambda_0 A_u^3 + g_0 \bar{\psi}_u \gamma_5 \psi_u$$
$$\left.\begin{aligned} &[A_u(x), \dot{A}_u(y)] = i\delta_3(x-y) \\ &\{\psi_{u\alpha}(x), \bar{\psi}_{u\beta}(y)\} = i\gamma^0_{\alpha\beta}\delta_3(x-y) \end{aligned}\right\} x_0 = y_0$$

The remaining (anti-) commutators vanish at equal times.

Renormalized quantities are introduced by

$$m^2 = m_0^2 + \delta m^2 \qquad M = M_0 + \delta M$$
$$A = Z_3^{-1/2} A_u \qquad \psi = Z_2^{-1/2}\psi_u$$
$$g = Z_1^{-1} Z_2 Z_3^{1/2} g_0 \qquad \lambda = Z_3^2 Z_4^{-1} \lambda_0$$

In terms of the renormalized quantities field equations and commutation relations become

$$(i\gamma\partial - M)\psi = Z_1 Z_2^{-1} g A\gamma_5\psi - \delta M\psi$$

$$-(\Box + m^2)A = Z_4 Z_3^{-1}\lambda A^3 + Z_1 Z_3^{-1} g\bar{\psi}\gamma_5\psi - \delta m^2 A$$

$$\left.\begin{aligned}[A(x), \dot{A}(y)] &= i Z_3^{-1}\delta_3(x-y)\\ \{\psi(x), \bar{\psi}(y)\} &= i Z_2^{-1}\gamma^0\delta_3(x-y)\end{aligned}\right\} x_0 = y_0 \qquad \text{(IV.54)}$$

We finally give the field equations for $T\bar{\psi}(x)\psi(y)$ and TA(x)A(y)

$$(i\gamma\partial - M)T\,\psi(x)\bar{\psi}(y) =$$

$$= Z_1 Z_2^{-1} g\,T A(x)\gamma_5\psi(x)\bar{\psi}(y) - \delta M\,T\psi(x)\bar{\psi}(y) + i Z_2^{-1}\delta(x-y)$$

$$-(\Box + m^2)\,T A(x) A(y) =$$

$$= Z_4 Z_3^{-1}\lambda\,T A(x)^3 A(y) + Z_1 Z_3^{-1} g\,T\bar{\psi}(x)\gamma_5\psi(x)A(y) +$$

$$-\delta m^2\,T A(x)A(y) + i Z_3^{-1}\delta(x-y)$$

$$\text{(IV.55)}$$

3b. Effective Lagrangian. Local Operator Products and Operator Product Expansions.

For the definition of the renormalized Green's functions in perturbation theory we use the effective Lagrangian

$$\mathcal{L} = \mathcal{L}_0 + \mathcal{L}_{INT} \qquad \mathcal{L}_0 = \mathcal{L}_A + \mathcal{L}_\psi \qquad \text{(IV.56)}$$

$$\mathcal{L}_{INT} = \mathcal{L}_1 + \mathcal{L}_2 \qquad \text{(IV.57)}$$

$$\mathcal{L}_A = \tfrac{1}{2}\partial_\mu A \partial^\mu A - \tfrac{1}{2}m^2 A^2$$
$$\mathcal{L}_\psi = \tfrac{1}{2} i \left(\bar\psi \gamma^\mu \partial_\mu \psi - \partial_\mu \bar\psi \gamma^\mu \psi\right) - M \bar\psi \psi \qquad (\text{IV}.58)$$

$$\mathcal{L}_1 = -\tfrac{1}{4!}\lambda A^4 - g \bar\psi \gamma_5 \psi A$$
$$\mathcal{L}_2 = \tfrac{1}{2}(\lambda a_1 + g a_2) A^2 + \tfrac{1}{2}(\lambda b_1 + g b_2)\mathcal{L}_A + \qquad (\text{IV}.59)$$
$$+ g a_3 \bar\psi \psi + g b_3 \mathcal{L}_\psi$$

The Green's functions are then given by the finite parts of the Gell-Mann Low series

$$\langle T O_1(x_1) \cdots O_n(x_n) \rangle =$$
$$= \text{Fin} \langle T e^{i \int dz \mathcal{L}^0_{INT}} O^0_1(x_1) \cdots O^0_n(x_n) \rangle_0 \qquad (\text{IV}.60)$$

where

$$O_j = \psi, \bar\psi, \text{ or } A$$

and O^0_j denotes the corresponding free field. $\mathcal{L}^0_{INT}$ is the interaction Lagrangian (IV.57) with free fields substituted for ψ, $\bar\psi$, or A.

On the basis of the Lagrangian (IV.57) the theory of local operator products and operator product expansions may be developed by analogy to the case of A^4-coupling treated in sections III.2-8. We restrict ourselves to stating the definitions and principal results.

We consider the product

$$Q(x) = \partial_{(\mu)_1} Q_1(x) \cdots \partial_{(\mu)_n} Q_n(x) \qquad (\text{IV}.61)$$

of operators $Q_j = \psi$, $\bar{\psi}$, or A and their derivatives

$$\partial_{(\mu)_j} Q_j = \partial_{\mu_{j_1}} \cdot\cdot \partial_{\mu_{j n(j)}} Q_j$$

$$(\mu)_j = (\mu_{j_1}, \cdots, \mu_{j n(j)})$$

The dimension d of (IV.61) is

$$d = \sum_{j=1}^{n} d(Q_j) + \sum_{j=1}^{n} \#(\mu)_j$$

$$d(\psi) = d(\bar{\psi}) = 3/2 \qquad d(A) = 1$$

$\#(\mu)_j = n(j)$ = degree of $\partial_{(\mu)_j}$

For $\delta \geq d$ we define the normal product

$$N_\delta \{ Q(x) \} \tag{IV.62}$$

by the Green's functions

$$\begin{aligned} &\langle T N_\delta \{ Q(x) \} O_1(y_1) \cdot\cdot O_N(y_N) \rangle = \\ &= \text{Fin} \langle T e^{i\int dz \mathcal{L}^0_{INT}} Q_0(x) O_1^0(x_1) \cdot\cdot O_N^0(y_N) \rangle_0 \end{aligned} \tag{IV.63}$$

Q_0 denotes the operator (IV.61) with free fields Q_j^0 substituted for Q_j. For $\delta \geq$ d the finite part (IV.60) is constructed with $c = \delta - d$ additional subtractions for proper parts containing x .
If the degree δ equals the dimension d we denote the normal product by

$$N_d \{ Q(x) \} = N \{ Q(x) \}$$

The difference

$$f = n_\psi - n_{\bar{\psi}}$$

of the number of ψ -operators and the number of $\bar{\psi}$ -operators occurring in (IV.61) is called the fermion number of the operator product.

In order to formulate the normalization condition for the operator (IV.62) we use the notation

$$\tilde{O}(o) = \tilde{O}_{1(\nu)_1}(o) \cdots \tilde{O}_{N(\nu)_N}(o)$$

where

$$\tilde{O}_{j(\nu)_j}(q) = \partial^{q}_{(\nu)_j} \tilde{O}_j(q)$$

$$O_j = \psi, \bar{\psi}, \text{ or } A$$

O_j^o **denotes the free field corresponding to** O_j. $\tilde{O}^o$ **means** :

$$\tilde{O}^o(o) = \tilde{O}^o_{1(\nu)_1}(o) \cdots \tilde{O}^o_{N(\nu)_N}(o)$$

With this notation definition (IV.63) leads to the following normalization conditions

$$\langle T N_\delta \{Q(o)\} \tilde{O}(o) \rangle^{PROP} = \langle T : Q_o(o) : \tilde{O}_o(o) \rangle^{PROP} \tag{IV.64}$$

provided

$$\sum_{j=1}^{N} d(O_j) + \sum_{j=1}^{N} \#(\nu)_j \leq \delta \tag{IV.65}$$

The constant appearing on the right hand side of the normalization condition (IV.64) is easily evaluated by using the following information. The definition (III.48) of the proper part implies that the proper function

$$\langle T : Q_o : O_o \rangle^{PROP}$$

of the free field products

$$Q_0 = \bar{\psi}_0(x_1) \cdot\cdot \bar{\psi}_0(x_a)\, \psi_0(y_1) \cdot\cdot \psi_0(y_b)\, A_0(z_1) \cdot\cdot A_0(z_c)$$

$$O_0 = \psi_0(x_1') \cdot\cdot \psi_0(x_{a'}')\, \bar{\psi}_0(y_1') \cdot\cdot \bar{\psi}_0(y_{b'}')\, A_0(z_1') \cdot\cdot A_0(z_{c'}')$$

is given by

$$\begin{aligned} \langle T : Q_0 \colon O_0 \rangle^{PROP} &= \delta_{aa'}\, \delta_{bb'}\, \delta_{cc'} \sum_P P \times \\ &\times \delta_{\alpha_1 \alpha'_{i_1}} \cdot\cdot \delta_{\alpha_a \alpha'_{i_a}}\, \delta(x_1 - x'_{i_1}) \cdot\cdot \delta(x_a - x'_{i_a}) \times \\ &\times \delta_{\beta_1 \beta'_{j_1}} \cdot\cdot \delta_{\beta_b \beta'_{j_b}}\, \delta(y_1 - y'_{j_1}) \cdot\cdot \delta(y_b - y'_{j_b}) \times \\ &\times \delta(z_1 - z'_{k_1}) \cdot\cdot \delta(z_c - z'_{k_c}) \end{aligned} \qquad \text{(IV.66)}$$

The sum extends over all permutations

$$(i_1 \cdot\cdot i_a)\,(j_1 \cdot\cdot j_b)\,(k_1 \cdot\cdot k_c)$$

of

$$(1 \cdot\cdot a)\,(1 \cdot\cdot b)\,(1 \quad c)$$

(IV.66) vanishes unless

$$a = a' \qquad b = b' \qquad c = c'$$

For an ordinary field operator $O = \psi,\ \bar{\psi}$, or A we have

$$\langle T\{O(x)\}\, O_1(y_1) \cdot\cdot O_N(y_N)\rangle^{PROP} = 0 \qquad \text{(IV.67)}$$

except for

$$\begin{aligned} \langle T\{\psi_\alpha(x)\}\, \bar{\psi}_\beta(y)\rangle^{PROP} &= \delta_{\alpha\beta}\, \delta(x-y) \\ \langle T\{\bar{\psi}_\alpha(x)\}\, \psi_\beta(y)\rangle^{PROP} &= \delta_{\alpha\beta}\, \delta(x-y) \\ \langle T\{A(x)\}\, A(y)\rangle^{PROP} &= \delta(x-y) \end{aligned} \qquad \text{(IV.68)}$$

Finally we state the results concerning Wilson's operator product expansion as far as needed for the work that follows. In perturbation theory the following expansions hold

$$T\,A(x+\xi)\,\psi(x-\xi) \sim \sum E_j(\xi)\,B_j(x) \qquad (\text{IV}.69)$$

$$T\,\bar{\psi}(x+\xi)\,\psi(x-\xi) \sim \sum E_{1j}(\xi)\,B_{1j}(x) \qquad (\text{IV}.70)$$

$$T\,A(x+\xi)\,A(x)\,A(x-\xi) \sim \sum E_{2j}(\xi)\,B_{2j}(x) \qquad (\text{IV}.71)$$

In (IV.69) the sum extends over all operator products (IV.61) of Fermion number one, degree

$$\delta = d(A) + d(\psi) = 5/2$$

and dimension $d \leq \delta$. In (IV.70-71) the sum extends over all local operator products (IV.61) of Fermion number 0, degree

$$\delta = d(\psi) + d(\bar{\psi}) = 3\,d(A) = 3$$

and dimension $d \leq \delta$. For

$$s = \sqrt{|\xi^2|} \rightarrow 0$$

the coefficients E behave no more singular than

$$\begin{aligned} E_j &\sim s^{-(5/2 - d(B_j))} \\ E_{1j} &\sim s^{-(3 - d(B_{1j}))} \\ E_{2j} &\sim s^{-(3 - d(B_{2j}))} \end{aligned} \qquad (\text{IV}.72)$$

3c. Field Equations and Renormalization Conditions

Definition (IV.63) leads to the following relations between the fields and the normal products of the interaction terms

$$(i\gamma\partial - M)\psi = g\left\{ N\{A\gamma_5\psi\} - a_3\psi - b_3(i\gamma\partial - M)\psi \right\} \qquad (\text{IV}.73)$$

$$-(\Box + m^2)A = \lambda\left\{ N\{A^3\} - a_1 A + b_1(\Box + m^2)A \right\} +$$
$$+ g\left\{ N\{\bar{\psi}\gamma_5\psi\} - a_2 A + b_2(\Box + m^2)A \right\} \qquad (\text{IV}.74)$$

Combined with the general principles of quantum field theory (IV.73-74) and the expansions (IV.69-71) give a description of the model in finite terms. For $T\psi(x)\bar{\psi}(y)$ and $TA(x)A(y)$ one obtains

$$(i\gamma\partial - M)T\psi(x)\bar{\psi}(y) =$$
$$= gT\gamma_5 N\{A(x)\psi(x)\}\bar{\psi}(y) - a_3 T\psi(x)\bar{\psi}(y) +$$
$$+ b_3(i\gamma\partial - M)T\psi(x)\bar{\psi}(y) + i\delta(x-y) \qquad (\text{IV}.75)$$

$$-(\Box + m^2)TA(x)A(y) =$$
$$= \lambda\left\{ TN\{A(x)^3\}A(y) - a_1 TA(x)A(y) + b_1(\Box + m^2)TA(x)A(y) \right\} +$$
$$+ g\left\{ TN\{\bar{\psi}(x)\gamma_5\psi(x)\}A(y) - a_2 TA(x)A(y) + b_2(\Box + m^2)TA(x)A(y) \right\} +$$
$$+ i\delta(x-y) \qquad (\text{IV}.76)$$

Similar relations hold for time ordered products of any number of operators.

For propagators and vertex functions we use the notations (III. 30), (IV. 17) and

$$\langle T\, \tilde{\Psi}(p)\, \tilde{\bar{\Psi}}(q)\, \tilde{A}(k)\rangle =$$

$$= \langle T\, \tilde{\Psi}(p)\, \tilde{\bar{\Psi}}(q)\, \tilde{A}(k)\rangle^{TRUNC}\, \delta(p+q+k)$$

$$\langle T\, \tilde{\Psi}(p)\, \tilde{\bar{\Psi}}(q)\, \tilde{A}(k)\rangle^{TRUNC} =$$

$$= \frac{1}{(2\pi)^2}\, \hat{S}_F'(p)\, \hat{\Delta}_F'(k)\, \Gamma(pqk)\, \hat{S}_F'(-q) \tag{IV.77}$$

In addition to the renormalization conditions (IV. 18), (IV. 20) we require

$$(\gamma p - M)\, \hat{S}_F'(p) = i \quad \text{at} \quad \gamma p = M \tag{IV.78}$$

and

$$\Gamma(00c) = i g \gamma_5 \tag{IV.79}$$

(IV.78) and (IV.79) are automatically satisfied due to the fact that **all subtractions are made at momentum zero.**

For the function Π we obtain from (IV. 76)

$$i\Pi = (2\pi)^2 \lambda \left(f_1 - a_1 - b_1 (p^2 - m^2)\right) +$$

$$+ (2\pi)^2 g \left(f_2 - a_2 - b_2 (p^2 - m^2)\right)$$

where

$$f_1(p^2) = \langle N\{A(0)^3\}\tilde{A}(p)\rangle$$

$$f_2(p^2) = \langle N\{\bar{\Psi}(0)\gamma_5\Psi(0)\}\tilde{A}(p)\rangle$$

(IV.20) and (IV.21) can be satisfied by setting

$$a_j = f_j(m^2) \qquad b_j = \left.\frac{\partial f_j}{\partial p^2}\right|_{p^2=m^2} \qquad (\text{IV}.80)$$

In order to satisfy the renormalization condition (IV.78) we introduce the function Σ by

$$S_F' = S_F + S_F \Sigma S_F'$$

Taking the Fourier transform of the vacuum expectation value of (IV.75) and dividing by S_F' we obtain for Σ the formula

$$i\Sigma = g\{h(p) - a_3 - b_3(\gamma p - M)\}$$

$$h(p) = (2\pi)^2 \langle TN\{A(0)\gamma_5\Psi(0)\}\bar{\Psi}(-p)\rangle$$

Setting

$$a_3 = h(p)\,,\quad b_3 = \frac{\partial h(p)}{\partial \gamma p} \quad \text{at } \gamma p = M \qquad (\text{IV}.81)$$

we obtain

$$\Sigma = 0\,,\quad \frac{\partial \Sigma}{\partial \gamma p} = 0 \quad \text{at } \gamma p = M \qquad (\text{IV}.82)$$

which implies the renormalization condition (IV.78).

With the new normal operators

$$\{A(x)\gamma_5\Psi(x)\} = N\{A(x)\gamma_5\Psi(x)\} - a_3\Psi - b_3(i\gamma\partial - M)\Psi \qquad (\text{IV}.83)$$

$$\{A(x)^3\} = N\{A(x)^3\} - a_1 A + b_1(\Box + m^2)A \qquad (\text{IV}.84)$$

$$\{\bar{\Psi}(x)\gamma_5\Psi(x)\} = N\{\bar{\Psi}(x)\gamma_5\Psi(x)\} - a_2 A + b_2(\Box + m^2)A \qquad (\text{IV}.85)$$

the field equations can be written in the form

$$(i\gamma\partial - M)\psi = g\{A\gamma_5\psi\} \qquad (IV.86)$$

$$-(\Box + m^2)A = \lambda\{A^3\} + g\{\bar{\psi}\gamma_5\psi\} \qquad (IV.87)$$

Moreover, time ordered products

$$T(\{A(x)\gamma_5\psi(x)\}C)$$

$$C = C_1(y_1)\cdots C_N(y_N) \qquad (IV.88)$$

$$C_j = \psi, \bar{\psi}, \text{ or } A$$

are defined by

$$T[\{A(x)\gamma_5\psi(x)\}C] = T[N\{A(x)\gamma_5\psi(x)\}C] +$$

$$-a_3 T[\psi(x)C] - b_3 T[(i\gamma\partial_x - M)\psi(x)C] \qquad (IV.89)$$

Similar definitions hold for

$$T[\{A(x)^3\}C], \quad T[\{\bar{\psi}(x)\gamma_5\psi(x)\}C]$$

$T\psi(x)\bar{\psi}(y)$ and $TA(x)A(y)$ then satisfy the relations

$$(i\gamma\partial - M)T\psi(x)\bar{\psi}(y) = \qquad (IV.90)$$

$$= gT\{A(x)\gamma_5\psi(x)\}\bar{\psi}(y) + i\delta(x-y)$$

$$-(\Box + m^2)TA(x)A(y) =$$

$$= \lambda T\{A(x)^3\}A(y) + gT\{\bar{\psi}(x)\gamma_5\psi(x)\}A(y) + i\delta(x-y) \qquad (IV.91)$$

(Similarly for products of an arbitrary number of operators.)

The new operators satisfy the normalization conditions

$$\left.\begin{array}{l} \langle T\{A(0)\gamma_5\psi(0)\}\tilde{A}(p)\rangle^{PROP} = 0 \\ \dfrac{\partial}{\partial \gamma p}\langle T\{A(0)\gamma_5\psi(0)\}\tilde{A}(p)\rangle^{PROP} = 0 \end{array}\right\} \text{ at } \gamma p = M \qquad (IV.92)$$

$$\left.\begin{array}{l} \langle T\{A(0)^3\}\tilde{A}(p)\rangle^{PROP} = 0 \\ \dfrac{\partial}{\partial p^2}\langle T\{A(0)^3\}\tilde{A}(p)\rangle^{PROP} = 0 \end{array}\right\} \text{ at } p^2 = m^2 \qquad (IV.93)$$

$$\left.\begin{array}{l} \langle T\{\bar{\psi}(0)\gamma_5\psi(0)\}\tilde{A}(p)\rangle^{PROP} = 0 \\ \dfrac{\partial}{\partial p^2}\langle T\{\bar{\psi}(0)\gamma_5\psi(0)\}\tilde{A}(p)\rangle^{PROP} = 0 \end{array}\right\} \text{ at } p^2 = m_1^2 \qquad (IV.94)$$

3d. Nucleon Field Equation and Current Operator

In order to obtain a field equation for ψ we must express the current operator $A\gamma_5\psi$ occurring in (IV.86) in terms of the fields A and ψ. We consider the expansion (IV.69)

$$A(y+\eta)\psi(y-\eta) \sim E_1(\eta)\psi(y) + E_{2\mu}(\eta)\partial^\mu\psi(y) + $$
$$+ E_3(\eta) N\{A(y)\psi(y)\}$$

Setting

$$y = x+\eta, \quad \eta = \tfrac{1}{2}\xi$$

we obtain the expansion

$$P(x\,\xi) = \frac{1}{2}\left(A(x+\xi) + A(x-\xi)\right)\psi(x) \sim$$

$$\sim F_1(\xi)\,\psi(x) + F_{2\mu}(\xi)\,\partial^\mu\psi(x) +$$

$$+ F_3(\xi)\,N\{A(x)\,\psi(x)\}$$

using the Taylor formula

$$\psi(y) \sim \psi(x) + \eta_\mu\,\partial^\mu\psi(x)$$

(up to terms of order s^2) and

$$N\{A(y)\psi(y)\} \sim N\{A(x)\,\psi(x)\}$$

(up to terms of order s). Expansions similar to (IV.95) hold for time ordered products

$$T\,P(x\,\xi)\,O_1(y_1)\cdots O_N(y_N)$$

We determine the coefficient F_3 by considering the expansion of

$$\langle T\,P(x\,\xi)\,\tilde{A}(c)\,\tilde{\bar{\psi}}(o)\rangle^{PROP}$$

Using the normalization conditions (IV.64) we get

$$F_3 \sim \gamma_5\,\gamma(\xi^2) =$$

$$= \frac{1}{(2\pi)^4}\langle P(o\,\xi)\,\tilde{A}(c)\,\tilde{\bar{\psi}}(o)\rangle^{PROP} \qquad \text{(IV.96)}$$

Replacing $\gamma_5 N\{A\psi\}$ by the new normal product $\{A\,\gamma_5\,\psi\}$ (see equ. (IV.83)) expansion (IV.95) becomes

$$P \sim F_1'\psi + F_{2\mu}'\,\partial^\mu\psi + \gamma\,\{A\,\gamma_5\,\psi\} \qquad \text{(IV.97)}$$

In order to determine F_1', F_2' we consider the expansion of

$$\langle T\, P(x\,\xi)\,\tilde{\tilde{\Psi}}(q)\rangle^{PROP} \sim F_1'\langle T\{\Psi(x)\}\,\tilde{\tilde{\Psi}}(q)\rangle^{PROP} +$$

$$+ F_{2\mu}'\,\partial_x^\mu \langle T\{\Psi(x)\}\,\tilde{\tilde{\Psi}}(q)\rangle^{PROP} + \gamma\langle T\{A\gamma_5\cdot\Psi(x)\}\,\tilde{\tilde{\Psi}}(q)\rangle^{PROP}$$

Using (IV.68) and setting x = 0

$$(2\pi)^2\langle T P(0\,\xi)\,\tilde{\tilde{\Psi}}(p)\rangle^{PROP} \sim F_1' + i p_\mu F_2'^{\mu} +$$

$$+ (2\pi)^2\gamma\langle T\{A(0)\gamma_5\cdot\Psi(0)\}\,\tilde{\tilde{\Psi}}(p)\rangle^{PROP} \qquad (IV.98)$$

The coefficient F_1' is found by setting $p = 0$

$$F_1'(\xi) \sim F_1''(\xi^2) =$$

$$= (2\pi)^2\langle T\, P(0\,\xi)\,\tilde{\tilde{\Psi}}(0)\rangle^{PROP} +$$

$$-(2\pi)^2\gamma\langle T\{A(0)\gamma_5\cdot\Psi(0)\}\,\tilde{\tilde{\Psi}}(0)\rangle^{PROP} \qquad (IV.99)$$

Differentiating with respect to p^μ we get

$$F_2'^{\mu} \sim (2\pi)^2\partial_p^\mu\langle T\, P(0\,\xi)\,\tilde{\tilde{\Psi}}(p)\rangle^{PROP}_{p=0} +$$

$$-(2\pi)^2\gamma\,\partial_p^\mu\langle T\{A(0)\gamma_5\cdot\Psi(0)\}\,\tilde{\tilde{\Psi}}(p)\rangle^{PROP}_{p=0}$$

The right hand side is of the general form

$$F_2'^{\mu} \sim \gamma^\mu F_2''(\xi^2) + \frac{\xi^\mu(\xi\gamma)}{\xi^2}\,\sigma(\xi^2) \qquad (IV.100)$$

Other terms permitted by Lorentz invariance are excluded since they are odd under the transformations $\xi \to P\xi$ or $\xi \to -\xi$

Inserting (IV.99) and (IV.100) back into (IV.98) we get

$$(2\pi)^2 \langle T\, P(c\,\xi)\, \tilde{\tilde{\Psi}}(p) \rangle^{PROP} \sim F_1''(\xi^2) + $$
$$+ (\gamma p) F_2''(\xi^2) + \frac{(p\xi)(\gamma\xi)}{\xi^2}\sigma(\xi^2) + \gamma(\xi^2) f(p^2) \qquad \text{(IV.101)}$$

$$f(p^2) = (2\pi)^2 \langle T \{ A(c)\gamma_5 \psi(c) \} \tilde{\tilde{\Psi}}(p) \rangle^{PROP}$$

This suggests defining the invariant part of (IV.98) by

$$(2\pi)^2\, \mathrm{Inv} \langle T\, P(c\,\xi)\, \tilde{\tilde{\Psi}}(p) \rangle^{PROP} =$$

$$= F_1''(\xi^2) + (\gamma p) F_2''(\xi^2) + \gamma(\xi^2) f(p^2) \qquad \text{(IV.102)}$$

which depends on ξ^2 and γp only. We introduce the renormalization functions

$$\alpha(\xi^2) = (2\pi)^2\, \mathrm{Inv} \langle T\, P(c\xi)\, \tilde{\tilde{\Psi}}(p) \rangle^{PROP}_{p=0} \qquad \text{(IV.103)}$$

$$\beta(\xi^2) = (2\pi)^2 \frac{\partial}{\partial(\gamma p)}\, \mathrm{Inv} \langle T\, P(c\xi)\, \tilde{\tilde{\Psi}}(p) \rangle^{PROP}_{p=0} \qquad \text{(IV.104)}$$

In terms of α and β the expansion (IV.97) becomes

$$P(x\,\xi) \sim \alpha\psi + \beta(i\gamma\partial - M)\psi - \sigma^{\mu\nu}\gamma_\mu \partial_\nu \psi +$$
$$+ \gamma\, \{A\gamma_5 \psi\} \qquad \text{(IV.105)}$$

$$\sigma^{\mu\nu} = \frac{\xi^\mu \xi^\nu}{\xi^2}\, \sigma(\xi^2)$$

We finally eliminate the kinetic term $(i\gamma\partial - M)\psi$ by inserting (IV.86) into (IV.105)

$$P(x\xi) \sim \alpha\psi - \sigma^{\mu\nu}\gamma_\mu\partial_\nu\psi + (g\beta+\gamma)\{A\gamma_5\psi\} \qquad (IV.106)$$

Dividing by $g\beta+\gamma$ and taking the limit $\xi \to 0$ the following formula for the current results

$$\{A(x)\gamma_5\psi(x)\} = \lim_{\xi\to 0} \frac{N(x\xi)}{\lambda(\xi)}$$

$$N(x\xi) = P(x\xi) - \alpha(\xi^2)\psi(x) + \sigma^{\mu\nu}(\xi)\gamma_\mu\partial_\nu\psi$$

$$\lambda(\xi^2) = g\beta(\xi^2) + \gamma(\xi^2)$$

$$\sigma^{\mu\nu}(\xi^2) = \frac{\xi^\mu\xi^\nu}{\xi^2}\sigma(\xi^2)$$

$$(IV.107)$$

Combining (IV.107) with (IV.86) we obtain the Dirac equation in the form

$$(i\gamma\partial - M)\psi = g\lim_{\xi\to 0}\frac{N(x\xi)}{\lambda(\xi^2)} \qquad (IV.108)$$

In a similar way we get the following field equation for $T\,\psi(x)\,\bar\psi(y)$

$$(i\gamma\partial_x - M)T\,\psi(x)\bar\psi(y) = g\lim_{\xi\to 0}\frac{N(x\xi y)}{\lambda(\xi^2)}$$

$$N(x\xi y) = T\,P(x\xi)\bar\psi(y) - \alpha(\xi^2)T\psi(x)\bar\psi(y) +$$

$$+\sigma^{\mu\nu}(\xi)\gamma_\mu\partial_\nu^x T\psi(x)\bar\psi(y) + \gamma(\xi^2)\,i\delta(x-y). \qquad (IV.109)$$

Analogous relations hold for arbitrary time ordered products

$$T\,\psi(x)\,O_1(y_1)\cdots O_N(y_N)$$

$$O_j = \psi,\ \overline{\psi} \text{ or } A$$

3e. Meson Field Equation and Current Operators

In the meson field equation (IV.87) appear the two normal products $\{\overline{\psi}\gamma_5\psi\}$ and $\{A^3\}$ which must be expressed by the fields. This will be accomplished by studying the expansions of the operators

$$P_1(x\xi) = \tfrac{1}{2}:\overline{\psi}(x+\xi)\gamma_5\psi(x-\xi): + \tfrac{1}{2}:\overline{\psi}(x-\xi)\gamma_5\psi(x+\xi): \qquad (IV.110)$$

and

$$P_2(x\xi) = :A(x+\xi)A(x)A(x-\xi): \qquad (IV.111)$$

in terms of normal products of degree $\delta = 3$ (see equ. (IV.70-71))

$$P_j \sim F_1^j A + F_2^{j\mu}\partial_\mu A + F_3^{j\mu\nu}\partial_\mu\partial_\nu A + + F_4^j N_3\{A^2\} + F_5^{j\mu} N\{A\partial_\mu A\} + + F_6^j N\{\overline{\psi}\psi\} + F_7^j N\{A^3\} \qquad (IV.112)$$

We determine the coefficients $F_4^j, \ldots, F_7^j$ by using the expansions of the Green's functions

$$\langle T\, P_j(0\xi)\,\tilde{A}(0)^2\rangle$$

$$\partial_\mu^p \langle T\, P_j(0\xi)\,\tilde{A}(0)\,\tilde{A}(p)\rangle_{p=0}$$

$$\langle T P_j(0\xi)\, \tilde{\psi}(0)\, \tilde{\bar{\psi}}(0)\rangle$$

$$\langle T P_j(0\xi)\, \tilde{A}(0)^3\rangle$$

and the normalization condition (IV.64)

$$F_i^{\,j} \sim F_i^{\prime\, j} \qquad i = 4,5,6,7 \quad j = 1,2$$

$$F_4^{\prime\, j} = \tfrac{1}{2}(2\pi)^4 \langle T P_j(0\xi)\, \tilde{A}(0)^2\rangle^{PROP}$$

$$F_5^{\prime\, j\mu} = -i\,(2\pi)^4\, \partial_q^\mu \langle T P_j(0\xi)\, \tilde{A}(0)\, \tilde{A}(q)\rangle^{PROP}_{q=0}$$

$$F_{6\alpha\beta}^{\prime\, j} = (2\pi)^4 \langle T P_j(0\xi)\, \tilde{\psi}_\alpha(0)\, \tilde{\bar{\psi}}_\beta(0)\rangle^{PROP}$$

$$F_7^{\prime\, j} = \tfrac{1}{6}(2\pi)^6 \langle T P_j(0\xi)\, \tilde{A}(0)^3\rangle^{PROP}$$

$F_5^{\prime\, j\mu}$ is of the form

$$\xi_\mu f(\xi^2)$$

and must vanish since even in ξ. $F_4^{\prime\, j}$ depends on ξ^2 and must vanish since odd under space inversion. For $F_7^{\prime\, j}$ we use the notation

$$F_7^{\prime\, j} = \gamma_{j2}(\xi^2)$$

$F_6^{\prime\, j}$ is of the form

$$F_6^{\prime\, j} = \gamma_5\, \gamma_{j1}(\xi^2)$$

Terms of the form $\gamma\xi$, $\gamma_5'\gamma\xi$ (times a function of ξ^2) are excluded since odd in ξ. A term $f(\xi^2)$ cannot occur since $F_i'^j$ transforms as

$$F_i'^j(\xi) = -\gamma_0 F_i'^j(P\xi)\gamma_0$$

under space inversion $\xi \to P\xi$

Using this information and replacing the operators $N\{\bar{\psi}\gamma_5\psi\}$ and $N\{A^3\}$ by $\{\bar{\psi}\gamma_5\psi\}$ and $\{A^3\}$ the expansion (IV.112) becomes

$$P_j \sim F_1'^j A + F_2'^{j\mu}\partial_\mu A + F_3'^{j\mu\nu}\partial_\mu\partial_\nu A +$$
$$+ \gamma_{j1}\{\bar{\psi}\gamma_5\psi\} + \gamma_{j2}\{A^3\} \qquad \text{(IV.114)}$$

The determination of the coefficients $F_1'^j, \ldots, F_3'^j$ follows the same pattern as in section IV, 2c. We give the final results only. The Green's functions

$$\langle T P_j(c\xi)\tilde{A}(p)\rangle^{PROP}$$

are of the general form

$$(2\pi)^2 \langle T P_j(c\xi)\tilde{A}(p)\rangle^{PROP} \sim$$
$$\sim (2\pi)^2 \, \mathrm{Inv} \langle T P_j(c\xi)\tilde{A}(p)\rangle^{PROP} +$$
$$+ \frac{(\xi p)^2}{\xi^2}\sigma_j(\xi^2) \qquad \text{(IV.115)}$$

where the invariant part is a function of ξ^2 and p^2 only. Defining

$$\alpha_j(\xi^2) = (2\pi)^2 \, \mathrm{Inv} \langle T P_j(c\xi)\tilde{A}(p)\rangle^{PROP}_{p^2=m^2} \qquad \text{(IV.116)}$$

$$\beta_j(\xi^2) = (2\pi)^2 \frac{\partial}{\partial p^2} I_{nv} \langle T P_j(0\xi) \tilde{A}(p) \rangle^{PROP}_{p^2=m^2} \qquad (IV.117)$$

we have

$$P_j \sim \alpha_j A - B_j (\Box + m^2) A +$$

$$- \sigma_j^{\mu\nu} \partial_\mu \partial_\nu A + \gamma_{j1} \{\bar\psi \gamma_5 \psi\} + \gamma_{j2} \{A^3\} \qquad (IV.118)$$

where

$$\sigma_j^{\mu\nu} = \frac{\xi^\mu \xi^\nu}{\xi^2} \sigma_j(\xi^2)$$

The kinetic term ($\Box$ + m^2) A can be eliminated from (IV. 118) by inserting (IV. 87). This leads to

$$P_1' \sim \gamma_{11}' \{\bar\psi \gamma_5 \psi\} + \gamma_{12}' \{A^3\}$$
$$P_2' \sim \gamma_{21}' \{\bar\psi \gamma_5 \psi\} + \gamma_{22}' \{A^3\} \qquad (IV.119)$$

$$\gamma_{11}' = g\beta_1 + \gamma_{11} \qquad \gamma_{21}' = g\beta_2 + \gamma_{21}$$
$$\gamma_{12}' = \lambda\beta_1 + \gamma_{12} \qquad \gamma_{22}' = \lambda\beta_2 + \gamma_{22} \qquad (IV.120)$$

where

$$P_j' = P_j + \sigma_j^{\mu\nu} \partial_\mu \partial_\nu A - \alpha_j A \qquad (IV.121)$$

Solving these equations for $\{\bar\psi\gamma_5\psi\}$ and $\{A^3\}$ we get the following expressions for the current operators

$$\{\bar\psi \gamma_5 \psi\} = \lim_{\xi\to 0} \frac{N_1}{\gamma'} \qquad \{A^3\} = \lim_{\xi\to 0} \frac{N_2}{\gamma'} \qquad (IV.122)$$

$$N_1 = \begin{vmatrix} P_1' & \gamma_{12}' \\ P_2' & \gamma_{22}' \end{vmatrix} \qquad N_2 = \begin{vmatrix} \gamma_{11}' & P_1' \\ \gamma_{21}' & P_2' \end{vmatrix} \qquad \gamma' = \begin{vmatrix} \gamma_{11}' & \gamma_{12}' \\ \gamma_{21}' & \gamma_{22}' \end{vmatrix}$$

(IV.122) combined with (IV.87)

$$-(\Box + m^2) A(x) = g\,\{\bar{\Psi}\gamma_5\Psi\} + \lambda\,\{A^3\}$$

represents the limit form of the meson equation.

We finally give the corresponding equations for the operator products $TP_j(x\xi)A(y)$ which we need for a comparison with the conventional renormalization constants. After elimination of the kinetic terms we have the expansions

$$\begin{aligned} T P_j(x\xi) A(y) \sim\ & \alpha_j\, T A(x) A(y) - \sigma_j^{\mu\nu}\, T A_{\mu\nu}(x) A(y) + \\ & + (g\beta_j + \gamma_{j1})\, T\{\bar{\Psi}\gamma_5\Psi(x)\} A(y) + \\ & + (\lambda\beta_j + \gamma_{j2})\, T\{A(x)^3\} A(y) + \\ & + \beta_j\, i\,\delta(x-y) \end{aligned} \qquad \text{(IV.123)}$$

or

$$P_j'(x\xi y) \sim \gamma_{j1}'\, T\{\bar{\Psi}(x)\gamma_5\Psi(x)\} A(y) + \gamma_{j2}'\, T\{A(x)^3\} A(y) \qquad \text{(IV.124)}$$

with

$$\begin{aligned} P_j' = {} & T P_j(x\xi) A(y) + \sigma_j^{\mu\nu}\, T A_{\mu\nu}(x) A(y) + \\ & - \alpha_j\, T A(x) A(y) - \beta_j\, i\,\delta(x-y) \end{aligned} \qquad \text{(IV.125)}$$

Solving these equations for $TP_j(x\,\xi\,)A\,(y)$ and taking the limit we get

$$T\{\bar{\Psi}(x)\gamma_5\Psi(x)\}\,A(y) = \lim_{\xi\to 0}\frac{N_1}{\gamma'}$$

$$T\{A(x)^3\}\,A(y) = \lim_{\xi\to 0}\frac{N_2}{\gamma'} \qquad (\text{IV}.126)$$

$$N_1 = \begin{vmatrix} P_1' & \gamma_{12}' \\ P_2' & \gamma_{22}' \end{vmatrix} \qquad N_2 = \begin{vmatrix} \gamma_{11}' & P_1' \\ \gamma_{21}' & P_2' \end{vmatrix}$$

3f. Comparison with Formal Field Equations

In order to write the field equations in a way resembling the formal field equations we introduce the following renormalization functions

$$Z_1^N = \gamma^{-1} \qquad Z_2 = 1 + g\,\frac{\beta}{\gamma}$$

$$\delta M = g\,\frac{\alpha}{g\beta+\gamma} \qquad (\text{IV}.127)$$

$$\tau_2^{\mu\nu} = \tau_2\,\frac{\xi^\mu\xi^\nu}{\xi^2} \qquad \tau_2 = \frac{\sigma}{g\beta+\gamma}$$

$$Z_3^{-1} = \frac{\gamma}{\gamma'} \qquad Z_4 = \frac{\lambda\gamma_{11} - g\,\gamma_{12}}{\lambda\gamma}$$

$$Z_1^{\pi} = \frac{\lambda \gamma_{21} - g\gamma_{22}}{g\gamma} \qquad \tau_3^{\mu\nu} = \tau_3 \frac{\xi^\mu \xi^\nu}{\xi^2}$$

$$\tau_3 = \frac{(g\gamma_{22} - \lambda\gamma_{21})\sigma_1 + (\lambda\gamma_{11} - g\gamma_{12})\sigma_2}{\gamma'}$$

$$\gamma = \begin{vmatrix} \gamma_{11} & \gamma_{12} \\ \gamma_{21} & \gamma_{22} \end{vmatrix} \qquad \gamma' = \begin{vmatrix} \gamma_{11}' & \gamma_{12}' \\ \gamma_{21}' & \gamma_{22}' \end{vmatrix}$$

With the exception of $\tau_2^{\mu\nu}$ and $\tau_3^{\mu\nu}$ all functions depend on ξ^2. The differential equations of the fields and propagators then take the form

$$(i\gamma\partial - M)\psi = \\ = \lim_{\xi\to 0} \left(Z_1^N Z_2^{-1} g P + \tau_2^{\mu\nu}\gamma_\mu \partial_\nu \psi - \delta M \psi \right) \qquad \text{(IV.128)}$$

$$-(\Box + m^2) A = \lim_{\xi \to 0} \left(Z_4 Z_3^{-1} g P_1 + Z_1^{\pi} Z_3^{-1} \lambda P_2 + \right. \\ \left. + \tau_3^{\mu\nu} \partial_\mu \partial_\nu A - \delta m^2 A \right) \qquad \text{(IV.129)}$$

$$(i\gamma\partial_x - M)\, T\psi(x)\bar\psi(y) = \lim_{\xi\to 0} R_\psi \qquad \text{(IV.130)}$$

$$-(\Box_x + m^2)\, TA(x)A(y) = \lim_{\xi \to 0} R_A \qquad \text{(IV.131)}$$

$$R_\psi = Z_1^N Z_2^{-1} g\, T\, P(x\xi)\, \bar{\psi}(y) +$$
$$+ \tau_2^{\mu\nu} \gamma_\mu \partial_\nu^x\, T\, \psi(x)\, \bar{\psi}(y) +$$
$$- \delta M\, T\, \psi(x)\bar{\psi}(y) + i Z_2^{-1} \delta(x-y)$$

$$R_A = Z_4 Z_3^{-1} g\, T P_1(x\xi)\, A(y) +$$
$$+ Z_1^\pi Z_3^{-1} \lambda\, T\, P_2(x\xi) A(y) +$$
$$+ \tau_3^{\mu\nu} \partial_\mu^x \partial_\nu^x\, T A(x)\, A(y) +$$
$$- \delta m^2\, T A(x) A(y) + i Z_3^{-1} \delta(x-y)$$

These equations correspond to the formal field equations (IV.54-55) with the renormalization constants replaced by functions of ξ .

Since in the formal approach $Z_1^N(0)$ and $Z_1^\pi(0)$ represent the same vertex renormalization constant Z_1 one should expect a relation

$$\lim_{\xi \to 0} \frac{Z_1^N(\xi^2)}{Z_1^\pi(\xi^2)} = 1$$

to hold.

4. Neutral Vector Meson Theory.

4a. Formal Lagrangian.

In this chapter we consider the model of a spin 1/2 field in interaction with a neutral vector meson field. Even from a formal point of view the renormalization of this model presents a number of problems. The most natural way of describing the interaction would certainly be to use the Proca Wentzel Lagrangian[55]

$$\begin{aligned} \mathcal{L} = & -\tfrac{1}{4} V_{\mu\rho\nu} V_{\mu}^{\rho\nu} + \tfrac{1}{2} m_0^2 V_{\mu\rho} V_{\mu}^{\rho} + \\ & + \tfrac{1}{2} i \left(\bar{\varphi}_{\mu} \gamma^{\rho} \partial_{\rho} \varphi_{\mu} - \partial_{\rho} \bar{\varphi}_{\mu} \gamma^{\rho} \varphi_{\mu} \right) - M_0 \bar{\varphi}_{\mu} \varphi_{\mu} + \\ & - e_0 \bar{\varphi}_{\mu} \gamma^{\rho} \varphi_{\mu} V_{\mu}^{\rho} , \end{aligned} \qquad (\text{IV}.132)$$

$$V_{\mu\rho\nu} = \partial_{\rho} V_{\mu\nu} - \partial_{\nu} V_{\mu\rho}$$

Unfortunately, however, the Feynman rules pertaining to this Lagrangian satisfy the criterion of a non-renormalizable theory, the reason being the two additional powers of momentum in the meson propagator

$$\frac{g_{\rho\nu} - \frac{k_{\rho} k_{\nu}}{m^2}}{k^2 - m^2}$$

A way out of this difficulty was suggested by Stueckelberg[72] who treated the model by analogy to electrodynamics with a Lagrangian which requires an indefinite metric in Hilbert space. It is not an entirely trivial matter, however, to write such a Lagrangian in a consistent manner. Symanzik[56] carried out the renormalization program for a neutral vector meson theory using the Lagrangian[75]

$$\mathcal{L} = -\frac{1}{4} A_{u\,\mu\nu} A_u^{\mu\nu} + \frac{1}{2} m_0^2 A_{u\mu} A_u^{\mu} - \frac{1}{2} \frac{m_0^2}{m^2} (\partial_\mu A_u^\mu)^2 +$$

$$+ \frac{1}{2} i (\bar{\psi}_u \gamma^\mu \partial_\mu \psi_u - \partial_\mu \bar{\psi}_u \gamma^\mu \psi_u) - M_0 \bar{\psi}_u \psi_u +$$

$$- e_0 \bar{\psi}_u \gamma^\mu \psi_u A_{u\mu} \quad . \qquad (\text{IV}.133)$$

$$A_u^{\mu\nu} = \partial^\mu A_u^\nu - \partial^\nu A_u^\mu$$

On the other hand de Calan[57] developed the renormalization theory of this model by generalizing the treatment of electrodynamics by Bogoliubov and Shirkov[58]. The Lagrangians of both methods are equivalent.

We shortly summarize the results. For (IV.133) the free propagator of the vector meson field is

$$\frac{g_{\mu\nu}}{k^2 - m^2 + i\epsilon}$$

hence the theory described by (IV.133) is renormalizable. Renormalized quantities are introduced by

$$A^\mu = Z_3^{-1/2} A_u^\mu \qquad \psi = Z_2^{-1/2} \psi_u \qquad e = Z_3^{1/2} e_0$$

$$m^2 = m_0^2 + \delta m^2 \qquad M = M_0 + \delta M$$

$$\bar{m}^2 = Z_3 m_0^2$$

The field equations are

$$(i\gamma\partial - M)\psi = e\gamma^\mu A_\mu \psi - \delta M \psi \qquad (\text{IV}.134)$$

$$-\partial_\nu A^{\mu\nu} = Z_3^{-1}\left(e j^\mu - \bar{m}^2 A^\mu - \frac{\bar{m}^2}{m^2}\partial^\mu \partial_\nu A^\nu\right)$$

$$j^\mu = Z_2^{-1}\,\bar{\psi}\gamma^\mu\psi \qquad (\text{IV}.135)$$

The current is conserved and $\partial_\mu A^\mu$ satisfies the Klein-Gordon equation

$$(\Box + m^2)\,\partial_\mu A^\mu = 0$$

The operator products $T\psi(x)\bar{\psi}(y)$ and $TA_\mu(x)A_\nu(y)$ satisfy

$$(i\gamma\partial_x - M)\,T\psi(x)\bar{\psi}(y) = e\,TA(x)\gamma_5\psi(x)\bar{\psi}(y) +$$

$$-\,\delta M\,T\psi(x)\bar{\psi}(y) + i\,Z_2^{-1}\,\delta(x-y) \qquad (\text{IV}.136)$$

$$-\,\partial^\rho\,TA_{\mu\rho}(x)A_\nu(y) = Z_3^{-1}\big(e\,Tj_\mu(x)A_\nu(y) +$$

$$-\,\bar{m}^2\,TV_\mu(x)A_\nu(y)\big) - g_{\mu\nu}\,i\,Z_3^{-1}\,\delta(x-y)$$

$$(\text{IV}.137)$$

Similar relations hold for arbitrary time-ordered products. The field $\partial_\mu A^\mu$ describes free ghost particles of mass m which require an indefinite metric in Hilbert space. A vector field describing the physical particles can be introduced by

$$V_\mu = A_\mu + \frac{1}{m^2}\,\partial_\mu \partial_\nu A^\nu \qquad (\text{IV}.138)$$

V_μ and $\partial_\mu A^\mu$ commute at all times

$$[V_\mu(x), \partial_\nu A^\nu(y)] = 0$$

The physical states are defined by the condition

$$(\partial_\mu A^\mu(x))^+ \Phi = 0$$

Changing the coefficient of $(\partial_\mu A^\mu)^2$ in (IV. 133) by a finite factor would lead to ghost particles of a different mass λ. The corresponding formalism describes the model in a different gauge. The renormalization constants Z_2, Z_3, δm^2 and δM can be chosen such that the perturbation expansions provide finite Green's functions of the renormalized fields.

$$\bar{m}^2 = Z_3 m_0^2 \qquad (IV.139)$$

is a finite constant depending on the finite parameters of the theory. It occurs in Johnson's sum rule[60]

$$\frac{1}{\bar{m}^2} = \frac{1}{m^2} + \int_{\kappa_0^2}^{\infty} d\kappa^2 \frac{\rho(\kappa^2)}{\kappa^2} \qquad (IV.140)$$

where ρ is the weight function of the meson propagator. The sum rule follows from the canonical commutation rules of the Lagrangian (IV. 133). (IV. 140) implies

$$\bar{m}^2 \leq m^2 \qquad (IV.141)$$

Finally it should be remarked that the Lagrangian (IV. 133) has only been shown to be correct if the limit is taken from a gauge invariant Pauli-Villars regularization.[57,58] For a more general regularization one needs additional counter terms in (IV. 133). This generalization was studied in detail by Symanzik.[56] Formally the field equations (IV.134-35) are invariant under the gauge transformations

$$A_\mu(x) \to A_\mu(x) + \partial_\mu \Lambda(x) \qquad \psi(x) \to e^{-ie\Lambda(x)}\psi(x)$$

provided

$$(\Box + m^2)\Lambda(x) = 0$$

4b. Effective Lagrangian and Renormalization Conditions

The theory will be based on the effective Lagrangian which is the finite version of the Lagrangian used by de Calan

$$\mathcal{L}_{EFF} = \mathcal{L}_0 + \mathcal{L}_{INT} \qquad \mathcal{L}_0 = \mathcal{L}_A + \mathcal{L}_\psi \qquad (IV.142)$$

$$\mathcal{L}_{INT} = \mathcal{L}_1 + \mathcal{L}_2 \qquad (IV.143)$$

$$\mathcal{L}_A = -\tfrac{1}{2}\partial_\mu A_\nu \partial^\mu A^\nu + \tfrac{1}{2}m^2 A_\mu A^\mu \qquad (IV.144)$$

$$\mathcal{L}_\psi = \tfrac{1}{2}i\left(\bar\psi\gamma^\mu\partial_\mu\psi - \partial_\mu\bar\psi\gamma^\mu\psi\right) - M\bar\psi\psi$$

$$\mathcal{L}_1 = e j^\mu A_\mu \qquad (IV.145)$$

$$\mathcal{L}_2 = e\tfrac{a}{2}\left(A_\mu A^\mu - \tfrac{1}{m^2}(\partial_\mu A^\mu)^2\right) - eb\mathcal{L}_A + ec\bar\psi\psi$$

We briefly comment on the special form of $\mathcal{L}_2$. A change of the coefficient of $(\partial_\mu A^\mu)^2$ would affect the mass of the ghost particles as was discussed in section 4a. With the present choice the ghost particles have the mass m of the vector mesons. For a more general form of the renormalization conditions one would have to add counter terms of the form

$$(1)\ \bar{\Psi}\gamma^{\mu}\partial_{\mu}\Psi \qquad (2)\ j^{\mu}A_{\mu} \qquad (3)\ (A_{\mu}A^{\mu})^{2}$$

We do not include these terms but adjust the vertex function and the four-meson vertex at momentum zero. The second subtraction of the fermion propagator is also taken at momentum zero. This "intermediate renormalization" was introduced by Bjorken and Drell.[59] In this renormalization method it is possible to take the limit to quantum electrodynamics without getting infrared divergencies for the Green's functions off the mass shell. While it is possible to include terms of type (1) and (2) a term of type (3) would be inconsistent since it leads to scattering between physical and ghost particles.

Most of the definitions and statements of section 3b may be taken over for the interaction Lagrangian (IV.143) replacing pseudoscalar fields A by vector fields A_{μ} . Instead of equ. (IV.66) we have

$$\begin{aligned} \langle T : Q_0 : O_0 \rangle^{PROP} &= \delta_{aa'}\,\delta_{bb'}\,\delta_{cc'} \sum_{P} P \times \\ &\times \delta_{\alpha_1\alpha'_{i_1}} \cdots \delta_{\alpha_a\alpha'_{i_a}}\, \delta(x_1 - x'_{i_1}) \cdots \delta(x_a - x'_{i_a}) \times \\ &\times \delta_{\beta_1\beta'_{j_1}} \cdots \delta_{\beta_b\beta'_{j_b}}\, \delta(y_1 - y'_{j_1}) \cdots \delta(y_b - y'_{j_b}) \times \\ &\times g_{\mu_1\mu'_{k_1}} \cdots g_{\mu_c\mu'_{k_c}}\, \delta(z_1 - z'_{k_1}) \cdots \delta(z_c - z'_{k_c}) \end{aligned} \qquad (IV.146)$$

where

$$Q_0 = \bar{\psi}_0(x_1)\cdots\bar{\psi}_0(x_a)\,\psi_0(y_1)\cdots\psi_0(y_b)\,A_{0\mu_1}(z_1)\cdots A_{0\mu_c}(z_c)$$

$$O_0 = \psi_0(x_1')\cdots\psi_0(x_{a'}')\,\bar{\psi}_0(y_1')\cdots\bar{\psi}_0(y_{b'}')\,A_{0\mu_1'}(z_1')\cdots A_{0\mu_{c'}'}(z_{c'}')$$

For a vector field the proper part of the propagator is

$$\langle T\{A_\mu(x)\}A_\nu(y)\rangle^{PROP} = g_{\mu\nu}\,\delta(x-y) \qquad (\text{IV } 147)$$

We have the following relations between the normal products of the interaction terms and the fields

$$(i\gamma\partial - M)\psi = e\,(N\{\gamma^\mu A_\mu\psi\} - c\psi) \qquad (\text{IV}.148)$$

$$(\Box + m^2)A_\mu = e\left(N\{j^\mu\} - aV^\mu + b(\Box + m^2)A_\mu\right) \qquad (\text{IV}.149)$$

$$V^\mu = A^\mu + \frac{1}{m^2}\partial^\mu\partial_\nu A^\nu$$

Similar relations hold for arbitrary time ordered products. We will only need the relations for two operators

$$(i\gamma\partial_x - M)\,T\,\psi(x)\bar{\psi}(y) = \qquad (\text{IV}.150)$$

$$= e\,TN\{\gamma^\mu A_\mu(x)\psi(x)\}\bar{\psi}(y) - ec\,T\psi(x)\bar{\psi}(y) + i\,\delta(x-y)$$

$$(\Box_x + m^2)\,T\,A_\mu(x)A_\nu(y) = e\,TN\{j_\mu(x)\}A_\nu(y) +$$

$$-eaTA_\mu(x)A_\nu(y) - \frac{ea}{m^2}\partial_\mu^x\partial_\rho^x TA^\rho(x)A_\nu(y) - ig_{\mu\nu}\delta(x-y) \quad (\text{IV}.151)$$

The vertex function is defined by

$$\langle T\tilde{\psi}(p)\tilde{A}_\mu(k)\tilde{\tilde{\bar{\psi}}}(q)\rangle^{\text{TRUNC}} =$$

$$= \frac{1}{(2\pi)^2}\hat{S}_F'(p)\hat{\Delta}'_{F\mu\nu}(k)\Gamma^\nu(pqk)\hat{S}_F'(q) \quad (\text{IV}.152)$$

The renormalization conditions are

$$(p^2-m^2)\hat{\Delta}_F' = i \quad \text{at } p^2=m^2 \quad (\text{IV}.153)$$

$$(\gamma p - M)\hat{S}_F' \neq 0 \quad \text{at } \gamma p = M \quad (\text{IV}.154)$$

$$\Gamma^\mu(000) = ie\gamma^\mu \quad (\text{IV}.155)$$

Here $\hat{\Delta}_F'$ is defined by

$$\hat{\Delta}_F'^{\mu\nu} = g^{\mu\nu}\hat{\Delta}_F'(k^2) + k^\mu k^\nu f(k^2)$$

The condition (IV. 155) is automatically satisfied. In the following we check that the conditions (IV. 153) and (IV. 154) can be satisfied by choosing appropriate constants a, b and c. We define $\Pi^{\mu\nu}$ by

$$\Delta'_{F\mu\nu} = \Delta_{F\mu\nu} + \Delta_{F\mu\rho}\Pi^{\rho\lambda}\Delta'_{F\lambda\nu} \quad (\text{IV}.156)$$

and decompose

$$\Pi^{\mu\nu} = g^{\mu\nu}\Pi(p^2) + p^\mu p^\nu\Phi(p^2)$$

(IV. 153) is then satisfied if

$$\Pi(m^2) = 0 \qquad \frac{d\Pi}{dp^2} = 0 \text{ at } p^2=m^2$$

From (IV.151) and (IV.155) one obtains

$$i\Pi_{\mu\nu} = e\,\{(2\pi)^2 \langle N\{j_\mu(0)\}\tilde{A}_\nu(p)\rangle^{PROP} - a g_{\mu\nu} + $$
$$- b(p^2 - m^2) g_{\mu\nu} + \frac{a}{m^2} p_\mu p_\nu\}$$

Setting

$$(2\pi)^2 \langle N\{j_\mu(0)\}\tilde{A}_\nu(p)\rangle^{PROP} = g_{\mu\nu} F(p^2) + p_\mu p_\nu \sigma(p^2)$$

we find

$$i\Pi = e(F - a - b(p^2 - m^2))$$

and the following relations for a and b

$$a = F, \quad b = \frac{dF}{dp^2} \quad \text{at} \quad p^2 = m^2$$

In case of the fermion propagator we have

$$i\Sigma = e\,\{(2\pi)^2 \langle N\{\gamma^\mu A_\mu(0)\psi(0)\}\tilde{\bar{\psi}}(-p)\rangle^{PROP} - c\}$$

Setting

$$c = (2\pi)^2 \langle N\{\gamma^\mu A_\mu(0)\psi(0)\}\tilde{\bar{\psi}}(-p)\rangle^{PROP}_{\gamma p = M}$$

it follows

$$\Sigma = 0 \quad \text{at} \quad \gamma p = M$$

which implies (IV.154). Since we did not introduce a second counter term we have

$$\partial_\mu \Sigma(p) = 0 \quad \text{at} \quad p = 0$$

For $m \neq 0$ it is possible to normalize the residue of the pole of S_F' by adding a counter term $ed\mathcal{L}_4$ to (IV.145). In the limit of quantum electrodynamics ($m \to 0$), however, this is not possible since

$$\frac{\partial \langle N\{\gamma^\mu A_\mu(0)\psi(0)\}\tilde{\bar{\psi}}(-p)\rangle^{PROP}}{\partial(\gamma p)}$$

becomes infrared divergent at $\gamma p = M$.

In the work that follows we will frequently make use of generalized Ward identities. It has been shown by Brandt that generalized Ward identities hold in renormalized perturbation theory of quantum electrodynamics. [13,14] His results can easily be extended to the case of a massive vector meson theory with intermediate renormalization.

4c. Dirac Equation

A special feature of the formal Dirac equation (IV.134) is the fact that the renormalized interaction term $\gamma^\mu A_\mu \psi$ is not multiplied by a divergent factor. This property is due to the relation $Z_1 = Z_2$. In this section it will be confirmed that the product

$$P_1(x\,\xi) = \tfrac{1}{2}\,T\left(\gamma_\mu A^\mu(x+\xi) + \gamma_\mu A^\mu(x-\xi)\right)\psi(x) \qquad \text{(IV.157)}$$

indeed requires only additive subtraction terms in the limit $\xi \to 0$.

We start out from the expansion of the operator

$$P_{1\mu}(x\,\xi) = \tfrac{1}{2}\,T\left(A_\mu(x+\xi) + A_\mu(x-\xi)\right)\psi(x) \qquad \text{(IV.158)}$$

$$P_1{}^\mu(x\xi) \sim E_1{}^\mu \psi + E_2{}^{\mu\nu}\partial_\nu\psi + E_3{}^{\mu\nu} N\{A_\nu \psi\}$$

The coefficients behave no more singular than

$$E_1^{\mu} \sim s^{-1} \lg^{\alpha} s \qquad E_2^{\mu\nu} \sim \lg^{\alpha} s \qquad E_3^{\mu\nu} \sim \lg^{\alpha} s \tag{IV.159}$$

for

$$s = \sqrt{|\xi^2|} \to 0$$

Similar relations hold for the time ordered products. The coefficients are determined by the normalization conditions (IV.64), (IV.146)

$$E_1^{\mu} \sim E_1'^{\mu} \qquad E_2^{\mu\nu} \sim E_2'^{\mu\nu} \qquad E_3^{\mu\nu} \sim E_3'^{\mu\nu}$$

$$E_1'^{\mu} = 2\pi \langle T\{A^{\mu}(\xi)\psi(0)\}\tilde{\bar{\psi}}(0)\rangle^{PROP}$$

$$E_2'^{\mu\nu} = (2\pi)^2 i \langle T\{A^{\mu}(\xi)\psi(0)\}\partial^{\nu}\tilde{\bar{\psi}}(0)\rangle^{PROP}$$

$$E_3'^{\mu\nu} = (2\pi)^4 \langle T\{A^{\mu}(\xi)\psi(0)\}\tilde{A}^{\nu}(0)\tilde{\bar{\psi}}(0)\rangle^{PROP} =$$

$$= (2\pi)^4 \langle T\{A^{\mu}(\xi)\psi(0)\}\tilde{A}^{\nu}(0)\tilde{\bar{\psi}}(0)\rangle^{EL} +$$

$$+ g^{\mu\nu} \tag{IV.160}$$

Setting $q = 0$ and taking the Fourier transform of the Ward identity

$$\partial_{\nu}^{q} \langle T\{\tilde{A}_{\mu}(k)\tilde{\psi}(0)\tilde{\bar{\psi}}(q)\rangle^{PROP} =$$

$$= -\frac{(2\pi)^2}{e} \langle TN\{\tilde{A}_{\mu}(k)\tilde{\psi}(0)\}\tilde{\bar{\psi}}(q)\tilde{A}_{\nu}(0)\rangle^{EL}$$

with respect to k we obtain

$$E_3'^{\mu\nu}(\xi) = i e E_2'^{\mu\nu}(\xi) + g^{\mu\nu} \tag{IV.161}$$

This relation is responsible for the absence of the multiplicative renormalization function in the Dirac equation.

The treatment of the directional dependent singularities is more complicated than in the case of the pseudoscalar interaction. We indicate the necessary modifications. The functions $E_1'^{\mu}$ and $E_2'^{\mu\nu}$ have the general form

$$E_1'^{\mu} = a_1 \gamma^{\mu} + a_2 \xi^{\mu} (\gamma \xi) \qquad \text{(IV.162)}$$

$$E_2'^{\mu\nu} = b_1 g^{\mu\nu} + b_2 \xi^{\mu} \xi^{\nu} + b_3 \sigma^{\mu\nu}$$

Other terms (permitted by Lorentz invariance) are excluded since they are odd in ξ or under space inversion. Inserting (IV. 161) into (IV. 158) we obtain

$$P_1^{\mu} \sim N\{A^{\mu}\psi\} + E_1'^{\mu}\psi + + E_2'^{\mu\nu}(\partial_{\nu}\psi + ie\, N\{A_{\nu}\psi\}) \qquad \text{(IV.163)}$$

and

$$P_1 = \gamma^{\mu} P_{1\mu} \sim \gamma^{\mu} N\{A_{\mu}\psi\} + E_1''\psi + + E_2''\gamma^{\mu}(\partial_{\mu}\psi + ie\, N\{A_{\mu}\psi\}) + + E_3''(\gamma\xi)\xi^{\mu}(\partial_{\mu}\psi + ie\, N\{A_{\mu}\psi\}) \qquad \text{(IV.164)}$$

where

$$E_i'' = E_i''(\xi^2)$$

In order to avoid implicit formulae for $N\{A^{\mu}\psi\}$ we have to eliminate the term

$$(\gamma\xi)\xi_{\mu}\; N\{A^{\mu}\psi\}$$

on the right hand side of (IV.164). To this end we return to the expression (IV.158) of $P_{1\mu}$ and form

$$\gamma_{\nu}\xi^{\nu}\xi^{\mu}P_{1\mu} \sim c_1\gamma_{\nu}\xi^{\nu}\xi^{\mu}N\{A_{\mu}\psi\} + c_2\psi + \\ + c_3\gamma_{\nu}\xi^{\nu}\xi^{\mu}\partial_{\mu}\psi \;\; + c_4\gamma^{\mu}(\partial_{\mu}\psi + ie N\{A_{\mu}\psi\})$$

$$c_j = c_j(\xi^2)$$

Solving this equation for $(\gamma\xi)\xi^{\mu}N\{A_{\mu}\psi\}$ and inserting this result into (IV.164) one finds

$$P_1 \sim \gamma^{\mu}N\{A_{\mu}\psi\} + E_1{}'''\psi + \\ + E_2{}'''\gamma^{\mu}(\partial_{\mu}\psi + ie N\{A_{\mu}\psi\}) + \\ + E_3{}'''(\gamma\xi)\xi^{\mu}(\partial_{\mu}\psi + ie P_{1\mu}) \qquad \text{(IV.165)}$$

Next we eliminate the kinetic term $(i\gamma\partial - M)\psi$ by using (IV.148)

$$P_1 \sim \gamma^{\mu}N\{A_{\mu}\psi\} + E_1{}'''\psi + i(M-c)E_2{}''' + \\ + E_3{}'''(\gamma\xi)\xi^{\mu}(\partial_{\mu}\psi + ie P_{1\mu})$$

Solving this equation for $\gamma^\mu N\{A_\mu \psi\}$ we find in the limit

$$\gamma^\mu N\{A_\mu(x)\psi(x)\} =$$

$$= \lim_{\xi \to 0} \left(P_1 + \mu_1 \psi + (\gamma\xi)\xi^\mu \mu_2 (\partial_\mu \psi + ie P_{1\mu})\right)$$

No multiplicative renormalization function appears in this formula for the interaction term of the Dirac equation

Setting

$$\mu_1(\xi^2) = c - \frac{\delta M(\xi^2)}{e}$$

$$\mu_2(\xi^2) = \frac{\mu(\xi^2)}{e\xi^2}$$

The final form of the Dirac equation becomes

$$(i\gamma^\mu \partial_\mu - M)\psi =$$

$$= \lim_{\xi \to 0} \left\{ eP_1 - \delta M \psi + \frac{(\gamma\xi)\xi^\mu}{\xi^2} \mu (\partial_\mu \psi + ie P_{1\mu}) \right\} \qquad \text{(IV.166)}$$

where the complete ξ-dependence is given by

$$P_{1\mu} = \tfrac{1}{2} T\left(A^\mu(x+\xi) + A^\mu(x-\xi)\right)\psi(x)$$

$$P_1 = \gamma^\mu P_{1\mu} \qquad \delta M = \delta M(\xi^2)$$

$$\mu = \mu(\xi^2) \qquad \text{(IV.167)}$$

4d. Vector Meson Field Equation and Brandt's Form of the Current Operator [14]

The field equation of the vector meson field (or the potential in electrodynamics) is particularly interesting because of the phenomenon of gauge invariance. Important work has been done by R. Brandt concerning gauge invariance and the structure of the electromagnetic current operator. [14] In this analysis it was shown that the directional dependent singularities of the current operator are closely related to the gauge invariance of the theory. To a large extent their form is determined by generalized Ward identities.

In this section it will be shown that for a massive vector meson field equations of the form

$$-\partial^{\rho} A_{\mu\rho}(x) = \lim_{\xi \to 0} Z_3^{-1}(\xi)\{Z_2(\xi)\, e\, j_\mu(x\xi) - \bar{m}^2 V_\mu(x)\}$$

$$-\partial^{\rho} T A_{\mu\rho}(x) A_\nu(y) =$$

$$= \lim_{\xi \to 0} \left[Z_3^{-1}(\xi)\{Z_2(\xi)\, e\, T j_\mu(x\xi) A_\nu(y) - \bar{m}^2 T V_\mu(x) A_\nu(y)\} + {} - g_{\mu\nu}\, i\, Z_3^{-1}(\xi)\, \delta(x-y) \right]$$

hold where $\bar{m}^2$ is a finite constant and $j_\mu(x\xi)$ has the form of Brandt's electromagnetic current. Brandt's field equation for the electromagnetic potential then follows by taking the limit $\bar{m}^2 \to 0$.

To simplify the discussion we assume that the vector ξ is spacelike. In the work that follows $\lim_{\xi \to 0}$ always means that the limit is taken with

$$\xi^2 < 0 \ , \quad \frac{\xi}{\sqrt{|\xi^2|}} \text{ bounded.}$$

We expand the current operator

$$P_{2\mu}(x\xi) = \tfrac{1}{2}\,\overline{\Psi}(x+\xi)\gamma_\mu \Psi(x-\xi) + \tfrac{1}{2}\,\overline{\Psi}(x-\xi)\gamma_\mu \Psi(x+\xi) \qquad \text{(IV.170)}$$

with respect to local operators of degree 3

$$\begin{aligned} P_{2\mu} \sim\ & E_1 + E_2^{\mu\nu} A_\nu + E_3^{\mu\nu\kappa}\partial_\kappa A_\nu + \\ & + E_4^{\mu\nu\kappa\lambda}\partial_\kappa\partial_\lambda A_\nu + E_5^{\mu\nu\kappa\lambda} N\{A_\nu A_\kappa A_\lambda\} + \\ & + E_6^{\mu} N\{\overline{\Psi}\Psi\} + E_7^{\mu\nu\kappa} N_3\{A_\nu A_\kappa\} + \\ & + E_8^{\mu\nu\kappa\lambda} N\{A_\nu(\partial_\lambda A_\kappa)\} \end{aligned} \qquad \text{(IV.171)}$$

In the last term we combined

$$\begin{aligned} & E_{81}^{\mu\nu\kappa\lambda} N\{(\partial_\lambda A_\nu)A_\kappa\} + E_{82}^{\mu\nu\kappa\lambda} N\{A_\nu \partial_\lambda A_\kappa\} = \\ & = E_8^{\mu\nu\kappa\lambda} N\{A_\nu \partial_\lambda A_\kappa\} \end{aligned} \qquad \text{(IV.172)}$$

$$E_8^{\mu\nu\kappa\lambda} = E_{81}^{\mu\kappa\nu\lambda} + E_{82}^{\mu\nu\kappa\lambda}$$

Hence E_8 is symmetric in κ and ν. We can also arrange that E_4 is symmetric in κ and λ, E_5 symmetric in ν, κ, λ and

E_7 symmetric in ν and κ. For the first term we have

$$E_1 \sim \langle P_{2\mu}(x\xi)\rangle = 0 \qquad \text{(IV.173)}$$

since $\langle P_{2\mu}\rangle$ is the even part of

$$J_\mu(\xi) = \langle T\,\overline{\psi}(\xi)\gamma_\mu\psi(-\xi)\rangle = \\ = \xi_\mu\, J(\xi^2) \qquad \text{(IV.174)}$$

The normalization conditions (IV.64), (IV.146) imply

$$E_j \sim E_j' \qquad \text{(IV.175)}$$

where the E_j represent certain Green's functions

$$E_2'^{\mu\nu} = (2\pi)^2 \langle T\, P_2^\mu(c\xi)\,\tilde{A}^\nu(0)\rangle^{PROP}$$

$$E_4'^{\mu\nu\kappa\lambda} = -\frac{1}{2}(2\pi)^2 \langle T\, P_2^\mu(0\xi)\,\partial^\kappa\partial^\lambda \tilde{A}^\nu(0)\rangle^{PROP}$$

$$E_5'^{\mu\nu\kappa\lambda} = \frac{1}{6}(2\pi)^6 \langle T\, P_2^\mu(c\xi)\,\tilde{A}^\nu(0)\,\tilde{A}^\kappa(0)\,\tilde{A}^\lambda(0)\rangle^{PROP}$$

$$E_8'^{\mu\nu\kappa\lambda} = -i(2\pi)^4 \langle T\, P_2^\mu(c\xi)\,\tilde{A}^\nu(0)\,\partial^\lambda \tilde{A}^\kappa(0)\rangle^{PROP}$$

Similar relations hold for E_3', E_6' and E_7'. For the time being we only need the information that the E' are even functions of ξ which transform covariantly under Lorentz transformations. Lorentz invariance implies

$$E_j'^{\mu\nu\kappa}(-\xi) = -E_j'^{\mu\nu\kappa}(\xi) \qquad j = 3,7$$

$$E_6^{i\mu}(\xi) = K_1 \gamma^\mu + K_2 \xi^\mu (\xi\gamma) + K_3 \xi^\mu + + K_4 \sigma^{\mu\nu} \xi_\nu \qquad (IV.177)$$

$$\sigma^{\mu\nu} = \frac{1}{2i} (\gamma^\mu \gamma^\nu - \gamma^\nu \gamma^\mu)$$

In (IV. 177) γ_5-terms were excluded using invariance under space inversion. Since all E_j are even functions of ξ it follows

$$E_3^{\mu\nu\kappa} = E_7^{\mu\nu\kappa} = 0$$

$$E_l^\mu = K^{\mu\nu} \gamma_\nu \qquad K^{\mu\nu} = K_1 g^{\mu\nu} + K_2 \xi^\mu \xi^\nu \qquad (IV.178)$$

With this information (IV. 171) simplifies to

$$P_2^\mu \sim E_2^{\mu\nu} A_\nu + E_4^{\mu\nu\kappa\lambda} \partial_\kappa \partial_\lambda A_\nu + + E_5^{\mu\nu\kappa\lambda} N\{A_\nu A_\kappa A_\lambda\} + K^{\mu\nu} N\{j_\nu\} + + E_8^{\mu\nu\kappa\lambda} N\{A_\nu \partial_\lambda A_\kappa\} \qquad (IV\ 179)$$

For spacelike ξ the operator $P_2^\mu(x\,\xi)$ transforms under charge conjugation as

$$P_2^\mu(x,\xi) \to -P_2^\mu(x,-\xi) = -P_2^\mu(x,\xi)$$

Since $N\{A_\nu \partial_\lambda A_\kappa\}$ is the only operator on the right hand side of (IV. 179) which transforms even under charge conjugation we have

$$E_8^{\prime\,\mu\nu\kappa\lambda} \sim 0$$

Hence

$$P_2^\mu \sim E_2^{\prime\,\mu\nu} A_\nu + E_4^{\prime\,\mu\nu\kappa\lambda} \partial_\kappa \partial_\lambda A_\nu +$$

$$+ E_5^{\prime\,\mu\nu\kappa\lambda} N\{A_\nu A_\kappa A_\lambda\} + K^{\mu\nu} N\{j_\nu\} \qquad \text{(IV. 180)}$$

For time ordered products one has corresponding formulae with $T(\cdots O_1(y_1)\cdots O_N(y_N))$ $\quad O_j = \psi, A_\mu$, or $\overline{\psi}$ taken on both sides of **(IV. 180).** Brandt has shown that E_2', E_5' and part of E_4' **are determined by the function** $J_\mu(\xi)$ which was defined in equ. **(IV. 174):**

$$E_2^{\mu\nu} = -2ie\,\xi^\nu J^\mu(\xi)$$

$$E_4^{\mu\nu\kappa\lambda} = -\tfrac{1}{3} ie\,\xi^\nu \xi^\kappa \xi^\lambda J^\mu(\xi) + E_4^{\prime\prime\,\mu\nu\kappa\lambda}$$

$$E_5^{\mu\nu\kappa\lambda} = \tfrac{1}{6} ie^3 2^3 \xi^\nu \xi^\kappa \xi^\lambda J^\mu(\xi) \qquad \text{(IV. 181)}$$

where

$$a_\nu a_\kappa a_\lambda E_4^{\prime\prime\,\mu\nu\kappa\lambda} = 0 \qquad \text{(IV. 182)}$$

for any 4-vector a_μ and

$$E_4^{\prime\prime\,\mu\nu\kappa\lambda} = E_4^{\prime\prime\,\mu\nu\lambda\kappa} \qquad \text{(IV. 183)}$$

These relations can be derived from generalized Ward identities In the following we check the first of equ. (IV.181). For the proof of the other relations we refer to Brandt's original paper. E_2 depends only on the difference of the coordinates of ψ and $\overline{\psi}$, hence

$$E_2^{\mu\nu}(\xi) = (2\pi)^2 \langle T\, \overline{\psi}(2\xi)\gamma^\mu \psi(0)\, \tilde{A}^\nu(0)\rangle^{PROP}$$

We write

$$C_2^{\mu\nu}(x) = E_2^{\mu\nu}(\xi) \quad \text{with } x = 2\xi$$

and take the Fourier transform

$$\tilde{C}_2^{\mu\nu}(p) = (2\pi)^2 \langle T\, \{\tilde{\overline{\psi}}(p)\gamma^\mu \psi(0)\}\, \tilde{A}^\nu(0)\rangle^{PROP}$$

$$= \frac{-1}{(2\pi)^2}\, \mathrm{tr}\, \gamma^\mu S_F'(p)\, \Gamma^\nu(p,-p,0)\, S_F'(p)$$

Applying the Ward identity

$$e\,\partial^\nu S_F'(p) = S_F'(p)\,\Gamma^\nu(p,-p,0)\, S_F'(p)$$

we find

$$\tilde{C}_2^{\mu\nu}(p) = \frac{e}{(2\pi)^2}\, \mathrm{tr}\, \gamma^\mu \partial^\nu S_F'(p)$$

or

$$C_2^{\mu\nu}(x) = -ig\, \mathrm{tr}\, \gamma^\mu x^\nu \langle T\, \psi(\tfrac{x}{2})\overline{\psi}(-\tfrac{x}{2})\rangle$$

and

$$E_2^{\mu\nu}(\xi) = -2ie\, \xi^\nu J^\mu(\xi)$$

Inserting (IV.181) into (IV.180) we obtain

$$P_2^\mu \sim J^\mu(\xi)\, Q(x\xi) + E_4''^{\,\mu\nu\kappa\lambda}\, \partial_\kappa \partial_\lambda A_\nu + \\ + K^{\mu\nu} N\{j_\nu\} \qquad \text{(IV.184)}$$

where

$$Q(x\xi) = -2ie\,\xi A - \tfrac{1}{3} ie\,(\xi\partial_x)^2(\xi A) + \\ + \tfrac{1}{6}\, ie^3 2^3 N\{(\xi A)^3\} \qquad \text{(IV.185)}$$

and $E_4''^{\,\mu\nu\kappa\lambda}$ satisfies (IV.182-83). Following Brandt we can give (IV.185) a particularly neat form by considering the expression

$$N\left\{ e^{-ie\int_{x-\xi}^{x+\xi} d\eta_\nu A^\nu(\eta)} \right\}$$

which is defined by taking the normal product of the power series expansion in ξ. Up to terms vanishing of order ξ^4 we have

$$N\left\{ e^{-ie\int_{x-\xi}^{x+\xi} d\eta_\nu A^\nu(\eta)} \right\} \sim$$

$$\sim 1 - 2ie\,\xi A - \tfrac{1}{3} ie\,(\xi\partial)^2\, \xi A + \\ - 2e^2 N(\xi A)^2 + \tfrac{1}{6} ie^3 2^3 N(\xi A^3)$$

The odd part of this expression becomes

$$\mathcal{E}(x\xi) = \tfrac{1}{2} N\left\{ e^{-ie\int_{x-\xi}^{x+\xi} d\eta_\nu A^\nu(\eta)} \right\} - \tfrac{1}{2}(\xi \to -\xi) \sim$$

$$\sim -2ie\,\xi A - \tfrac{1}{3} ie\,(\xi\partial)^2\, \xi A + \tfrac{1}{6} ie^3 2^3 N\{(\xi A)^3\} \qquad \text{(IV.186)}$$

Since the leading singularities of $J_\mu(\xi)$ have dimension 3 it follows

$$J_\mu(\xi)\, Q(x\xi) \sim J_\mu(\xi)\, \mathcal{E}(x\xi) \qquad \text{(IV.187)}$$

up to terms which vanish for $\xi \to 0$.

Next we study the term

$$E_4''^{\,\mu\nu\kappa\lambda}(\xi)\, \partial_\kappa \partial_\lambda A_\nu$$

The most general tensor satisfying (IV.182) and (IV.183) is

$$E_4''^{\,\mu\nu\kappa\lambda}(\xi) = F_1(\xi^2)\left(\tfrac{1}{2} g^{\mu\kappa} g^{\nu\lambda} + \tfrac{1}{2} g^{\mu\lambda} g^{\nu\kappa} - g^{\mu\nu} g^{\kappa\lambda}\right) +$$

$$+ F_2(\xi^2)\left(\tfrac{1}{2} g^{\mu\kappa} \xi^\nu \xi^\lambda + \tfrac{1}{2} g^{\mu\lambda} \xi^\nu \xi^\kappa - g^{\mu\nu} \xi^\kappa \xi^\lambda\right) +$$

$$+ F_3(\xi^2)\left(\tfrac{1}{2} \xi^\mu \xi^\kappa g^{\lambda\nu} + \tfrac{1}{2} \xi^\mu \xi^\lambda g^{\kappa\nu} - \xi^\mu \xi^\nu g^{\lambda\kappa}\right)$$

This implies

$$E_4''^{\,\mu\nu\kappa\lambda} \partial_\kappa \partial_\lambda A_\nu = F_1\, \partial_\nu A^{\mu\nu} + F_2\, \xi_\nu \xi_\lambda \partial^\lambda A^{\mu\nu} +$$

$$+ F_3\, \xi^\mu \xi^\kappa \partial^\lambda A_{\kappa\lambda} \qquad \text{(IV.188)}$$

Combining (IV. 178), and (IV. 186-88) we get

$$P_2^{\,\mu} \sim J^\mu \mathcal{E} + F_1\, \partial_\nu A^{\mu\nu} + F_2\, \xi_\nu \xi_\lambda \partial^\lambda A^{\mu\nu} +$$

$$+ F_3\, \xi^\mu \xi^\kappa \partial^\lambda A_{\kappa\lambda} + K_1 N\{j^\mu\} + K_2\, \xi^\mu \xi^\nu N\{j_\nu\} \qquad \text{(IV.189)}$$

Finally the term $\xi^\mu \xi^\nu N\{j_\nu\}$ should be eliminated on the right hand side. To this end we multiply (IV.189) by $\xi^\rho \xi^\mu$ and solve the equation for $\xi^\nu N\{j_\nu\}$ Inserting the result into (IV.189) we obtain

$$P_2^\mu(x\xi) \sim K N\{j^\mu\} - L\partial_\kappa A^{\mu\kappa} + J^\mu \mathcal{E} +$$
$$+ \mathcal{N}_1 \xi_\kappa \xi_\lambda \partial^\lambda A^{\mu\kappa} + \mathcal{N}_2 \xi^\mu \xi^\kappa \partial^\lambda A_{\kappa\lambda} +$$
$$+ \mathcal{N}_3 \xi^\mu \xi^\kappa j_\kappa(x\xi) \qquad \text{(IV.190)}$$

Similarly

$$T P_2^\mu(x\xi) A_\nu(y) \sim K T N\{j^\mu(x)\} A^\nu(y) + J^\mu T \mathcal{E}(x\xi) A^\nu(y) +$$
$$- L\partial_\kappa T A^{\mu\kappa}(x) A^\nu(y) + \mathcal{N}_1 \xi_\kappa \xi_\lambda \partial^\lambda T A^{\mu\kappa}(x) A^\nu(y) +$$
$$+ \mathcal{N}_2 \xi^\mu \xi^\kappa \partial^\lambda T A_{\kappa\lambda}(x) A^\nu(y) +$$
$$+ \mathcal{N}_3 \xi^\mu \xi^\kappa T P_{2\kappa}(x\xi) A^\nu(y) \qquad \text{(IV.191)}$$

In order to construct a field equation in limit form we proceed differently from the previous cases. We first express Nj^μ and $TNj^\mu(x) A^\nu(y)$ by A^μ and $TA^\mu(x)A^\nu(y)$ and their derivatives using (IV.149), (IV.151):

$$N\{j_\mu\} = -\, s\,\partial^\rho A_{\mu\rho} + t V_\mu \qquad \text{(IV.192)}$$

$$T\, N\{j_\mu(x)\}\, A_\nu(y) =$$

$$= -s\,\partial^\rho T A_{\mu\rho}(x) A_\nu(y) + t\, T V_\mu(x) A_\nu(y) + $$

$$+ \frac{i}{e}\, g_{\mu\nu}\, \delta(x-y) \qquad \text{(IV. 193)}$$

s and t are finite constants

$$s = \frac{1-eb}{e} \qquad t = a + \frac{m^2}{e}(1-eb)$$

representing power series in e with finite coefficients. Inserting (IV. 192) into (IV. 190), solving for $\partial^\rho A_{\mu\rho}$ and taking the limit $\xi \to 0$ we obtain

$$\partial^\rho A_{\mu\rho} = \lim_{\xi\to 0} \frac{1}{Ks+L}\left(j_\mu(x\xi) - K t V_\mu \right)$$

where

$$j_\mu(x\xi) = P_\mu(x\xi) - J_\mu(\xi)\, \mathcal{E}(x\xi) +$$

$$-\nu_1\, \xi_\kappa \xi_\lambda\, \partial^\lambda A^{\mu\kappa} - \nu_2\, \xi^\mu \xi^\kappa\, \partial^\lambda A_{\kappa\lambda} +$$

$$-\nu_3\, \xi^\mu \xi^\kappa\, P_{2\kappa}(x\xi)$$

Similarly from (IV. 191) and (IV. 193)

$$-\partial^\rho T A_{\mu\rho}(x) A_\nu(y) =$$

$$= \lim_{\xi\to 0}\left[\frac{1}{Ks+L}\left(T j_\mu(x\xi) A_\nu(y) - K t\, T V_\mu(x) A_\nu(y) \right) + \right.$$

$$\left. -i\, \frac{K}{e(Ks+L)}\, g_{\mu\nu}\, \delta(x-y) \right]$$

The final result we write in the form (IV. 168-69) setting

$$Z_3^{-1} = \frac{K}{e(K_3+L)} \qquad \bar{m}^2 = et \qquad Z_2 = K^{-1}$$

$j_\mu(x\xi)$ has Brandt's form of the electromagnetic current (IV. 194) with

$$P_{2\mu}(x\xi) = \tfrac{1}{2}\bar{\Psi}(x+\xi)\gamma_\mu\Psi(x-\xi) + (\xi\to-\xi)$$

$$J_\mu(\xi) = \langle T\bar{\Psi}(x+\xi)\gamma_\mu\Psi(x-\xi)\rangle$$

$$\mathcal{E}(x\xi) = \tfrac{1}{2}N\left\{e^{-ie\int_{x-\xi}^{x+\xi}d\eta_\nu A^\nu(\eta)}\right\} - (\xi\to-\xi)$$

$$\sim 2ie\xi A - \tfrac{1}{3}ie(\xi\partial_x)^2\xi A + \tfrac{1}{6}ie^3 2^3 N\{(\xi A)^3\}$$

•

We summarize the main results of this section. It was shown that the vector field A_μ satisfies a field equation of the form (IV. 168). While the renormalization function Z_3^{-1} is logarithmically divergent at $\xi = 0$ the constant $\bar{m}^2$ is a power series with finite coefficients. Since formally

$$\bar{m}^2 = Z_3 m_0^2$$

it follows that the bare meson mass m_0 is logarithmically divergent despite the quadratic divergence of meson self-energy diagrams. This is not surprising since the current operator $j_\mu(x\xi)$ contains a term linear in A_μ with quadratically divergent coefficient in the limit $\xi \to 0$. If this term were combined with $Z_3^{-1}\,\bar{m}^2\,A_\mu$ the total bare meson mass would of course be quadratically divergent in perturbation theory. It is entirely a matter of convenience (related to gauge invariance, see below) to include the quadratically divergent self-energy term in the definition of the current operator $j_\mu(x\,\xi)$.

Apart from direction dependent terms Brandt's current operator $j_\mu(x\,\xi)$ contains the characteristic subtraction term

$$J_\mu(\xi)\, N\left\{ e^{-ie\int_{x-\xi}^{x+\xi} d\eta_\nu A^\nu(\eta)} \right\} \qquad (IV.197)$$

The exponential of the line integral is familiar from Valatin's and Schwinger's work [61] on external field problems. The original motivation for this expression was that

$$\bar{\psi}(x+\xi)\, e^{-ie\int_{x-\xi}^{x+\xi} d\eta_\nu A^\nu(\eta)}\, \psi(x-\xi)$$

is invariant under the gauge transformation

$$A_\mu(x) \longrightarrow A_\mu(x) + \partial_\mu \Lambda(x)$$

$$\psi(x) \longrightarrow e^{-ie\Lambda(x)}\,\psi(x)$$

already before the limit $\xi \to 0$ is taken. The current (IV.194) does not meet this requirement, but it is already an important result that the contribution (IV.197) to the current can rigorously be justified. [76]

While not necessary, it would certainly be preferable to have a current operator which is gauge invariant for $\xi \neq 0$. Some work has been done in that direction, but so far no check in higher orders of renormalized perturbation theory has been given.[62]

Another problem is that the current (IV.194) contains normal products of operators A_μ which have not been defined other than by their perturbative expansions. To eliminate all normal products from the meson field equation would require to simultaneously discuss the Wilson expansions of

$$\bar{\psi}(x+\xi)\gamma_\mu\psi(x-\xi)$$

and

$$A_\mu(x+\xi)A_\nu(x)A_\rho(x-\xi)$$

similar to the treatment of the pseudoscalar meson equation given in section 3e.

5. Quantum Electrodynamics

In this section we briefly discuss quantum electrodynamics as limit of a vector meson theory by letting the renormalized meson mass approach zero.[13]

We consider the field equations (IV.166), (IV.168) which, however, should be interpreted as weak limit equations for suitable state vectors other than the asymptotic incoming or outgoing states (IV.141) implies

$$\lim_{m\to 0} \bar{m} = 0$$

Since

$$\int d\kappa^2 \, \frac{\rho(\kappa^2)}{\kappa^2}$$

stays finite in the limit $m \to 0$ we obtain from

$$\frac{m^2}{\bar{m}^2} = 1 + m^2 \int d\kappa^2 \, \frac{\rho(\kappa^2)}{\kappa^2}$$

the relation

$$\lim_{m \to 0} \frac{m^2}{\bar{m}^2} = 1$$

Applying the limit $m \to 0$ to the field equation (IV.168) we get

$$-\partial_\nu A^{\mu\nu}(x) = \lim_{\xi \to 0} Z_3^{-1}(\xi)\left\{e\, j^\mu(x\xi) - \partial^\mu \partial_\nu A^\nu(x)\right\}$$

with the current given by (IV.194-96). It was assumed here that the limit $m \to 0$ exists and can be interchanged with the limit $\xi \to 0$. (IV.198) represents Brandt's field equation for the electromagnetic potential. The form of the Dirac equation (IV.166) does not change in the limit $m \to 0$.

6. Further Problems Concerning Current Operators in Renormalized Perturbation Theory.

Our discussion has been restricted to current operators which are associated with local field equations of renormalizable models. Except for commutation relations the properties of such operators seem to be fairly understood in perturbation theory.[62] However, relatively little has been done concerning the rigorous treatment of current operators in general. In a renormalizable theory it is certainly possible to define local polynomials of field operators which are finite in every order of perturbation theory. But the problem is to find specific current operators related to symmetries or broken symmetries of the system which have certain properties like commutation relations, divergence relations, etc. [74]

A particular important case is the construction of an energy-momentum tensor which is finite in perturbation theory. In this connection interesting work has been done by Callan, Coleman and Jackiw which seems to indicate that finite energy-momentum tensors should indeed exist for renormalizable models.[63,64] The problem will be to express this tensor in terms of the renormalized field operators (using a suitable limiting process) and to show that the space integrals of the zero components equal the generators of the translation group representation.

Acknowledgments

I am indebted to Dr. Brandt for many stimulating and clarifying discussions. My thanks go to Drs. Deser, Grisaru and Pendleton for the perfect organization of the Summer Institute. In particular, I am grateful to Dr. Pendleton for the great help he provided in editing this manuscript.

Footnotes and References

1. H. Lehmann, Nuovo Cimento 11, 342 (1954).

2. L. Garding and A. Wightman, Arkiv Fys. 28 (1965). This paper contains further references.

3. K. Wilson, On Products of Quantum Field Operators at Short Distances, Cornell Report (1964).

4. K. Wilson, Phys. Rev. 179, 1499.

5. J. Wess, Nuovo Cimento 18, 1086 (1960).

6. W. Thirring, Ann. Phys. 3, 91 (1958),
K. Johnson, Nuovo Cimento 20, 773 (1961),
B. Klaiber, Boulder Lectures (1967), Lectures in Theoretical Physics, Gordon and Breach, New York.

7. J. Lowenstein, Normal Products in the Thirring Model, preprint University of Sao Paulo (1969),
K. Wilson, Operator Product Expansions and Anomalous Dimensions in the Thirring Model.

8. More precisely a and b are power series in λ with finite coefficients.

9. J. Valatin, Proc. Roy. Soc. A, 225, 535 and 226, 254 (1954).

10. For recent work on field equations in limit form see ref. [12 - 17].

11. Following is a list of papers which investigate field equations in limit form for solvable models (see also ref. [7]).
R. Haag and G. Luzzato, Nuovo Cimento 13, 415 (1959),
P. Federbush, Progr. Theor. Phys. 26, 148 (1961),
K. Johnson, Nuovo Cimento 20, 773 (1961),
C. Sommerfield, Ann. Phys. (N.Y.) 26, 1 (1963),
F. Schwabl, W. Thirring and J. Wess, Ann. Phys. (N.Y.) 44, 200 (1967),
D. A. Dubin and J. Tarski, Ann. Phys. (N.Y.) 43, 263 (1967).

12. R. Brandt, Ann. Phys. 44, 221 (1967), (Neutral pseudoscalar meson theory, Dirac equation).

13. R. Brandt, Ann. Phys. 52, 122 (1969).

14. R. Brandt, Maryland Reports No. 673 (1967) and No. 887 (1968), to be published in Fortschritte der Physik (Quantum Electrodynamics).

15. W. Zimmermann, Comm. Math. Phys. 6, 161 (1967), 10, 325 (1968) (A^4-Coupling).

16. W. Zimmermann, Comm. Math. Phys. 8, 66 (1968) (Consequences of infinite mass renormalization).

17. Th. A. J. Maris, D. Dillenburg and G. Jacob, Nuovo Cimento 53 A, 823 (1968) and preprint 1969, Universidade Federal do Rio Grande do Sul, Porto Alegre, Brasil (1969), (Quantum Electrodynamics of Mass Zero Fermions).

18. In this chapter Bogoliubov's combinatorial technique is applied to define the finite part of Feynman integrals in momentum space. The method is thus an extension of the original work of Dyson [19] and Salam [20]. Following is a partial list of papers on renormalization (ref. [19] - [37]).

19. F. J. Dyson, Phys. Rev. 75, 1736 (1949).

20. A. Salam, Phys. Rev. 82, 217 and 84, 426 (1951).

21. J. C. Ward, Proc. Phys. Soc. London A 64, 54 (1951).

22. N. N. Bogoliubov and D. W. Shirkov, Uspechi fiz. Nauk 55, 149 (1955), translated into German in Fortschr. Phys. 4, 438 (1956),

N. N. Bogoliubov and O. Parasiuk, Acta Math. 97, 227 (1957)

N. N. Bogoliubov and D. W. Shirkov, Introduction to the Theory of Quantized Fields, New York, Interscience Publ. 1959.

23. K. Symanzik, Lectures on High Energy Physics, edited by B. Jakšic, Zagreb (1961), Gordon and Breach, New York (1966).

24. T. T. Wu, Phys. Rev. 125, 1436 (1962).

25. J. G. Taylor, Suppl. al Nuovo Cimento 1, 857 (1963).

26. R. L. Mills and C. N. Yang, Progr. Theor. Phys., Suppl., 37, 495 (1966).

27. O. Steinmann, Ann. Phys. 29, 76 (1964) and 36, 267 (1966).

28. K. Hepp, Comm. Math. Phys. 2 , 301 (1966).

29. E. Speer, J. Math. Phys. 9 , 1404 (1968) and Annals of Mathematical Studies, No. 62, Princeton University Press 1969.

30. R. J. Johnson, University of Syracuse preprint (1968).

31. P. Breitenlohner and H. Mitter, Nucl. Phys. B7, 443 (1968).

32. E. R. Caianiello, R. F. Guerra, M. Marinaro, Nuovo Cimento 60 , 713 (1969).

33. T. Appelquist, Ann. of Physics, 54 , 27 (1969).

34. P. K. Kuo and D. R. Yennie, Ann. Phys. 51 , 496 (1969).

35. W. Zimmermann, Comm. Math. Phys. 15 , 208 (1969).

36. H. Epstein and V. Glaser, CERN-preprint No. TH. 1156 (1970).

37. This is always possible, see ref. [35], p. 213.

38. S. Weinberg, Phys. Rev. 118, 838 (1960).

39. Y. Hahn and W. Zimmermann, Comm. Math. Phys. 10, 330 (1968),
W. Zimmermann, Comm. Math. Phys. 11 , 1 (1969).

40. See ref. [28], p. 321.

41. K. Hepp and W. Zimmermann, unpublished.

42. W. Pauli and F. Villars, Rev. Mod. Phys. 21 , 434 (1949).

43. M. Gell-mann and F. Low, Phys. Rev. 84, 350 (1951). In this formula it is understood that differentiations are always taken in front of the time ordering symbol T.

44. A. S. Wightman, Phys. Rev. 101, 860 (1965).

45. Note that the limit $\epsilon \rightarrow 0^{+}$ has already been performed.

46. Here the limits $\epsilon \rightarrow 0^{+}$ and $H \rightarrow \infty$ are performed in reverse order, which requires justification.

47. The unitarity relations of the renormalized S-matrix elements were first derived in perturbation theory by T. T. Wu for A^4-coupling using parametrized Feynman integrals. (Phys. Rev. 125, 1436 (1962)). In this paper also the proof of (II. 82-85) is sketched. The method was independently developed by M. Veltman, Physica 29, 186 (1963).

48. Partial results concerning the rigorization of this method have been obtained by K. Osterwalder (private communication).

49. It is still an unresolved mathematical problem how to define Feynman amplitudes on the mass shell in the proper sense of distribution theory. This difficulty does not seem to occur in Steinmann's approach [27].

50. Similar identities were derived by Hepp [28] for the renormalized integrands in coordinate space. In Brandt's work [12-14] these identities were used in the derivation of local field equations.

51. A different method of deriving operator product expansions in perturbation theory has been developed by K. Wilson (private communication).

52. R. Brandt has developed an interesting method which allows to deduce the light cone singularities from the singularities near the origin, independently from perturbation theory. R. Brandt and G. Preparata, Rockefeller University preprint (1970), R. Brandt, preprint in preparation.

53. The material of this chapter is based on ref. [3-4], [12-15].

54. Unless otherwise noted $\lim\limits_{\xi \to 0}$ denotes the limit taken with the restriction that

$$\frac{\xi_\mu}{\sqrt{|\xi^2|}}$$

be bounded.

55. J. Proca, J. Phys. Radium 7, 347 (1936), G. Wentzel, Quantum Theory of Fields, Interscience Publishers, Inc., New York (1949).

56. K. Symanzik, Lectures of the University of Islamabad, 1969.

57. C. de Calan, Doctoral Thesis, University of Paris, 1968.

58. N. N. Bogoliubov and D. V. Shirkov, introduction to the Theory of Quantized Fields (New York, 1959).

59. J. D. Bjorken and S. D. Drell, Relativistic Quantum Fields (New York, 1965).

60. K. Johnson, Nucl. Phys. 25, 435 (1961).

61. J. Valatin, Proc. Roy. Soc. A, 222, 93, 228 (1954) and [7], J. Schwinger, Phys. Rev. Letters 3, 296 (1959).

62. In this connection see ref. [16-17].

63. C. Callan, S. Coleman and R. Jackiw, to be published in Ann. Phys. (N.Y.).

64. Further papers relevant for the problem of defining an energy-momentum tensor are:

C. A. Orzalesi, J. Sucher and C. H. Woo, Phys. Rev. Letters 21, 1550 (1968),

C. Callan, Caltec-preprint (1970),

R. Jackiw, preprint, Harvard University (1970),

K. Symanzik, DESY-preprint (1970).

65. In general, the principles of quantum field theory and the local field equations in limit form do not uniquely determine the perturbative solutions. This question is being investigated by G. Dell'Antonio for two-dimensional models (private communication). Using some additional assumptions Brandt [12] has shown how the renormalized expansion of the Green's functions can be generated starting from a local field equation in limit form.

66. In (II. 23-24) the differential operator t_λ only acts on those factors of I_Γ which are associated with lines of the subdiagram λ .

67. The following abbreviation is used here:

$$\langle T A_{(\mu)_1}(x_1)\cdots A_{(\mu)_n}(x_n) A(y_1)\cdots A(y_N)\rangle =$$

$$= \partial^{x_1}_{(\mu)_1}\cdots\partial^{x_n}_{(\mu)_n}\langle T A(x_1)\cdots A(x_n)A(y_1)\cdots A(y_N)\rangle$$

68. Here and in the work that follows we use the abbreviation:

$$\langle T N_\delta\{A_{(\mu)_1}(x_1)\cdots A_{(\mu)_n}(x_n)\}\tilde{A}_{(\nu)_1}(q_1)\cdots\tilde{A}_{(\nu)_N}(q_N)\rangle =$$

$$= \partial^{q_1}_{(\nu)_1}\cdots\partial^{q_N}_{(\nu)_N}\langle T N_\delta\{A_{(\mu)_1}(x_1)\cdots A_{(\mu)_n}(x_n)\}\tilde{A}(q_1)\cdots\tilde{A}(q_N)\rangle$$

69. V. Glaser, H. Lehmann and W. Zimmermann, Nuovo Cimento 6, 1122 (1957).

70. O. Steinmann, Comm. Math. Phys. 10, 245 (1968).

71. R. de Mottoni and H. Genz, Field Products at Short Distances, preprint, University of Hamburg (1970).

72. E. C. G. Stueckelberg, Helv. Phys. Acta 11, 225 and 299 (1938).

73. C. de Calan, R. Stora and W. Zimmermann, Lettere al Nuovo Cimento, 1, 877 (1969).

74. The following papers deal with the rigorous definition of renormalized current operators and their commutation relations in perturbation theory:
R. Brandt, Phys. Rev. 166, 1795 (1968) and 180, 1490 (1969),
B. Schroer and P. Stichel, Comm. Math. Phys. 8, 327 (1968),
P. Otterson, University of Maryland thesis, to be published.
These papers contain further references.

75. This Lagrangian was suggested by O. V. I. Ogievetskii and I. V. Polubarinov, Zurn. Eksp. Teor. Fiz. 41, 247 (1961), Sov. Phys. JETP, 14, 179 (1962), G. Feldman and P. T. Matthews, Phys. Rev. 130, 1633 (1963).

76. The current is of course gauge invariant in the limit $\xi \to 0$ as has been discussed by Brandt [14].

INDEX

www.ingramcontent.com/pod-product-compliance
Lightning Source LLC
Chambersburg PA
CBHW060418220326
41598CB00021BA/2216

* 9 7 8 0 2 6 2 5 4 0 1 3 1 *